FOUNDATIONS OF SET THEORY

STUDIES IN LOGIC

AND

THE FOUNDATIONS OF MATHEMATICS

VOLUME 67

NORTH-HOLLAND PUBLISHING COMPANY
AMSTERDAM • LONDON

FOUNDATIONS OF SET THEORY

SECOND REVISED EDITION

ABRAHAM A. FRAENKEL
Late Professor of Mathematics, The Hebrew University of Jerusalem

YEHOSHUA BAR-HILLEL
Professor of Logic and Philosophy of Science, The Hebrew University of Jerusalem

AZRIEL LEVY
Professor of Mathematics, The Hebrew University of Jerusalem

with the collaboration of

DIRK VAN DALEN
University of Utrecht

1973

NORTH-HOLLAND PUBLISHING COMPANY
AMSTERDAM · LONDON

Library of Congress Catalog Card Number 68-54505
ISBN 0-7204-2270-1

First edition 1958
Second revised edition 1973

PUBLISHER
NORTH-HOLLAND PUBLISHING COMPANY — AMSTERDAM • LONDON

PRINTED IN THE NETHERLANDS

CONTENTS

PREFACE TO THE SECOND EDITION

The first edition of this book was published exactly half a century after the year (1908) in which set theory in its original, naive form created by Cantor, gravely shaken by the antinomies, underwent a thorough reconstruction in the hands of Brouwer, Russell, and Zermelo. During the thirteen years that have passed since the publication of that edition there have been many developments in most of the topics that were discussed in it. But only in the axiomatic foundations have there been such extensive, almost revolutionary, developments as to warrant the almost complete rewriting of Chapter II by one of the authors (A.L.). The senior author, A.A. Fraenkel, had left upon his death in 1966, extensive notes for the updating of Chapter IV on the intuitionistic conceptions of mathematics, but it was deemed necessary to invoke further help, since neither of the other authors felt himself on firm enough ground in this respect. We would like to thank Dr. D. van Dalen for his readiness to take upon himself, on short notice, this task, in connection with which Professor Heyting's helpful comments should gratefully be acknowledged.

In Chapter III, only Sections 3 and 4, dealing with Quine's systems, were rewritten, while the remaining sections were left more or less intact. Similarly, Chapter V remained essentially untouched, with the exception of Section 7 which was considerably updated. Section 8 would have deserved to be completely revamped and enlarged, but this task had to be postponed for another occasion, in order not to delay publication of this edition still further.

We tried to avoid discussing in detail those topics which would have required heavy technical machinery, while describing the major results obtained in their treatment if these results could be stated in relatively non-technical terms. Thus the notions of Gödel-constructibility and Cohen-forcing were not defined but many of the results obtained by means of these notions were discussed in considerable detail. Similarly, we did not remark on a treatment of the highly important and rapidly developing area of very large cardinals, refrained from even defining such notions as compact, Ramsey, measurable, supercompact, and extendable cardinals, and restricted ourselves to provide references to the literature.

The book contains repetitions of more than average frequency. The authors thought that many readers would prefer them to constant back-references; they have no illusions as to having hit on the right proportions.

The bibliography is now self-contained. By dropping many older items, we were able to keep its size unchanged, in spite of many additions. No claims to any kind of exhaustiveness are now made.

In 1966, a Russian translation of the first edition was published in Moscow. This translation, by Yu.A. Gastev, contains some 30 pages of additional bibliography, carefully prepared by the translator and updated till approximately 1965.

The main intention of the present edition is to serve as a first reference for those who would like to get acquainted with the state of the art in the foundations of set theory.

Y.B.-H.
A.L.

CHAPTER I

THE ANTINOMIES

§1. HISTORICAL INTRODUCTION

In *Abstract Set Theory* [1]) the elements of the theory of sets were presented in a chiefly *genetic* way: the fundamental concepts were defined and theorems were derived from these definitions by customary deductive methods. To be sure, some quasi-axiomatic ingredients were inserted there in the form of seven Principles, whose main purpose was the delimitation of the notion of set. The precise significance of these Principles will be discussed in detail in Chapter II of the present book.

At a few places in *Theory* (pp. 11, 98, 201, 218), special precautionary measures had to be taken in order to avoid certain *contradictions* that would have otherwise evolved. These contradictions, arising mainly in connection with a natural unrestricted use of the notions of set, cardinal number, ordinal, and Aleph, have been called *antinomies,* or *paradoxes,* of set theory. In general, we say that a certain theory contains an antinomy when each of two contradictory statements, or else one single compound statement having the form of an equivalence between two contradictory statements, has been proved within this theory, though the axioms of the theory seem to be true and the rules of inference valid.

Before we go into a detailed systematic investigation of the ways in which antinomies threaten the foundations of set theory, a few historical remarks are appropriate. Cantor's discoveries, starting around 1873 and slowly expanding to an autonomous branch of mathematics, had at first met with distrust and even with open antagonism on the part of most mathematicians and with indifference on the part of almost all philosophers. It was only in the early nineties that set theory became fashionable and began, rather suddenly, to be widely applied in analysis and geometry. But at this very

[1]) A.A. Fraenkel, *Abstract Set Theory,* Amsterdam, 3rd ed., 1966, or 2nd ed., 1961. This book will henceforth be referred to as *Theory* or as *T.*

moment, when Cantor's daring vision seemed finally to have reached its triumphant climax, when his achievements had just received their final systematic touch, he met the first of those antinomies. This happened in 1895. The antinomy was not published immediately. Two years later, Burali-Forti rediscovered it. Though neither Cantor nor Burali-Forti were able at the time to offer a solution, the matter was not considered to be very serious; this first antinomy emerged in a rather technical region of the theory of well-ordered sets, and it was apparently hoped that some slight revision in the proofs of the theorems belonging to this region would remedy the situation, as had happened so often before in similar circumstances.

This optimism however was radically shattered when Bertrand Russell in 1902 surprised the philosophical and mathematical public with the presentation of an antinomy (see § 2) lying at the very first steps of set theory and indicating that something was rotten in the foundations of this discipline. But not only was the basis of set theory shaken by Rusell's antinomy; logic itself was endangered. Only a slight shift in the formulation was required in order to turn Russell's antinomy into a contradiction that could be formulated in terms of most basic logical concepts.

To be sure, Russell's antinomy was not the first one to appear in a basic philosophical discipline. From Zenon of Elea up to Kant and the dialectic philosophy of the 19th century, epistemological contradictions awakened quite a few thinkers from their dogmatic slumber and induced them to refine their theories in order to meet these threats. But never before had an antinomy arisen at such an elementary level, involving so strongly the most fundamental notions of the two most "exact" sciences, logic and mathematics.

Russell's antinomy came as a veritable shock to those few thinkers who occupied themselves with foundational problems at the turn of the century. Dedekind, in his profound essay on the nature and purpose of the numbers [1]), had based number theory on the membership relation – his method of "chains" may even be taken as a basis for the theory of well-ordered sets (cf. *T*, p. 230) – and had utilized the notion of a set in its full Cantorian sense for the proof of the existence of an infinite ("reflexive", cf. p. 45) set. Under the impact of Russell's antinomy, he stopped for some time the publication of his essay, the fundaments of which he regarded as shattered [2]). Still more tragic was Frege's fate: he had just put the final touches on his chief

[1]) Dedekind 1888.

[2]) See the preface of the 3rd ed. of Dedekind 1888; cf. Dedekind 30–32, p. 449.

work [1]), after decades of tiresome effort, when Russell wrote him about his discovery. In the first sentence of the appendix, Frege admits that one of the foundations of his edifice had been shaken by Russell [2]).

It is not surprising that many mathematicians who had just begun to accept set theory as a full-fledged member of the community of mathematical disciplines reversed their attitude. This reversal is typically illustrated by the leading mathematician of the time, Poincaré, who himself had contributed to the propagation and application of set theory. For some years after 1902 he met Russell's own proposals for a rehabilitation of set theory (see Chapter III) with an air of mockery [3]).

Cantor himself, to be sure, did not for a moment lose faith in his theory in its full "naive" extent though he was unable to meet the challenge of Russell's antinomy. Other scholars professed not to be especially disturbed by this and other antinomies and, distinguishing between "Cantorism" and "Russellism" [4]), warned against attributing to the "artificially constructed" antinomies any decisive significance. It is however difficult to defend this attitude. Even if Burali-Forti's antinomy does not appear so long as one restricts himself to the ordinals of a few number-classes (*T*, p. 216), this cannot release the serious thinker from the obligation of scrutinizing the theorems that involve the *general* concept of an ordinal; and the contemptuous reference to the "artificial" character of many antinomies should be no more convincing than the claim, say, that every continuous function has a derivative since continuous functions without derivatives are "artificial". It may be safely stated that, on the contrary, throughout mathematics – and other disciplines – the investigation of the most general notions, in all their unrestricted generality, has often proved to be of extreme value for the advancement of research. To think that difficulties could be overcome simply by disregarding the general case [5]) is somewhat naive. Finally, to draw a sharp line between mathematics (which is fine) and logic (which a self-conscious mathematician should shun for the benefit of his soul) is less than useless:

[1]) Frege 1893–1903.

[2]) But he set out immediately to repair the damage, though without success; cf. Frege 1893–03 II, pp. 253 ff, Geach-Black 52 (Preface and pp. 234 ff, especially note o on p. 243), Sobociński 49–50 (pp. 220 ff), Quine 55.

[3]) Poincaré 08, book 2.

[4]) See e.g., Schoenfliess 00–07 II, p. 7; 11, pp. 250–255.

[5]) Or by saying that one is going to disregard it; see the witty exposition in Jourdain 18, pp. 75 ff. As Russell (08, p. 226) says: "One might as well, in talking to a man with a long nose, say 'when I speak of noses, I except such as are inordinately long' which would not be a very successful effort to avoid a painful topic".

logic is constantly applied in mathematics, though this use is not often brought into the open and explicitly taken into account, and if one wishes to put restrictions on this application, as some intuitionists do (see Chapter IV), it is better to formulate these restrictions openly and clearly rather than leaving them in the dark.

It is true that the field of mathematical activity proper, both in analysis and in geometry, is not directly affected by the antinomies. They appear chiefly in a region of extreme generalization, beyond the domain in which the concepts of these disciplines are actually used. It is in general not difficult to take precautionary measures in order to avoid the dangerous region. This is the main reason why many mathematicians recoiled so quickly from the initial shock caused by the appearance of the antinomies. The very fact that one continued to speak of *paradoxes,* or *antinomies,* rather than of *contradictions* serves as an indication that deep in their heart most modern mathematicians did not want to be expelled from the paradise into which Cantor's discoveries had led them.

Nevertheless, even today the psychological effect of the antinomies on many mathematicians should not be underestimated. In 1946, almost half a century after the despairing gestures of Dedekind and Frege, one of the outstanding scholars of our times made the following confession:

> We are less certain than ever about the ultimate foundations of (logic and) mathematics. Like everybody and everything in the world to-day, we have our "crisis". We have had it for nearly fifty years. Outwardly it does not seem to hamper our daily work, and yet I for one confess that it has had a considerable practical influence on my mathematical life: it directed my interests to fields I considered relatively "safe", and has been a constant drain on the enthusiasm and determination with which I pursued my research work [1]).

Though the present book is officially dedicated to the treatment of the foundations of set theory alone, the fact that set theory is one, and according to some even the only [2]), fundamental discipline of the whole of mathematics on the one hand, as well as part and parcel of logic on the other hand, will force us to interpret our topic very liberally and often go into a discussion of the foundations of logic on the whole and of mathematics on the whole. It is well known that many thinkers are at a loss to delimit the borderline between these disciplines. It is often said that set theory belongs to them simultaneously and forms their common link. We shall be in a better position to discuss this view later on.

[1]) Weyl 46.

[2]) See, e.g., Bourbaki 49, p. 7.

Having decided that a treatment of the logico-mathematical antinomies is a task that cannot be dodged, we shall proceed, in the subsequent sections of this chapter, to classify the known antinomies as well as to exhibit some of the most significant ones; we shall then present a preliminary and informal analysis of these specimens and conclude this chapter with some remarks on the role of antinomies in the foundations of mathematics and on crises in the foundations of mathematics in general.

§ 2. LOGICAL ANTINOMIES

Since Ramsey [1]) it has become customary to distinguish between *logical* and *semantic* (sometimes also called *syntactic* or *epistemological*) antinomies. The significance of this distinction will become clear in the following section. In this section, we shall present three antinomies of the first kind, viz., the antinomies named respectively after Russell, Cantor, and Burali-Forti.

1. Russell's Antinomy

In 1903, Russell [2]) published the antinomy he had discovered two years before and communicated to Frege by letter. The same antinomy was simultaneously and independently discussed in Göttingen by Zermelo and his circle without however reaching the stage of publication.

It seems to make perfect sense to inquire, for any given set, whether it is a member of itself or not. For certain sets one would hardly hesitate to commit himself to saying that they are not members of themselves: the set of planets, e.g., is certainly not a planet itself, hence not a member of itself. For other sets, one would as little hesitate to regard them as being members of themselves; the set of all sets is an obvious example. Therefore it seems to make perfect sense to ask the same question with regard to *the set of all sets that are not members of themselves.* The answer to this question, however, is alarming: denoting the set under scrutiny by 'S', we see quickly that if S is a member of S, it belongs to the set of all sets that are *not* members of themselves, i.e. it is not a member of itself, but also that if S is not a member of S, it does *not* belong to the set of all sets that are not members of themselves, hence is a member of itself; taken together, we convince ourselves that S is a member of S if and only if S is not a member of S, a glaring

[1]) Ramsey 26.
[2]) Russell 03 (in particular § 78 and Chapter X).

contradiction, derived from most plausible assumptions by a chain of seemingly unquestionable inferences.

The careful reader might perhaps have felt that the case was overstated. He might object that the contradiction was derived, among other premises, also from the assumption that there exists such a thing as the set of all sets that are not members of themselves, the set we called 'S', hence that we are entitled only to derive that *if S exists,* then S is a member of S if and only if S is not a member of S, from which would only follow the falsity of the antecedent, by *reductio ad absurdum,* hence only that S does not exist or, in colloquial terms, that there just ain't no such animal as S.

Though this objection is valid (and will be taken into consideration later in Chapter III), it does but little to reduce the paradoxical character of the result arrived at. That there should not exist the set containing all those objects that satisfy a certain seemingly precisely delimited condition – viz., that of not containing itself as a member – is probably not less repugnant to common sense than a plain contradiction.

A similar objection against another would-be logical paradox is not only valid but also conclusive. It is worthwhile to deal with this paradox, since it has not been generally recognized that there is indeed a decisive difference between this paradox and Russell's which takes the sting completely out of it. In short, one considers the man who, supposedly, shaves all and only those inhabitants of a certain village who do not shave themselves. Abbreviating the expression 'that inhabitant of the village who shaves all and only those inhabitants of the village who do not shave themselves' by 'b', we arrive, by an argument which is completely analogous to that occurring in Russell's antinomy, at the conclusion that b shaves b if and only if b does not shave b. Noticing, however, that we are only entitled to infer that *if b exists,* then b shaves b if and only if b does not shave b, we could only derive that b does not exist, i.e. that there is no such inhabitant of a village who shaves all and only those inhabitants of that village who do not shave themselves, a result which – though perhaps somewhat surprising to the unaware bystander – is no more paradoxical than, say, the fact that there is no inhabitant of a village who is both more and less than fifty years old.

The condition which the luckless "village barber" was supposed to satisfy simply turned out to be self-contradictory, hence unsatisfiable. (This fact was masked by the circumstance that the insertion of just one inconspicuous word would have made the condition a perfectly satisfiable one: '... all *other* inhabitants ...'.) The condition occurring in Russell's antinomy, on the other hand, does not seem at all to be self-contradictory; the non-existence of a corresponding set is, consequently, a disturbing and unfamiliar result.

The same careful reader who was supposed to make the above-mentioned and partially sustained objection might have also asked himself – recalling the contents of *Theory* – how the emergence of Russell's antinomy can there be obviated. Trying to *prove* the existence of the paradoxical set, he will have noticed that this proof is not forthcoming: the relevant Principle of Subsets (*T*, p. 16) only enables him to prove the existence of a set satisfying a given condition *if this set is a subset of a set already secured.* Russell's paradoxical set, however, cannot be proven to fulfil this additional condition.

Let it be very clearly stated at the outset that there was absolutely nothing in the traditional treatments of logic and mathematics that could serve as a basis for the elimination of this antinomy. We think that all attempts to handle the situation without any departure from traditional, i.e. pre-20th century, ways of thinking have completely failed so far and are misguided as to their aim. Some departure from the customary ways of thinking is definitely indicated, though it is by no means clearly determined where this departure should take place. Indeed, 20th century research into the foundations of logic and mathematics can be fruitfully classified in terms of the place of departure from the Cantorian approach. This attitude will be adopted in the following chapters.

For the sake of historical completeness it should be admitted that certain misgivings as to the status of "self-referential" concepts, of which being-a-member-of-itself is an obvious specimen, were already voiced in the middle ages [1]). These misgivings, however, never were given the form of a clear proposal for a revision of the customary ways of thinking and expression.

Certain "philosophical" doubts as to the validity of the *tertium non datur,* the logical law of the excluded middle, of which free use was made in the derivation of Russell's antinomy, had already been uttered prior to the intuitionists (see Chapter IV), but again these doubts were nowhere formulated in anything approaching the way they have been expressed in the 20th century, and never until then was a non-Aristotelian logic developed to any tolerable degree of completeness and responsibility.

In order to show that Russell's antinomy is not a specifically mathematical one, depending perhaps on some out-of-the-way peculiarities of the concept of set, we shall briefly reformulate it in purely logical terms. It seems to make perfect sense to inquire of a property whether it applies to itself or not. The property of being red, for instance, does not apply to itself since red is surely not red, whereas (the property of being) abstract, being itself abstract, applies to itself. Calling the property of not applying to itself 'impredicable', we arrive at the paradoxical consequence that impredicable is impredicable if and only if impredicable is not impredicable. The property-theoretical (logical) variant is as paradoxical as the set-theoretical (mathematical) one [2]).

2. Cantor's Antinomy

According to Cantor's theorem (*T*, p. 70), the set ***Cs*** of all the subsets of any given set *s* has a greater cardinal than has *s* itself. Consider now the set of *all* sets, call it *U.* Its "power-set" *CU*, i.e. the set of all subsets of *U*, has then a greater cardinal than *U* itself, which is paradoxical in view of the fact that *U* by definition is the most inclusive set of sets.

This antinomy was known to Cantor himself in 1899 though – ironically enough – it was published only in 1932 [3]). In June 1901, it came to the

[1]) Cf. Bocheński 56, § 35.

[2]) Russell 03, p. 102.

[3]) Cantor 32.

attention of Russell who under its stimulation proceeded to construct his own antinomy which is of course much more elementary – at least superficially so – since it makes no allusion to such technical concepts as subset and power-set.

The strong connection between Cantor's antinomy and Russell's antinomy should be clear to all who recall the proof given for Cantor's theorem in *Theory* (pp. 70–71).

3. **Burali-Forti's Antinomy**

As the last antinomy of this group – the qualifier 'logical' is in this case rather misleading – we shall mention the historically earliest one. It is named after Burali-Forti who published it in 1897 [1]). Cantor himself, however, discussed it as early as in 1895 and communicated it to Hilbert in 1896.

The formulation of this antinomy is extremely simple: according to the *Theorem 7* (*T*, p. 201), the well-ordered set W of *all* ordinals has an ordinal which is greater than any member of W, hence greater than any ordinal.

Again in the development of set theory, as presented in *Theory*, neither Cantor's nor Burali-Forti's antinomy are forthcoming since the existence of the relevant sets, viz. the set of all sets and the set of all ordinals, cannot be proven on the basis of the principles laid down there. The reader who guessed that the formulation given there to the Principle of Subsets was intended to obviate the emergence of these and other logical antinomies guessed correctly.

§3. SEMANTICAL ANTINOMIES

A few years after the appearance of the antinomies mentioned in the previous section, antinomies of a somewhat different kind made their debut. Again we shall treat here only a few of the more important ones.

1. **Richard's Antinomy**

This antinomy, published by Richard in 1905 [2]), is of special significance since it is a sort of caricature of Cantor's diagonal method (*T*, p. 52). Many variants of this antinomy are known; the following is one of the simpler ones.

Let us consider all those real numbers between 0 and 1 that can be uniquely characterized by sequences of English words of any finite (but unbounded)

[1]) Burali-Forti 1897.
[2]) Richard 05, 07.

length, e.g. 'point eight', 'the positive square root of point zero seven four', 'the smallest number satisfying the condition that the sum of the square of this number and its product by point one equals point three'. Clearly there are only denumerably many such numbers. Let R be their set. R can then be enumerated. Consider any such enumeration. We now characterize a real number r as *that real number between 0 and 1 whose n-th digit after the decimal point is the cyclic sequent of the n-th digit of the n-th number in the enumeration under consideration* (where '1' is the cyclic sequent of '0', ..., and '0' the cyclic sequent of '9'). From an argument that is almost entirely analogous to that presented in *T*, pp. 52–53, it follows that r is different from all the members of R and is therefore not uniquely characterizable by a finite sequence of English words, in plain contradiction to the fact that r has just been characterized in this fashion, viz. by the italicized sequence of English words in the preceding sentence.

Berry's antinomy [1]), essentially only an instructive and ingenious simplification of Richard's antinomy, will not be discussed here since it has no additional theoretical interest and lacks the straightforward connection with the diagonal method that makes Richard's antinomy so especially embarrassing.

2. **Grelling's Antinomy**

In 1908, Grelling and Nelson [2]) called attention to the following antinomy which they regarded as only a variant of Russell's antinomy but which turned out to be essentially different from, though still remarkably analogous to, the paradox regarding the property of being impredicable.

Grelling's antinomy can be formulated very simply: A few English adjectives, such as 'English' and 'polysyllabic', have the very same property that they denote, e.g. the adjective 'English' is English and the adjective 'polysyllabic' is polysyllabic, while the vast majority, such as 'French', 'monosyllabic', 'blue' and 'hot', do not. Calling the adjectives of the second kind *heterological,* we immediately discover to our dismay that the adjective *'heterological'* is heterological if and only if it is not heterological.

3. **The Liar**

Of this antinomy very many versions are known, among them quite a few that are not truly paradoxical at all. Some of these versions go back to antiquity, to the time when the Megaric philosophers used them to tease the

[1]) Published for the first time in Russell 06.

[2]) Grelling-Nelson 08.

members of Plato's academy [1]). We shall present here one of the more recent versions.

Assume that John Doe utters on December 1st, 1970 the following English sentence and nothing else all day: "The only sentence uttered by John Doe on December 1st, 1970 is false". Since this sentence is a declarative sentence, with nothing elliptical (like "The only sentence uttered by John Doe on December 1st is false") or context-dependent (like "The only sentence uttered by him on December 1st, 1970 is false") about it, one seems entitled to inquire whether this sentence is true or false. However, one realizes before long that the sentence is true if and only if it is false.

Against this antinomy one might raise the objection that it is based on a factual assumption, viz. that John Doe did utter a certain sentence, and nothing else, on a certain day. This is true enough but does little to diminish the paradoxical result. Besides, it has been shown that an analogous antinomy can be constructed which does not rely on any factual assumptions [2]).

§4. GENERAL REMARKS

We have had no intention of presenting an exhaustive description of all the antinomies that have turned up in foundational research during the last seventy years [3]). Among those not treated here so far, the most important is Skolem's paradox because of its basic significance in axiomatic set theory. But just for this reason its exposition and discussion will be postponed to Chapter V, §5.

We have already remarked that only few mathematicians were seriously disturbed by the appearance of the antinomies. But even among those mathematicians who were alert to the crisis in the foundations of their discipline, brought about by the emergence of the antinomies, the great majority shared Peano's opinion that *Exemplo de Richard non pertine ad mathematica, sed ad linguistica,* from which fact they concluded that qua mathematicians they need not bother about Richard's antinomy and the semantic antinomies in general. Indeed, semantic terms like 'denote', 'characterize', or 'true' are necessary ingredients of these antinomies, and these are not terms about which an ordinary mathematician will feel obliged to think very hard. Howev-

[1]) Cf. Bocheński 56, § 23.

[2]) See Tarski 44, note 11.

[3]) For an enumeration and careful description of some twelve antinomies, including the six treated here, see Beth 51, Ch. 17.

er in one of the most interesting developments in modern foundational research it became clear that the problem presented by the semantic antinomies was not just a methodological one of at most indirect relevance to mathematics proper, but rather served as the starting point for investigations of immense direct impact on modern mathematics. How this came about will be discussed in Chapter V.

The literature dealing with the antinomies is very extensive. Whereas for the first few years after the publication of Russell's antinomy they were discussed chiefly by mathematicians, they later began to attract the attention of logicians, methodologists, and philosophers at large in an ever increasing measure. Much of this literature is concerned with piecemeal solutions of the various antinomies, exhibiting no general methodological insight and often contradicting each other. Some of them are based on misunderstandings and errors, others lose themselves in epistemological or metaphysical considerations far from the point. On the whole it seems that though a piecemeal solution might occasionally be appropriate with regard to antinomies emerging in the context of natural languages [1]), insofar as they refer to language systems nothing short of the profound investigations described in the following chapters will do.

All the antinomies, whether logical or semantic, share a common feature that might be roughly and loosely described as *self-reference*. In all of them the crucial entity is defined, or characterized, with the help of a totality to which it belongs itself. There seems to be involved a kind of circularity in all the argumentations leading up to the antinomies, and it is obvious that attempts should have been made to see therein the culprit. However a wholesale exclusion of all reasonings involving any kind of self-reference is certainly too strong a medicine and would throw away the baby with the bath-water. There are innumerably many ordinary ways of expression that are self-referential but still perfectly harmless and useful [2]). To characterize someone as the tallest man on a certain team is doubtless utterly innocuous as well as effective, in spite of the fact that the characterization is performed on the basis of a totality to which the man himself belongs. And many a crucial concept in mathematics – as in every other discipline – is formed in a similar fashion. Not all self-reference leads to contradiction, and some self-reference seems to be an indispensable tool in science as in everyday life.

Since the wholesale exclusion of self-referential concept formation is then

[1]) A recent attempt for a solution of Liar-type antinomies in the framework of natural languages was made by Bar-Hillel 57a, 66.

[2]) For a witty and persuasive defence of self-referential reasoning, see Popper 54.

apparently not feasible, many authors looked for an additional criterium that would separate the sheep from the goats. We shall deal with some of these attempts in Chapter III. Here we shall mention only one such proposal. It amounts, in essence, to disqualifying those would-be concepts whose elimination, on the basis of their definition, would lead to infinite regress [1]); in positive terms, to accepting into the community of scientifically legitimate concepts only those applicants for which finite eliminability can be shown. Without entering here into a detailed discussion, it should be remarked that this proposal, even if effective in overcoming all known antinomies, suffers from the following defect: the proof of finite eliminability, though often extremely tedious, will nevertheless have to be produced from scratch for *every single newly introduced concept.* It is doubtful whether mathematics could stand such a severe imposition. It is therefore understandable that this proposal could not dissuade other authors from looking for more efficient and practicable remedies.

For those mathematicians who believe in the essential soundness of classical mathematics, the task posed by the antinomies is that of constructing a system in which all the notions of classical mathematics can be defined and all (or essentially all) the theorems of mathematics up to and including analysis can be derived but such that its consistency can be proved or, short of this, such that the argumentations leading to the known kinds of antinomies are effectively excluded. It seems that the achievement of this task will require some radical changes in the "naive" attitude that is still prevalent among many mathematicians. It might not be necessary to abandon the belief in the essential soundness of classical analysis – as the intuitionists would advice us to do – but one might be persuaded to leave the paradise into which Cantor has led the mathematicians and to withdraw into a less opulent but more secure habitat. Those unwilling to do this might perhaps prefer to stay in the realm of plenty and build walls around it to keep away the beastly antinomies without, however, being certain that some of these beasts were not walled in themselves. This theme will be developed in a more prosaic way in the following chapters.

§5. THE THREE CRISES

The twentieth century is not the first period in which mathematics underwent a foundational crisis. It might add to the perspective in which con-

[1]) This is "Behmann's solution", proposed in Behmann 31.

temporary antinomies should be looked upon if prior crises are, if only briefly, sketched.

In the fifth century B.C., only a short time after mankind attained one of the most brilliant achievements in its history, viz. the development of geometry as a rigorous deductive science, two discoveries were made that were extremely paradoxical: the first was that not all geometrical entities of the same kind were commensurable with each other, so that, for instance, the diagonal of a given square could not be measured by an aliquot part of its side [1]) (in modern terms, that the square root of 2 is not a rational number); the other were the paradoxes of the Eleatic school (Zenon and his circle) developing with many variations the theme of the non-constructibility of finite magnitudes out of infinitely small parts [2]).

This crisis shocked the Greek mathematicians into obtaining two more brilliant achievements [3]): the *theory of proportions,* as contained in books 5 and 10 of Euclid's *Elements*, and the *method of exhaustion,* as invented by Archimedes, that was nothing less than a strict, though not sufficiently general, forerunner of modern theories of integration. Their theory of proportions should have enabled the Greeks to define *irrational number* and develop, accordingly, an arithmetical theory of the *continuum*; somehow they did not quite make it.

The Greek theory of proportions was soon forgotten – so much so that when rigorous arithmetical theories of irrational numbers were constructed in the second half of the 19th century, one was not at first aware of the fact that these methods were not in principle much different from those already in the possession of the Greek mathematicians two thousand years earlier. Before that, in the 17th and 18th centuries, the great power and fruitfulness of the newly invented calculus led most mathematicians of those times into feverish applications of the new ideas without caring much for the solidity of the basis upon which the calculus was founded [4]). However, the shakiness of this basis became clear at the beginning of the 19th century, constituting *the second crisis in the foundations of mathematics.*

In order to overcome this crisis, Cauchy, in the eighteen thirties, showed how to replace the irresponsible use of infinitesimals by a careful use of

[1]) The profound impression made by this discovery may be gathered from Plato's report in *Theaitetos* that Theodoros had proved the irrationality of the square roots of 3, 5, ... 17, a result that was later generalized by his pupil Theaitetos; cf. Reidemeister 49.

[2]) See *T*, p. 7, footnote 1 and Grünbaum 67.

[3]) Cf. van der Waerden 54.

[4]) Typical for this attitude is d'Alembert's famous dictum: *Allez en avant, et la foi vous viendra.*

limits, whereas Weierstrass and others, in the sixties and seventies, demonstrated how all of analysis and function theory could be "arithmetized". This solidification of the foundations was so successful that Poincaré, in an address delivered in 1900 before the Second International Congress of Mathematicians on the role of intuition and logic in mathematics, could proudly claim that mathematics had by then acquired a completely solid and sound basis. In his own words: "Today there remain in analysis only integers and finite or infinite systems of integers ... Mathematics ... has been arithmetized ... We may say today that absolute rigor has been obtained" [1]).

Ironically enough, at the very same time that Poincaré made his proud claim, it had already turned out that the theory of the "infinite systems of integers" – nothing else but a part of set theory – was very far from having obtained absolute security of foundations. More than the mere appearance of antinomies in the basis of set theory, and thereby of analysis, it is the fact that the various attempts to overcome these antinomies, to be dealt with in the subsequent chapters, revealed a far-going and surprising divergence of opinions and conceptions on the most fundamental mathematical notions, such as set and number themselves, which induces us to speak of the *third foundational crisis* that mathematics is still undergoing [2]).

[1]) Poincaré 02.

[2]) For a rather extensive bibliography on antinomies, up to 1956, see Ch. I, § 6 of Fraenkel-Bar-Hillel 58..

CHAPTER II

AXIOMATIC FOUNDATIONS OF SET THEORY

§1. INTRODUCTION

The discovery of the antinomies led to major changes in set theory affecting both its contents and its methodology. Cantor's definition of the concept of set[1]) reads (translated from German): "A set is a collection into a whole of definite distinct objects of our intuition or of our thought. The objects are called the elements (members) of the set". The occurrence of the antinomies showed that the naive concept of set as appearing in Cantor's "definition" of set, and in the most general conclusions derivable from it, cannot form a satisfactory basis for set theory, much less for mathematics as a whole [2]). One may compare this function of the antinomies as controlling and restricting the deductive systems of logic and mathematics to the function of experiments as controlling and modifying the semi-deductive systems of sciences like physics and astronomy [3]). The discovery of the antinomies called therefore for a re-examination of the concept of set or, rather, of the way this concept was handled. This re-examination resulted in a great divergence in the diagnosis of the ills of Cantor's set theory, and, naturally, different diagnoses led to the recommendation of different cures.

Most of the various diagnoses and cures which came forth since the beginning of the present century can be classified into three main groups each of which divides into several subgroups, viz. the *axiomatic,* the *logicistic,* and the *intuitionistic* attitudes. This order of arrangement may be considered to proceed from more conservative to rather revolutionary attitudes, though the logicistic frame comprises widely different degrees of radicalism. The arrange-

[1]) Given in Cantor 1895–97 I, p. 481, at the start of the final exposition of his lifework in set theory. For earlier attempts to define this concept, see Cantor 1879–84 III, pp. 114 ff., and V, p. 587.

[2]) Cantor himself recognized this fact after having concluded his work; see his letters to Dedekind of 1899 (Cantor 32, pp. 443–448) where he speaks of *inkonsistenten Mengen.*

[3]) Cf. Bourbaki 49.

ment is, however, not a historical one; curiously enough, the first and decisive steps in each of these three directions were taken simultaneously and independently during the years 1906–1908. The present and the two following chapters exhibit some main features of the attitudes mentioned, in the above order.

The various systems of set theory which emerged after the discovery of the antinomies differ greatly in their contents. Not only that certain statements concerning sets are truths in one system and at the same time are falsehoods in another system, but in many cases different systems use different languages, and there is not always a natural translation of the statements in the language of one system to statements in the language of another system. The picture is quite different, as we shall see, with respect to the methodological basis of set theory. When those various systems of set theory were introduced they differed also considerably in their methodology. However, during the half-century which followed the discovery of the antinomies these differences have almost completely disappeared and a high degree of unanimity has been attained.

The methodological basis of almost all branches of mathematics is the *axiomatic method*. It emerged with great perfection in Euclid's *Elements* (c. 300 B.C.), was revived only in the course of the 19th century (again in geometry), and has developed impetuously since the beginning of the present century; most fields of mathematics and logic and some other scientific theories have since been axiomatized. However, the axiomatization of the various fields of mathematics is usually based, explicitly or tacitly, on some fragment of set theory. For example, the axiom of induction of number theory is "If P is any set of natural numbers which contains 0 and which, for every natural number n contained in P, contains also $n+1$, then P contains all natural numbers"; rational numbers are defined as pairs (or sets of pairs) of integers; real numbers are defined as sets (or sets of sequences) of rational numbers, etc. Since axiomatization of a mathematical theory meant deductive development of the theory within, or with the aid of, some fragment of set theory, when set theory itself emerged as a mathematical theory towards the end of the 19th century, it did not seem to be a natural candidate for axiomatization.

Until the discovery of the antinomies, set theory was a branch of mathematics of a methodological status somewhat similar to that of a natural science, as is evident from Cantor's definition of the notion of set. As will be seen later, it is the contents of Cantor's naive set theory, not its methodological status, that bears the blame for the logical antinomies. However, many of the amendments of set theory, or of mathematics in general, which were brought forth in order to avoid the logical antinomies, especially those along the logicistic and, to a considerable extent, also the intuitionistic lines of thought, required a much stricter methodological basis for set theory. Since, as mentioned above, axiomatization of set theory in the traditional sense could not conform to strict methodological requirements, it was advocated, mostly by the proponents of the logicistic attitude, to base set theory directly on logic, i.e., either to consider set theory, and mathematics in general, as part of logic and to obtain the set-theoretical truths as logical truths, or, what turned out to be more adequate, to introduce some of the set-theoretical truths as axioms and deduce from them other set-theoretical truths by means of logic only. This point of view was gradually adopted also by the proponents of the other attitudes [1]) (the proponents of the intuitionistic attitude usually use a system of logic different from the standard one – but it is still a system of logic). This growing adoption of logic as the methodological basis of set theory, and through it of mathematics in general, brought logic into the limelight of foundational investigation in mathematics and was a major, if not the main, stimulant for the subsequent rapid development of mathematical logic and metamathematics. However, modern mathematical logic can by no means be defined as the study of the methodological basis of set theory; it is now an active and interesting branch of mathematics in its own right.

A general description of the treatment of formal mathematical theories based on logic, and of the problems connected with it, is given in Chapter V. The second part of the present section contains a superficial sketch which is sufficient for the understanding of the axiomatic systems described in the present chapter.

The most important directions taken in axiomatic set theory are, on one hand, that of Zermelo and his early successors, later taken up, with new and different approaches, by von Neumann, Bernays, Gödel and Ackermann, and,

[1]) In the beginning of Mostowski 55 the following illuminating sentences are found regarding the foundations, set theory, and the antinomies. "The present stage of investigations on the foundations of mathematics opened at the time when the theory of sets was introduced. The abstractness of that theory and its departure from the traditional stock of notions which are accessible to experience, as well as the possibility of applying many of its results to concrete classical problems, made it necessary to analyze its epistemological foundations. This necessity became all the more urgent at the moment when antinomies were discovered. However, there is no doubt that the problem of establishing the foundations of the theory of sets would have been formulated and discussed even if no antinomy had appeared in the set theory."

on the other hand, that of Quine (*New Foundations* and *Mathematical Logic*). An exposition of the former direction is given in the present chapter while Quine's methods are explained in Chapter III which deals with the logicistic foundation of set theory.

The axiomatic attitude towards set theory and the foundations of mathematics differs from the logicistic and the intuitionistic attitudes not in that the latter attitudes are less strict in demanding a rigorous development of set theory, but in that it believes in the soundness of logic as used in mathematics throughout the ages and views the logical antinomies not as a failure of logic but only as a failure of Cantor's basic assumption about sets as expressed in his "definition" of set. Therefore the cure advocated by the axiomatic attitude is to formulate new basic assumptions, in other words, new axioms, concerning sets in a way which will, at least apparently, avoid the occurrence of antinomies.

We begin the exposition of axiomatic set theory with a detailed development of a *modified form of Zermelo's system* in §§2–6. In §7 we shall treat modified versions of the systems of von Neumann and Bernays and related systems, as well as the system of Ackermann. We shall consider in detail the common features and the disparities of these systems. Special attention will be paid to the nature and the implications of the *axiom of choice* which is common to all these systems and which has, throughout the first half of the present century, formed a focus of discussions.

As we shall see in the present chapter, Zermelo's system and the systems of von Neumann and Bernays have so much in common that they can be regarded as different variants of the same theory. Zermelo's system, with certain modifications in various directions, and with or without the classes added by von Neumann and Bernays, is used today by almost all mathematicians as the basis for set theory and the whole of mathematics [1]).

Before presenting Zermelo's system a few explanations regarding the axiomatic method in general are in order; a more extensive treatment will be given in Chapter V.

Every axiomatic theory (with the exception of axiomatic theories of logic itself) is constructed by adding to a certain *basic discipline* – usually some system of logic (with or without a set theory) but sometimes also a system of arithmetic – new terms and axioms, the *specific* undefined terms and axioms of the theory under consideration. However, mathematicians are in general not used to making the underlying basic discipline explicit. They assume that the interpretation of the "logical" words and phrases they employ, such as

[1]) See Bourbaki 49 and 54.

'not', 'and', 'if...then', 'all', etc., as well as their performance within deduction, is well known and not in need of special discussion. Yet, this happy-go-lucky attitude towards the basic discipline is not quite safe with respect to axiomatic set theory where antinomies are always lurking in the background. Therefore, an explicit taking into account of the basic discipline is now almost universally accepted. This may be done in various degrees of depth and rigor. A complete exposition of the discipline presupposed in our further treatment of axiomatic set theory is out of question if only for reasons of space. We shall employ a somewhat uneasy compromise and describe the basic discipline in general terms only, referring the reader who is interested in details to the ample literature in existence [1]).

In addition to being formalized, the language in which the axiomatic theory is formulated may also be symbolized, i.e. artificial symbols may be used instead of the words of a natural language. A complete symbolization – in contradistinction to a partial symbolization to which every mathematician is accustomed in his daily work – though certainly involving a further increase in rigor and facility of mechanically checking proffered proofs and derivations, is something which for its effective use requires a preliminary training that can be neither presupposed nor required of the average reader, and which makes reading much more difficult even for the reader who has the necessary training. The use of logical symbolism in the main body of this book will therefore be restricted to a minimum, mostly for the formulation of the axioms and of some definitions. (A more extensive use of such symbolism will be made in the remarks and discussions printed in petit, the reading and understanding of which is not necessary for grasping the argument presented in the main body.)

For our purpose of constructing an axiomatic set theory the basic discipline – unless otherwise stated – is assumed to be the so-called *first-order predicate calculus*. We mention here only that this calculus contains a set of *connectives* sufficient for expressing negation, conjunction, disjunction, conditional, and biconditional, and two *quantifiers* denoting universal and existential quantification. For these notions we shall use the symbols '$\neg$', '$\wedge$', '$\vee$', '$\rightarrow$', '$\leftrightarrow$', '$\forall$' and '$\exists$', respectively. These symbols are used in the language as follows, where $\mathfrak{A}$ and $\mathfrak{B}$ are arbitrary statements and 'x' is an arbitrary variable: $\neg\mathfrak{A}$, $\mathfrak{A}\wedge\mathfrak{B}$, $\mathfrak{A}\vee\mathfrak{B}$, $\mathfrak{A}\rightarrow\mathfrak{B}$, $\mathfrak{A}\leftrightarrow\mathfrak{B}$, $\forall x\mathfrak{A}$, $\exists x\mathfrak{A}$. These statements are read "it is not the case that $\mathfrak{A}$", "$\mathfrak{A}$ and $\mathfrak{B}$", "$\mathfrak{A}$ or $\mathfrak{B}$", "if $\mathfrak{A}$ then $\mathfrak{B}$", "$\mathfrak{A}$ if and only if $\mathfrak{B}$", "for all x, $\mathfrak{A}$", "there exists an x such that $\mathfrak{A}$", respectively. The statement $\mathfrak{A}$ in $\forall x\mathfrak{A}$ and in $\exists x\mathfrak{A}$ usually does assert something about x, i.e., in technical terms, 'x' occurs free in $\mathfrak{A}$.

The language in which one deals with the *expressions* of a given theory

[1]) See, for instance, Rosser 53, Church 56, Quine 50, Kleene 67, Mendelson 64, or Shoenfield 67. In Church's terminology we are proceeding according to the informal axiomatic method rather than according to the formal axiomatic method; cf. Church 56, p. 57.

(not with the entities denoted by these expressions!) is called the *metalanguage* of this theory. In our case the metalanguage will be ordinary English, supplemented by a few symbols and some rules governing their use. The language in which the theory itself is formulated is called the *object-language* of this theory. In our case the object-language is a certain extremely restricted sub-language of ordinary English, again supplemented by a few symbols and their rules.

As stated above, the object-language of a given theory is sometimes an artificial symbolic language. Only in very rare cases, however, when an extraordinarily high degree of precision and rigor is indicated or for certain very special purposes is the metalanguage itself taken to be a symbolic language, in which case its metalanguage, the meta-metalanguage of the theory, is still some natural language.

In order to refer in the metalanguage to particular expressions of the theory under discussion, names or other designations of these expressions have to be used. This can be done in various ways [1]). One of the simplest is to employ quotation marks. Often, however, particular signs of the metalanguage are utilized and even more often those expressions themselves are used for this purpose. This last method in which some expressions are doing double duty, first as normal signs for something different from themselves and second as autonymous signs for themselves, is not without dangers. Since this method is, however, the one favored by almost all mathematicians, we shall use it when the other more exact methods would look pedantic and when no misunderstanding will be likely to arise.

The situation is somewhat more complicated when reference has to be made not to particular expressions of the theory but to classes of such expressions, e.g. to all expressions of a certain kind. In order to do this in a rigorous fashion metalinguistic variables have to be used. (A name of the object-linguistic variable 'x', such as ''x'' or 'the last-but-two letter of the English alphabet', is of course a metalinguistic constant and not a variable itself.) Various rigorous methods have been proposed for handling this situation [2]), but we decide, again, not to use those methods but to rely on the context on the one hand, and on the common sense and good-will of the reader on the other. There will be a certain amount of inconsistency in all these matters, but this is probably to be preferred to a usage that would appear overpedantic to most readers.

Within the framework of the first-order predicate calculus we have a (potentially) infinite list of *individual variables* x,y,z,w,x',y',z',w', etc. We

[1]) For an extensive treatment of this question see Carnap 37, §41.
[2]) See Carnap 37, Church 56 and Quine 51.

shall use these variables also as metamathematical variables which range over the variables. E.g., when we shall speak of the set of all statements $x \in y$ (or all statements *of the form* $x \in y$) we shall mean thereby the set which contains, in addition to the statement $x \in y$, also the statements $y \in z, z \in z, u \in y'$, etc. In most cases when we use variables as metamathematical variables for variables we shall assume, tacitly, that different variables stand for different variables. E.g., the set of all statements $\forall z(z \in x \leftrightarrow z \in y)$ is also assumed to contain the statement $\forall u(u \in w \leftrightarrow u \in v)$, but not to contain the statement $\forall x(x \in x \leftrightarrow x \in y)$. In some other cases we do not insist that different variables stand for different variables. E.g., when we mention the set of all statements $x \in y$ we usually mean the set which contains also the statement $x \in x$. In most cases the intended meaning will be obvious from the context; in those cases where there might be some doubt we shall explicitly say what we mean.

A statement is said to be *closed* if it contains no free variables. It is said to be *open* if it contains at least one free occurrence of a variable. E.g., the statements, "Every set x is a member of itself" and "0 is the least natural number", are closed statements, whereas "x is a set", "There exist a z such that $x<y<z$", and "If $x=z$ then $x=y$" are open statements.

When we deal with a symbolized system we often use '(well-formed) formula' instead of 'statement'.

In the formula '$\exists y(x \in y)$' the variable 'y' is bound by the existential quantifier '$\exists y$' but the variable 'x' is free; the formula is therefore open. In the formula '$\forall x \exists y(x \in y)$' all the variables are bound, 'y' as before and 'x' by the universal quantifier '$\forall x$'; the formula is therefore closed.

An open statement in which the variable 'x' is free will also be called, according to mathematical custom, a *condition on* x. A condition on x can also have free variables other than 'x'; those additional variables will be called *parameters*. For example, the statement "x is a member of y" can be regarded as a condition on x and y with no parameters, or as a condition on x with the parameter y, or as a condition on y with the parameter x, just as in ordinary mathematical usage the function $ax+b$ can be regarded as a function of x with a and b as parameters, or as a function of x, a, b with no parameters, etc.

An axiomatic system is in general constructed in order to axiomatize a certain scientific discipline previously given in a pre-systematic, "naive", or "genetic" form. The primitive, undefined terms of the system are meant to denote some of the concepts treated in this discipline while terms denoting the remaining concepts are introduced into the system by definition. The axioms of the system are meant to stand for some of the facts about these concepts while other facts are expressed by the theorems, i.e. the statements that can be derived from the axioms on the basis of the underlying discipline.

If a scientific discipline is axiomatized this discipline forms an *interpretation* or a *model* of the axiom system. In general, however, the axiom system can be interpreted in many additional ways; in that case the original scientific discipline forms the intended or principal interpretation [1]).

§2. SOME BASIC NOTIONS, EQUALITY AND EXTENSIONALITY

We shall now present two variants of Zermelo's set theory, which we denote with ZF [2]) and ZFC. In the first part of the present section we shall describe the language in which ZF and ZFC are formulated. In this and in the next three sections we shall list the axioms of ZF and ZFC. The axioms of ZF will be the axioms known as the axioms of extensionality (I), pairing (II), union (III), power-set (IV), subsets (V), infinity (VI), replacement (VII), and foundation (IX). ZFC will be the system obtained from ZF by adding to it the axiom of choice (VIII) as an additional axiom. We do not wish thereby to brand the axiom of choice as an axiom of set theory less legitimate than the others; on the contrary, as far as we are concerned, ZFC is a better system of set theory than ZF. We still segregate the axiom of choice because, as we shall see in §4, it is of a different nature than the other axioms and plays a special role in set theory and therefore it is desirable for our purposes to know exactly in which parts of our discussion the axiom of choice plays a significant role.

As stated above, the discipline underlying ZF will be the first-order predicate calculus. The primitive symbols of this set theory, taken from logic, are the connectives, quantifiers and variables mentioned above and, possibly, also the symbol of equality (discussed below), as well as such auxiliary symbols as commas, parentheses and brackets. The only specific set-theoretical primitive

[1]) See Church 56, p. 56; also Carnap 39 who uses a somewhat different terminology. The notion of an interpretation is discussed in detail in Chapter V, §3.

[2]) This system is mostly referred to in the literature as the Zermelo–Fraenkel set theory. Some authors use the more appropriate name Zermelo–Fraenkel–Skolem set theory. It rests mainly on Zermelo 08a; cf. Zermelo 30. The principal modifications inserted in the following exposition are contained in Fraenkel 22, 22a, 26 and in Skolem 23, 29. Cf. the formalizations in Wang–McNaughton 53 (pp. 15–18), Carnap 54 (§ 43), Suppes 60. A general survey of Zermelo's intension is given in Weyl 46, pp. 10 f.

While the criticism of Poincaré 13 (Chapter IV) is unjustified in many respects, the criticism of Skolem 23 is incorporated in the following exposition.

symbol of ZF will be the binary predicate $\in$ which denotes the membership relation [1]). We shall read $x \in y$ as "x is a member of y", and, synonymously, as "x belongs to y", "x is contained in y", "y contains x (as a member)" [2]). The *atomic* formulae of ZF are the formulae of the form $x \in y$, and possibly also $x = y$; all other formulae are obtained from these formulae by means of the connectives and the quantitifiers.

The range of the individual variables, the so called *universe of discourse*, consists of *objects*. Since we are dealing with set theory, it is natural to assume that each of these objects is a member of some set (which is again an object). This is in accordance with one of the tacit principles of Cantor's naive set theory that every object can serve as a building block for a set [3]). This set can be, for example, a set which contains just this single object. Some of the systems of set theory discussed in §7 abandon this principle. However, as far as ZF is concerned, this principle remains valid; it will be implemented by the axiom of pairing. Throughout the present chapter we shall mean by *element* an object which is a member of some object. In ZF the term 'element' is synonymous with the term 'object', yet we shall prefer to use the former so as to facilitate the comparison of ZF with those systems of set theory discussed in §7 in which not all objects are elements.

Let us refer to those elements which have members as *sets* [4]), and to those elements which have no members as *individuals* [5]). When we develop a system of set theory we have to make up our mind as to how many individuals we want to have. The same question arises also with respect to the sets, i.e., we have to make up our mind as to "how many" sets we want to have. The latter question is indeed the central question of set theory, and in answering it we are guided mostly by the idea of preserving the intuitive rules for "construction" of sets available in Cantor's naive set theory, to the extent that they do not lead to contradictions. In contrast, the former question is of much less significance. In our answer to that question we cannot rely much on intuitive arguments since there are no ways of "constructing" individuals and, as a con-

[1]) Schoenfliess 21 and Wegel 56 take as the primitive concept the part-whole relation ("x is a part, or a proper subset, of y") instead of the membership relation. Their systems are quite cumbersome and all they yield is a theory of magnitude which is far from being a full-scale set theory.

[2]) We avoid the phrase "x is an element of y" since we shall use the term "element" in a different meaning.

[3]) This follows from the axiom of comprehension – see §3.1.

[4]) The null-set, which has no members, will also be called a set – see below.

[5]) This is not the standard meaning of "individual" in logic. Our individuals were called "Urelemente" by Zermelo.

sequence, there is nothing to tell us how many individuals to admit [1]). Therefore we shall be guided here by arguments of simplicity and elegance rather than by deep insights into the nature of the mathematical universe.

The existence of at least one individual is called for by both philosophical and practical reasons. An individual is needed in order to serve as the *foundation* of the universe. Once we have an individual a, we can construct a set b whose only member is a, a set c whose only member is b, a set d whose only members are a and b, and so on. The way in which the universe of set theory is constructed, starting with a single individual, will be discussed at length in §5. The practical reasons which call for the existence of an individual are as follows. When we define the intersection of two sets r and s to be the set t which consists of those elements which belong to both r and s, we want the intersection to be defined even in the case where r and s have no members in common. In this case the intersection t has to be a memberless element, i.e., an individual. There are also many other examples where the existence of a memberless element makes things simpler. The same practical reasons which call for the existence of such an element also call for using always the same element for the intersection of any two sets r and s with no members in common, and for referring to this element as a set. Therefore we shall call this element the *null-set* and our sets are, from now on, the elements which have members as well as the null set. Let us, however, stress at this point that whereas the existence of at least one individual is required for serious philosophical reasons, referring to one of the individuals as the null-set is done only for reasons of convenience and simplicity, and can be regarded as a mere notational convention.

Having decided that we need an individual we now face the question of whether we need *more than one* individual. It turns out that for mathematical purposes there seems to be no real need for individuals other than the null-set [2]). Therefore we shall indeed not admit any such individuals in ZF. Thus all our elements are either sets which have members or the null-set. Hence, the terms 'set' and 'element' are synonymous in ZF. Yet, mostly for metamathematical purposes, there is also considerable interest in systems of set theory which admit individuals other than the null-set. Therefore we shall formulate the verbal versions of the axioms in such a way that they will serve, with

[1]) In fact, if ZF is consistent then so are corresponding systems of set theory with any prescribed number, finite or infinite, of individuals (in addition to the null-set discussed below) – Mostowski 39, A. Levy 64. One can also prove this by a method similar to that of the index model of Rieger 57.

[2]) Fraenkel 22, p. 234, and 25. This attitude was adopted, among others, by von Neumann 25 and Bourbaki 54.

possible minor modifications which we shall point out as we go along, also as axioms for a corresponding system of set theory which admits individuals [1]). Thus we shall distinguish between the term 'element', which in such a system refers also to the individuals, and the term 'set'. Also, from now on we shall use the term 'individual' only for individuals other than the null-set; thus every element is a set or an individual, but nothing is both a set and an individual.

One of the most fundamental notions of mathematics is the notion of equality. One can adopt any one of the following three attitudes towards equality.

a) The equality symbol is understood to denote identity and is thus regarded as *belonging to the underlying logic*. In our case the underlying discipline is taken to be the *first-order predicate calculus with equality* [2]). The basic properties of equality, which from the point of view of the present attitude are logical truths, are as follows.

(*i*) Reflexivity: (For every x) $x = x$.

(*ii*) Symmetry: If $x = y$ then $y = x$.

(*iii*) Transitivity: If $x = y$ and $y = z$ then $x = z$.

(*iv*) Substitutivity: For every statement $\mathfrak{P}(x)$, if $\mathfrak{P}(x)$ holds and $x = x'$ then $\mathfrak{P}(x')$ also holds.

It is enough to require (*iv*) only for two particular statements $\mathfrak{P}(x)$ as follows.

(*iv'*) If $x \in y$ and $x = x'$ then also $x' \in y$; if $y \in x$ and $x = x'$ then also $y \in x'$. In the presence of (*i*)–(*iii*), all of (*iv*) follows from (*iv'*); this is a particular case of the general fact that it is enough to postulate (*iv*) for atomic statements $\mathfrak{P}(x)$ only, and to infer from this all other cases of (*iv*) [3]).

This attitude seems to have been adopted in essence by Zermelo [4]). He regards x and y as equal when "they denote the same thing", exhibiting there-

[1]) In such an axiom system we need another primitive notion in addition to membership. This primitive notion can be taken be O – an individual constant denoting the null-set – or the unary predicate $S(x)$, to be read "x is a set". If we take O as the additional primitive notion, we define x to be a set if it is O or it has members, and we adopt, in addition to the axioms listed in §§2–5, also the axiom "O *has no members*". If we take $S(x)$ as the additional primitive notion then we define the null-set O as in §3.4 below, and we adopt the axiom "*If x is a member of y then y is a set*". For axiom systems which admit individuals other than the null-set, see Suppes 60, Borgers 49, and Mostowski 39.

[2]) Cf. Church 56, §48, or Mendelson 64, Ch. 2, §8.

[3]) See Church 56, Exercise 48.4, and Mendelson 64, Prop. 2.25.

[4]) Zermelo 08a. Whenever in the present Chapter Zermelo is mentioned without additional reference, this paper is meant; it is not only fundamental for our exposition in general but also contains many details appearing in the following sections.

by, incidentally, a confusion between use and mention of symbols [1]). Eliminating this confusion one winds up saying that x and y are said to be equal when they are the same thing. This is also the attitude which we choose to adopt officially throughout this chapter. However, our treatment will cover the other attitudes as well.

b) Equality is regarded as *one of the primitive relations of the system*, on a par with the others. In our case the equality symbol can be regarded as a second primitive binary predicate. (*i*)–(*iii*) and (*iv*′) are now taken to be axioms of our system. All the instances of (*iv*), for all different statements $\mathfrak{P}(x)$, now become theorems of the system. For all practical considerations the system obtained by adopting this attitude is the same as that obtained by adopting attitude (a), since in both cases exactly the same statements are theorems of the system.

c) Equality is introduced by a *definition* [2]). In this case the definition must be such that (*i*)–(*iv*) become provable, either by arguments of logic only or by arguments which make also use of the axioms of set theory. As we shall see later there are several ways of defining equality in set theory.

After these preliminary remarks we start establishing our system ZF. In general, no symbolism beyond the customary set-theoretic symbols '$\in$', '$\subseteq$' etc. will be used. Only the axioms and some definitions will be fully symbolized, in addition to their semi-symbolic formulation. Instead of $\neg x=y$ and $\neg x \in y$ we shall usually write $x \neq y$ and $x \notin y$, respectively. When $x \neq y$ we say that x is *different* from y.

DEFINITION I (Relation of inclusion). If y and z are sets such that for all x, if $x \in y$ then $x \in z$, we shall write $y \subseteq z$ and say that y is a *subset* of z (y is *included* in z); if, in addition, there is at least one w such that $w \in z$ but $w \notin y$, we write $y \subset z$ and say that y is a *proper subset* of z (or y is *properly included* in z).

THEOREM 1. Every set is a subset of itself ($x \subseteq x$); if $x \subseteq y$ and $y \subseteq z$ then $x \subseteq z$. In other words, the relation $\subseteq$ is reflexive and transitive. The relation $\subset$, on the other hand, is irreflexive, asymmetric, and transitive. (*T*, p. 129.)

One has to distinguish clearly between the relations $\in$ and $\subseteq$ (of which, in the present exposition, the first is primitive and the second derived). The confusion between them, enhanced by equivocations of English and other languages – the copula 'is' of Aristotelian fame is used in both these senses (and many others) – had disastrous consequences in the early development of logic. Frege seems to have been the first logician to point out the necessity

[1]) See, e.g., Quine 51, §4.
[2]) Fraenkel 27a; cf. A. Robinson 39, Hailperin 54.

of this distinction; nowadays only beginners are prone to fall prey to the confusion between $\in$ and $\subseteq$. In our terminology, while a set always *includes* itself and its subsets, it *contains*, in general, neither itself nor its subsets.

Having in mind attitude (c) towards equality let us present the following definition.

DEFINITION II. x is said to be *membership-congruent* to y ($x =_{\mathrm{m}} y$) if for all z, $x \in z$ if and only if $y \in z$ and also for all u, $u \in x$ if and only if $u \in y$; in other words, $x =_{\mathrm{m}} y$ if every set which contains one of them contains also the other and every element contained in one of them is also contained in the other.

In symbols, $x =_{\mathrm{m}} y =_{\mathrm{Df}} \forall z(x \in z \leftrightarrow y \in z) \wedge \forall u(u \in x \leftrightarrow u \in y)$ [1]).

It is easily seen that the relation of membership-congruence is reflexive, symmetric, transitive, and substitutive with respect to the atomic statements $x \in y$, i.e., (*i*)–(*iii*) and (*iv'*) hold for $=_{\mathrm{m}}$. If we adopt attitude (c), then the atomic statements $x \in y$ are the only atomic statements of set theory (since $x = y$ is introduced by a definition and can be regarded as an abbreviation of a non-atomic statement) and therefore we get that (*iv*), too, holds for $=_{\mathrm{m}}$ [2]). The proof of (*i*)–(*iv*) for $=_{\mathrm{m}}$ does not use any of the axioms of set theory, but only the definition of $x =_{\mathrm{m}} y$. Now, it can be easily seen that any defined relation $x = y$ which satisfies requirements (*i*) and (*iv*) must coincide with the relation $=_{\mathrm{m}}$, i.e., we get that for all x and y, $x = y$ if and only if $x =_{\mathrm{m}} y$. Since for any relation of equality which we may define, $x = y$ is equivalent to $x =_{\mathrm{m}} y$, we lose nothing by *defining* $x = y$ as $x =_{\mathrm{m}} y$ [3]).

Our first axiom is

AXIOM (I) OF EXTENSIONALITY. **If $x \subseteq y$ and $y \subseteq x$, then $x = y$; in** other words, sets containing the same members are equal.

In symbols, $\forall x \forall y [\forall z(z \in x \leftrightarrow z \in y) \rightarrow x = y]$.

This axiom affirms the extensional nature of the sets, i.e., that each set is completely determined by its members. The antonym of the present meaning of 'extensional' is 'intensional'. Sets would be of an intensional character if identity of sets would depend not only on their extension (i.e., on their members) but also on the way they are presented ("defined"). Thus, from an intensional point of view the set of all non-nega-

[1]) '$=_{\mathrm{Df}}$' is short for 'is short, by definition, for', this symbol is of course a meta-linguistic symbol which does not belong to the object language of ZF.

[2]) As in p. 25 above.

[3]) Thiele 55, Quine 63. In essentially the same way we can define equality in any first-order theory with finitely many primitive symbols – Hailperin 54. This is in line with a tradition which goes back at least as far as Leibniz (identitas indiscernibilium) – Cf. for instance Grelling 37.

tive real numbers and the set of all squares of real numbers are not necessarily identical, even though they have the same extension. The purely extensional notion of set is chosen to be the basic notion of set theory, rather than any intensional notion, for the following reasons. First, the extensional notion of set is simpler and clearer than any possible intensional notion of set. Second, whereas there is just one extensional notion of set, there may be many intensional notions of set, depending on the purpose for which those sets are needed [1]); so if we wanted to base set theory on some intensional notion of set, we would have to choose among the various intensional notions of set in a way which is bound to be at least somewhat arbitrary. Third, as we shall see, starting with the simple notion of extensional set we shall obtain, by means of the axioms, a system of set theory in which much more complicated notions can be constructed. In particular, we shall be able to construct intensional notions of set within our system [2]).

If we adopt one of the attitudes (a) or (b) towards equality then we know that if $x=y$ then also $x =_{\mathrm{m}} y$. As a consequence of the axiom of extensionality we also get the converse that if $x =_{\mathrm{m}} y$ then $x=y$. Thus, even if equality is taken to be primitive rather than defined, membership-congruence coincides with equality [3]).

Axiom I establishes the relation of having the same members as a characteristic property of equality in set theory. There is also a "dual" characterization of equality, which is given by the following principle.

(*) $x=y$ if and only if every set z which contains one of the elements x and y contains also the other.

In symbols, $\forall z(x \in z \leftrightarrow y \in z) \leftrightarrow x=y$.

This principle cannot be proved from Axiom I alone. However, it will follow from the axiom of pairing (Axiom II below) that for every set x there is a set u whose only member is x, i.e., for every element v, v is a member of u if and only if $v=x$. Therefore, u is a set which contains x but does not contain any y which is not equal to x [4]).

One can use each of these two characteristic properties of equality in set

[1]) E.g., since the intension of a set is determined also by its definition we can choose between regarding verbally different, but logically equivalent, definitions as different definitions or as the same definition.

[2]) We shall, actually, deal with a particular such notion in §3.5.

[3]) In order to get that membership-congruence coincides with equality it is enough to know that the statement $x=y$ is equivalent in set theory to *some* statement which does not involve equality.

[4]) This proves the even stronger principle that if y is a member of every set which contains x then $y=x$.

theory as a definition of equality, instead of membership-congruence. In each case one has to make sure that it will indeed follow from the axioms of the system that equality satisfies requirement (*iv*), i.e., that $x = y$ implies $x =_{\mathrm{m}} y$. In both cases it follows trivially from the definitions of $x = y$ and $x =_{\mathrm{m}} y$ that if $x =_{\mathrm{m}} y$ then $x = y$.

If we define that $x = y$ if they are members of exactly the same sets, then it will follow from the power-set axiom (Axiom IV below) or, alternatively, from the axioms of pairing and subsets (Axioms II and V below, respectively) that if $x = y$, then x and y have the same extension (i.e., for every element z, $z \in x$ if and only if $z \in y$) [1]), hence, by Axiom I, $x = y$ if and only if x and y are membership-congruent.

Now suppose we define that $x = y$ if x and y have the same extension. Then Axiom I becomes a tautology, and in order to be able to prove that $x = y$ implies that x and y are membership-congruent we have, essentially, to assume it as an axiom, i.e., we have to adopt the following [2]):

AXIOM I* . **If** $x \in z$ **and** $y = x$ **then also** $y \in z$ [3]).

The former definition of equality is not appropriate for the systems of set theory considered in §7, in which there are many different objects which are not members of any object, since by that definition any two such objects are equal. On the other hand, in a system of set theory which admits individuals other than the null-set the latter definition is inappropriate, since it implies that all the individuals are equal to the null-set. Interestingly enough, by a slight deviation from ordinary usage, viz. by treating individuals as a special kind of sets (sets which contain themselves as their only member), Quine [4]) succeeds in introducing equality by means of the latter definition without having to renounce individuals in his ontology [5]). As he points out himself,

[1]) Thiele 55, pp. 176f.

[2]) Cf. A. Robinson 39.

[3]) We can also use the following version of the axiom of extensionality: *If x and y have exactly the same members then they are members of exactly the same sets.* This version can serve as the axiom of extensionality independently of whichever attitude we adopted among (a), (b), (c), and of whichever definition of equality we use if we adopt attitude (c) – Scott 61; As we saw above in the case of attitudes (a) and (b), it follows from the axiom of pairing that if x and y are members of exactly the same sets then $x=y$. The converse of the present version of the axiom of extensionality, viz. that if x and y are members of exactly the same sets then they have exactly the same members, follows from the power-set axiom or, alternatively, from the axiom of pairing and subsets, as mentioned above.

[4]) Quine 63, pp. 31–33.

[5]) The assumption of the existence of such individuals does not introduce any contradictions in a system like ZF – Cf. Bernays 37–54 VII and Rieger 57.

the situation can be described alternatively by saying that the $\in$-relation is to be interpreted as "is a member of, or is equal to, according to whether the right-hand object is a set or not" [1]).

Whichever of the various methods discussed above one chooses for introducing equality in set theory, the *intended* interpretation of '$x = y$' is that the objects denoted by 'x' and 'y' are identical. E.g., the direct way to say, in the language of set theory as given above, that a set z has exactly one member is to say that there is a member x of z such that every member y of z is equal to x; if equality is not intended to be necessarily identity, then such a set z can contain two or more members equal to one another.

Since a set, according to the axiom of extensionality, is fully determined by its members, we may denote the (finite or infinite) set which consists of the members $a, b, c, \ldots$ by

$$\{a, b, c, \ldots\} ,$$

where the order in which the members are written does not matter.

DEFINITION III. Two sets which contain no common member are said to be *disjoint*. If the members of a set s are pairwise disjoint, s is said to be a *disjointed set* [2]).

§3. AXIOMS OF COMPREHENSION AND INFINITY

3.1. **The Axiom Schema of Comprehension**. Having determined, by means of Axiom I, one of the basic properties of the notion of set, our next task is to introduce axioms which guarantee the existence of sufficiently many sets, at least as many as are needed for the development of arithmetic and analysis. Let us pretend to be unaware of the logical antinomies and try to set up an axiom system by adapting Cantor's "definition" of set (in §1, p. 15) to our present rigorous setup. According to that "definition" every collection of

[1]) This hardly differs from using equality as a primitive relation for individuals only and considering two *sets* to be equal if they have exactly the same members. One may doubt if the technical economy achieved by Quine's method suffices to compensate for its susceptibility of misinterpretation (such a misinterpretation occurs in Quine 63, p. 285, where the axiom of foundation is said to clash with the existence of Quine's individuals; the usual formulation of this axiom – Axiom IX – becomes now equivalent to "every non-void set x contains an individual or contains a set y such that $x \cap y = 0$"; the latter statement does not clash with the assumption of existence of individuals).

[2]) A set which contains no member or a single member is, trivially, disjointed.

elements is a set; therefore, for every rule or process by means of which a collection of elements is obtained there is a set which contains exactly the elements which conform to the rule, or are obtained in the process, respectively. The simplest general axiom in this direction is the following

AXIOM OF COMPREHENSION[1]). **For any condition $\mathfrak{P}(x)$ on x there exists a set which contains exactly those elements x which fulfil this condition.**

In symbols, this axiom reads $\forall z_1 \ldots \forall z_n \exists y \forall x(x \in y \leftrightarrow \mathfrak{B}(x))$, where $z_1, \ldots, z_n$ are the free variables of $\mathfrak{B}(x)$ other than x, and y is not a free variable of $\mathfrak{B}(x)$.

On this occasion, let us remark that Axiom I, without any additional axioms, obviously implies that for every condition $\mathfrak{P}(x)$ on x there exists *at most* one set y which contains exactly those elements x which fulfil the condition $\mathfrak{P}(x)$; in other words, if y_1 and y_2 are two sets each of which contains exactly those elements x which fulfil the condition $\mathfrak{P}(x)$, then y_1 and y_2 are equal.

Since we were guided by Cantor's naive notion of set in formulating the axiom of comprehension, we can hardly be surprised when this axiom turns out to be inconsistent (i.e., it implies a logical falsehood). Russell's antinomy can easily be derived from the axiom of comprehension as follows. We prove first, with use of logic only:

THEOREM 2. There exists no set (element) which contains exactly those elements which do not contain themselves (in symbols: $\neg \exists y \forall x(x \in y \leftrightarrow x \notin x)$).

Proof. By contradiction. Assume that y is a set such that for every element x, $x \in y$ if and only if $x \notin x$. For $x = y$, we have $y \in y$ if and only if $y \notin y$. Since, obviously, $y \in y$ or $y \notin y$, and, as we saw, each of $y \in y$ and $y \notin y$ implies the other statement, we have both $y \in y$ and $y \notin y$, which is a contradiction.

Theorem 2, which poses as a theorem of set theory, is really a theorem of logic (i.e., a logical truth) – it remains true with the membership relation replaced by any other binary relation. Theorem 2 directly contradicts the statement which is obtained from the axiom of comprehension by taking '$x \notin x$' for $\mathfrak{P}(x)$; therefore the latter statement is a logical falsehood.

The axiom of comprehension is not a single statement of the object-language of ZF; by taking different conditions $\mathfrak{P}(x)$ in the axiom of comprehension we get different statements. Therefore the axiom of comprehension is said to be an *axiom schema*. Each statement obtained from the axiom

[1]) The first explicit use of this axiom seems to be in Frege 1893–1903 (I, § 9) (where it is given in a somewhat different form).

of comprehension by taking a particular condition $\mathfrak{P}(x)$ is said to be an *instance* of this axiom schema, or an *axiom of comprehension*. As we saw above, the contradiction was obtained from the single instance "There exists a set y which contains exactly those elements x which fulfil the condition $x \notin x$" of the axiom schema of comprehension.

The axiom of comprehension turned out to be inconsistent and therefore cannot be used as an axiom of set theory. However, since this axiom is so close to our intuitive concept of set we shall try to retain a considerable number of instances of this axiom schema. The instance which we used here to get a contradiction is by no means the only contradictory instance of the axiom schema of comprehension; moreover, there are non-contradictory instances of this axiom schema which contradict each other [1]). Therefore, the decision as to which instances to keep is not an easy one; different decisions made at this point lead to different systems of set theory, as we shall see. Our guiding principle, for the system ZF, will be to admit only those instances of the axiom schema of comprehension which assert the existence of sets which are not too "big" compared to sets which we already have. We shall call this principle *the limitation of size doctrine* [2]).

As pointed out above, the set whose existence is asserted by a given axiom of comprehension is uniquely determined by that axiom. Therefore, in the verbal formulation of an axiom of comprehension we shall use the definite article ('there exists *the* set of all ...').

3.2. **The Axiom of Pairing. Ordered Pairs.** We shall start with sets of very modest size by means of the

AXIOM (II) OF PAIRING. **For any two elements** a **and** b **there exists the set** y **which contains just** a **and** b (i.e., a and b and no different member).

In symbols: $\forall a \forall b \exists y \forall x [x \in y \leftrightarrow (x = a \vee x = b)]$ [3]).

The set which contains just a and b is called the *pair* of a and b and is

[1]) See Quine 63, pp. 66–68. Additional examples are easy to obtain once one notices that for every closed formula φ, the instance $\exists y \forall x(x \in y \leftrightarrow x \notin x \wedge \neg \varphi)$ of the schema of comprehension is logically equivalent to $\varphi \wedge \exists y \forall x(x \notin y)$, cf. Putnam 57.

[2]) This principle, implicit in Cantor 32 (p. 444), was first stated by Russell 06; see also pp. 135–136.

[3]) Using the other axioms we can replace Axiom II by any of the following axioms, "For any *different* sets a and b there exists a set which contains just a and b" (Fraenkel–Bar-Hillel 58), "For any sets a and b there exists their union" (Kuratowski 25), "For any sets a and b there exists a set which contains exactly all the members of a and the set b itself" (Bernays 37–54 I). When we shall introduce Axiom VII (of replacement) it will be shown that it implies Axiom II.

denoted by '$\{a, b\}$' or, synonymously, by '$\{b, a\}$'. If in $\{a, b\}$ $a=b$ then a is the only member of the pair $\{a, a\}$, which is denoted also by $\{a\}$ and which is called the *singleton*, or *unit-set*, of a.

Incidentally, it is only by means of Axiom IX that we shall be able to prove that the pair $\{a, b\}$ is different from a and from b [1]).

Given the elements a, b, c, and d, repeated application of Axiom II allows us to build various sets from these elements, for instance $\{\{a, b\}, \{c, d\}\}$, $\{\{a, b\}, c\}$, $\{\{c, d\}\}$, ... However, all sets obtained in this way have just one or two members.

A simple notion which is extremely useful in mathematics is the notion of an ordered pair. The *ordered pair* $\langle a, b\rangle$ is an element which corresponds to a and b (taken in that order) such that

(*i*) For all a, b, c, d, if $\langle a, b\rangle = \langle c, d\rangle$ then $a=c$ and $b=d$.

As suggested by Wiener and Kuratowski such a notion can be defined in ZF by

DEFINITION IV. $\langle a, b\rangle = \{\{a\}, \{a, b\}\}$ [2]).

(*i*) is established by means of Definition IV as follows. Let $\langle a,b\rangle = \langle c,d\rangle$; then $\{a\} \in \{\{a\}, \{a,b\}\} = \{\{c\}, \{c,d\}\}$; hence $\{a\} = \{c\}$ or $\{a\} = \{c,d\}$. We shall now deal with two cases. Case (a): $\{a\} = \{c\}$ and $\{a\} \neq \{c, d\}$. Since $\{a\} = \{c\}, a=c$. $\{c, d\} \in \{\{c\}, \{c, d\}\} = \{\{a\}, \{a, b\}\}$, hence, by $\{c, d\} \neq \{a\}$, we must have $\{c, d\} = \{a, b\}$. By $\{a\} \neq \{c, d\}$ and $a=c$ we have $d \neq a$, thus $d \in \{c, d\} = \{a, b\}$ implies $d=b$. Case (b): $\{a\} = \{c, d\}$. Then $c=d=a$. $\{a, b\} \in \{\{a\}, \{a, b\}\} = \{\{c\}, \{c, d\}\} = \{\{a\}, \{a, a\}\} = \{\{a\}\}$, hence $\{a, b\} = \{a\}$ and $b=a=d$.

Almost all applications of the notion of an ordered pair to mathematics make use only of (*i*) and not of Definition IV. For example, the notion of an *ordered triple*, and in general, the notion of an *ordered n-tuple*, for $n \geq 3$, are defined by

DEFINITION V. $\langle a, b, c\rangle =_{\mathrm{Df}} \langle a, \langle b, c\rangle\rangle$. $\langle a_1, \ldots, a_n\rangle =_{\mathrm{Df}} \langle a_1, \langle a_2, \langle a_3, \ldots \ldots, \langle a_{n-1}, a_n\rangle\rangle \ldots\rangle\rangle$ [3]). It follows immediately from (*i*) that the notion of an ordered n-tuple has the property corresponding to (*i*), namely that if $\langle a_1, \ldots, a_n\rangle = \langle b_1, \ldots, b_n\rangle$ then $a_i = b_i$ for $1 \leq i \leq n$.

3.3. **The Axioms of Union and Power-Set.** To obtain sets of more than two members we have to look for another procedure. The operation of union is

[1]) Cf. Rieger 57.

[2]) Bourbaki 54 (and, to some extent, also von Neumann 25) introduces the pairing operation as a primitive notion and assumes (*i*) as an axiom.

[3]) The use of natural numbers in this definition is only superficial. To define the notion of an ordered n-tuple for a particular n, say 17, it is enough to have 17 different variables. For a different definition of ordered n-tuples see Skolem 57.

one of the simplest set-theoretical operations. According to our program of not introducing "too large" sets, the sets whose union is to be formed will not be taken arbitrarily – they must be members of a single given set. Thus we have

AXIOM (III) OF UNION (or *Sum-set*). **For any set a there exists the set whose members are just the members of the members of a.**

This set is called the *union* of the members of a, or the *union-set* (or *sum-set*) of a; it is denoted by '$\bigcup a$'. Hence, $x \in \bigcup a$ holds if and only if there is a $z \in a$ (at least one z) such that $x \in z$.

In symbols: $\forall a \exists y \forall x [x \in y \leftrightarrow \exists z (x \in z \wedge z \in a)]$.

Roughly speaking, if the set a contains the members $t, u, v, \ldots$ then just the members of $t, u, v, \ldots$ are contained in $\bigcup a$. Therefore, we shall sometimes denote the union-set of a by $t \cup u \cup v \cup \ldots$ [1]), where the order of the terms is insignificant.

If a and b are sets, their union $a \cup b$ certainly exists. For by the axiom of pairing the pair $\{a, b\} = p$ exists, and by the axiom of union also $\bigcup p = a \cup b$ exists.

The axioms of pairing and union, taken together, enable us to "construct" sets of various kinds. For instance, using the notation introduced in §2, we know what the expressions '$\{a,b,c,d\}$' and '$\{a_1, \ldots, a_n\}$' denote – $\{a,b,c,d\}$ is the set, if there exists such, whose members are exactly *a, b, c,* and *d,* and similarly for '$\{a_1, \ldots, a_n\}$'; by means of the axioms of pairing and union we can prove the existence of these sets. By the axiom of pairing, the set $\{\{a,b\},\{c,d\}\}$ exists, hence, by the axiom of union, $\bigcup\{\{a,b\},\{c,d\}\} = \{a,b,c,d\}$ exists. The existence of $\{a_1, \ldots, a_n\}$ is shown similarly; the number of applications of the axioms in this proof depends, naturally, on n.

$\bigcup\{a, b, c\}$ is also denoted, as mentioned above, by $a \cup b \cup c$. Precisely as in *T*, p. 85, we can prove that the sets $a \cup b \cup c$, $(a \cup b) \cup c$ and $a \cup (b \cup c)$ contain the same members and are, hence, equal. The general associativity of the union operation can be shown in a similar way.

$a \cup b \cup c$ contains the members of three sets a, b, c. Proceeding further in the same or in a similar way – provided we have at our disposal more than three different sets – we may obtain more and more comprehensive sets. But in spite of the considerable strength of the axiom of union, a glance at Cantor's theory shows that the axioms of pairing and of union do not give us sufficient liberty in forming new sets, even if we make fairly strong assumptions about the existence of initial sets. In fact, let us assume that there exist

[1]) Some authors write $b + c$ for $b \cup c$ and Σa for $\bigcup a$ and call it, accordingly, the *sum* of b and c and the *sum-set* of a, respectively.

infinite [1]) sets of the kind called denumerable, and even denumerably many different such sets. Not even with this assumption would the axioms of pairing and union be strong enough to guarantee the existence of a more-than-denumerable set; for instance, the existence of a continuum [2]).

Cantor's (second and principal) tool for reaching sets with a higher cardinality was (transfinite) multiplication, in particular exponentiation. We shall see that for this purpose the power-set (*T*, Theorem 2 on p. 70, the so-called 'Cantor's Theorem', and Theorem 2 on p. 112) is sufficient; hence we formulate:

AXIOM (IV) OF POWER-SET. **For any set a there exists the set whose members are just all the subsets of a.**

In symbols: $\forall a \exists y \forall x (x \in y \leftrightarrow x \subseteq a)$.

The set of all subsets of a is called the *power-set* of a and is denoted by $\mathbf{P}a$ [3]). $x \in \mathbf{P}a$ is true if and only if $x \subseteq a$, i.e. if $\forall z (z \in x \to z \in a)$.

3.4. **The Axiom Schema of Subsets.** In the axiomatic system, Axiom IV fulfils a decisive task, for without it we are not able to form comprehensive enough sets. Yet Axiom IV cannot be utilized for this purpose at the present stage of our axiomatization. For Axiom IV permits only the use of those subsets *whose existence has previously been established.* Now Definition I (p. 26) does not enable us to *form* subsets of a given set s but merely, given a certain set, to ascertain whether it is a subset of s, and the axioms of pairing and union allow the construction of very special subsets only. Hence Axiom IV by itself is not a tool anywise comparable to Cantor's power-set. To use Axiom IV as an instrument for obtaining comprehensive sets another axiom is needed, apt to yield subsets of a given set in a general way.

To realize this we start with a set S which contains at least two members s_1 and s_2; by the axiom of pairing the pair $p = \{s_1, s_2\}$ exists and is, by Definition I, a subset of S. Hence the power-set $\mathbf{P}S$ at any rate contains the member p. If S contains more than two members we may do the same with any two members of S, thus obtaining new members of $\mathbf{P}S$. By means of the axiom of union we obtain more general subsets of S, but only such as are

[1]) 'Infinite' and 'denumerable' are formally defined only in § 3.6. We assume that the reader has an informal knowledge of these notions; we use them here only for a heuristic argument.

[2]) Cf. Bernays 37–54 VI, where it is shown, in essence, that Axioms I–III, and V–IX do indeed hold in a "model" which consists only of finite and denumerable sets.

[3]) The notation for the power-set is by no means uniform in the literature. In Fraenkel–Bar-Hillel 58 and in *T* the power set of a was denoted by $\mathbf{C}a$; $\mathbf{P}$ was used there to denote the Cartesian (outer) product.

finite in the naive sense. For instance, if s_1, s_2, s_3 are members of S, then, as we saw above, the set $\{s_1, s_2, s_3\}$ exists; it is obviously a subset of S, i.e., a member of **PS**.

Yet by such methods we fail to obtain infinite subsets of an infinite set S (other than S itself). If S contains all natural numbers we cannot even guarantee, at the present stage, the existence of the subset of all numbers greater than 1, let alone the subset of all even, or prime, numbers. Therefore, as yet we are not able to prove that **PS** has a greater cardinality than S.

What we want may very loosely be described as follows. The axioms of pairing, of union, and of power-set have an *expansive* function inasmuch as they yield the existence of sets which, when compared to the sets to which the axioms were applied, turn out, in general, to contain something new. Now we are in need of a *restrictive* operation in order to obtain sets whose extent is less than that of the given set; viz., subsets of it. Therefore we add the

AXIOM SCHEMA (V) OF SUBSETS (or *Separation*) [1]). **For any set a and any condition $\mathfrak{P}(x)$ on x there exists the set that contains just those members x of a which fulfil the condition $\mathfrak{P}(x)$.**

This set, which is, clearly, a subset of a, is denoted by $a_{\mathfrak{P}}$. In symbols axiom schema V reads

$$\forall z_1 \ldots \forall z_n \forall a \exists y \forall x [x \in y \leftrightarrow x \in a \wedge \mathfrak{P}(x)] \ ,$$

where $z_1, \ldots, z_n$ are the free variables of $\mathfrak{P}(x)$ other than x, and y is not a free variable of $\mathfrak{P}(x)$ [2]).

Axiom V plays, in many respects, a central role in ZF. This axiom schema (together with the axiom schema of replacement below) contains whatever is left of the general comprehension axiom in ZF, as distinguished from the particular cases of Axioms II–IV. Evidently, Axiom V admits general comprehension only for members x *of a given set*.

The history of the axiom of subsets is quite interesting and we shall review it here briefly.

In 1908 Zermelo formulated this axiom as follows (translated from German).

If the statement $\mathfrak{E}(x)$ is *definite* for all members of a set M, then M has always a

[1]) Zermelo's term is *Axiom der Aussonderung* (axiom of 'separating', 'singling out', 'sifting', or 'selecting', viz. selecting those members of a which fulfil the condition $\mathfrak{P}(x)$.

[2]) Axiom schema V can be weakened by admitting only conditions $\mathfrak{P}(x)$ with no parameters. From the weakened form of Axiom V one cannot prove directly the existence of the intersection of two sets (see Theorem 4 below). However, the weakened form of Axiom V in conjunction with Axioms II–IV implies the full Axiom V.

subset $M_{\mathfrak{C}}$ which contains those members of M for which $\mathfrak{C}(x)$ is true, and only those members.

This sounds more similar to Axiom V than it really is. The difference between Zermelo's axiom of subsets and Axiom V is that the notion of a 'condition $\mathfrak{P}(x)$ on x' in Axiom V is a well-defined notion, since in the beginning of § 2 we described explicitly what our object language is and at the end of § 1 we said that a condition $\mathfrak{P}(x)$ on x is an open statement (of our object language) in which the variable x is free; on the other hand, Zermelo did not have any particular object language in mind and therefore his notion of a statement $\mathfrak{C}(x)$ was quite vague.

This vagueness of Zermelo's notion of a statement threatened his system with appearence of antinomies of the semantical type. To avoid these antinomies Zermelo included in his formulation of the axiom of subsets the requirement that the statement $\mathfrak{C}(x)$ should be *definite* for all members of M. The concept of definiteness was explained by Zermelo as follows.

A question or statement $\mathfrak{C}$ is said to be 'definite' if the primitive relations of the system [1]), by means of the axioms and the general laws of logic, determine without ambiguity whether $\mathfrak{C}$ is true or not. Likewise, a statement $\mathfrak{C}(x)$ whose variable x ranges over all members of a class $\mathfrak{K}$ is said to be definite if this statement is definite for each member x of the class $\mathfrak{K}$. For instance, the question whether $a \in b$, as well as whether $M \subseteq N$, is always definite.

What Zermelo meant by saying that the truth or falsity of $\mathfrak{C}$ is determined by the primitive relations of the system is not that there is a procedure which leads to a decision whether $\mathfrak{C}$ is true or false in a finite number of steps, but that once the primitive relation of the system (namely, the membership relation) is "given" then the very meaning of $\mathfrak{C}$ makes it either true or false. Using modern terminology we can say that $\mathfrak{C}$ is definite if it belongs to a formal system with an interpretation which makes $\mathfrak{C}$ true or false; likewise, $\mathfrak{C}(x)$ is definite for a class $\mathfrak{K}$ of objects of the system if $\mathfrak{C}(x)$ belongs to an interpreted formal system which makes $\mathfrak{C}(x)$ true or false for every member of the class $\mathfrak{K}$ [2]).

Zermelo's vague notion of a definite statement did not live up to the standard of rigor customary in mathematics. This would be considered a serious shortcoming in the case of any axiomatic theory, let alone an axiomatic system of set theory, which is always viewed more suspiciously because of the antinomy ridden past of set theory. In 1921/22, independently and almost simultaneously, two different methods [3]) were offered for replacing in the axiom of subsets the vague notion of a definite statement by a well-defined, and therefore much more restricted, notion of a statement.

The first method, proposed by Fraenkel, uses a certain notion of function [4]) which is defined by the operations of Axioms II–V. Only statements of the form $f(x) \in g(x)$

[1]) What Zermelo meant by 'system' is a domain of objects on which the binary membership relation is defined. If the system satisfies the axioms it is called 'set theory'.

[2]) This explanation is similar to, but not identical with, Zermelo's explanation of the notion of definiteness in Zermelo 29, p. 341.

[3]) Fraenkel 22a and 25; Skolem 23 and 29 (§ 2).

[4]) Cf. also Fraenkel 27 (pp. 103–115) and the important supplement given in von Neumann 28a.

and $f(x) \notin g(x)$, where f and g are functions, are allowed in the axiom of subsets. This method seems to be more special than that of Skolem (which will be discussed immediately) under which it may be subsumed, but it is sufficient for the purpose of developing general set theory [1]).

The second method, proposed by Skolem and, by now, universally accepted because of its simplicity and generality, is the method adopted in our Axiom V, where the notion of a definite statement is replaced by that of a condition on x, i.e., a well-formed formula of the first-order predicate calculus with the free variable x, built up from atomic $\in$-statements.

Zermelo, while later admitting the need of formalizing his loose notion of definiteness, rejected both methods just described [2]), in particular because in his view they implicitly involve the notion of finite cardinal (natural number) which should be based on set theory. Therefore, within the frame of his axiomatic system he introduces a special axiomatization of the notion of definiteness. His system became thus somewhat similar to that of Skolem, but it has certain serious shortcomings which render it undesirable.

Axiom V has the awkward property of being *impredicative*. (A definition of a set is called impredicative if it contains a reference to a totality to which the set itself belongs. One may also say that a definition written in symbols is impredicative if it defines an object which is one of the values of a bound variable occurring in the defining expression.) The significance and the riskiness of impredicative definitions and procedures in mathematics, as well as various attempts made since Poincaré and Russell to eliminate them or to render them harmless, will be discussed in Chapters III and V. Here just the special case of impredicativity involved in Axiom V shall be exhibited. Whenever the condition $\mathfrak{P}(x)$, used in Axiom V to produce the subset $a_{\mathfrak{P}}$ of a given set a, essentially refers to the power-set $\mathbf{P}a$ or to all sets, a *particular* subset of a is determined by the totality of *all* subsets of a, or even by the totality of all sets – which is just the procedure against which Russell's *vicious circle principle* was directed. Naturally a Platonistic attitude would judge this situation quite differently than a constructive attitude [3]).

Axiom V is distinguished from the preceding axioms in that it is an axiom schema, i.e., it consists of infinitely many instances. A very natural problem is the question whether Axiom V can be replaced by a finite number of (single) axioms. The answer to this question is negative [4]), and the reason for

[1]) See Fraenkel 25 and – for the theory of ordered and well-ordered sets, not covered by Zermelo – Fraenkel 26 and 32. Special existence theorems, e.g., those concerning ordinal numbers, need the supplement provided by von Neumann 28a.

[2]) Zermelo 29. Cf. the (justified) criticism of this essay in Skolem 30.

[3]) Cf., for instance, Scholz 50; also Bernays 35.

[4]) Proved by Mostowski (cf. Montague 57); for improved results see Montague 61 and Kreisel–Levy 68. The latter paper proves that Axiom V is not implied even by an

this negative answer can be attributed to the existence of instances of Axiom schema V of an "arbitrarily high degree of impredicativity". On the other hand, the system of von Neumann–Bernays discussed in §7, as well as that of Quine's *New Foundations* (see Chapter III) [1]), can be presented by means of a finite number of single axioms.

Finally we draw a few simple conclusions from Axioms I–V. We first define:

DEFINITION. A set n which contains no member (i.e. for which $\neg\exists x(x \in n)$) shall be called a *null-set*.

THEOREM 3. There exists just one null-set.

Proof. Take for $\mathfrak{P}(x)$ in Axiom V a self-contradictory condition on x, for instance '$x \neq x$'. Then for any a [2]) we obtain a subset $y = a_{\mathfrak{P}}$ which contains no member, i.e. a null-set, and its uniqueness follows from extensionality.

The null-set will be denoted by 'O' [3]). According to Definition I on p. 26 O is a subset of every set.

THEOREM 4. For every two sets a and b there exists the set of the members that belong to both a and b. More generally, for every non-empty set t there exists the set of the members common to *all* members of t.

These sets are called the *intersection* (or meet) of *a and b*, in symbols $a \cap b$ [4]), and the *intersection of the members of t*, in symbols $\bigcap t$. (If there is no x common to all members of t we have $\bigcap t = \mathrm{O}$).

Proof. $a \cap b$ may be defined as that subset of a which by Axiom V corresponds to the condition $x \in b$. As to $\bigcap t$, since we assumed that t is non-empty

infinite consistent set Γ of axioms, if the number of quantifiers in the axioms of Γ is bounded. Nevertheless, by adding new symbols to the language any axiom schema can be implied by a finite set of closed formulas (Kleene 52a). For Axiom V this was done in a special way that is much simpler than the general way of Kleene, by von Neumann and Bernays (§7.2). Mostowski 55 (p. 20) gives a weaker form of Axiom V, due to Tarski, which makes the system finitizable.

[1]) See the proof in Hailperin 44. The result is rather surprising since Quine's original comprehension schema is impredicative.

[2]) That there exists a set (or an object) at all does not follow directly from Axioms I–V, which only assert, at most, that if some set, or sets, exist then some other set exists. The existence of at least one set is a tacit assumption made at the point where it was decided to base set theory on the first-order predicate calculus and let the variables range over the sets. The first-order predicate calculus assumes non-emptiness of the range of values of the variables (the "universe of discourse") by allowing to infer "there is an x such that..." from "for all x...". In addition, Axiom VI (of infinity) below asserts explicitly that some set exists.

[3]) Some authors use Λ, $\emptyset$ or 0.

[4]) Some authors use $a \cdot b$.

it has at least one member; let c be some member of t. Let $\mathfrak{P}(x)$ be the condition "x is contained in *each* member of t". By Axiom V the subset $c_{\mathfrak{P}}$ of c exists and its members are, obviously, just those common to all members of t, i.e., $c_{\mathfrak{P}} = \bigcap t$.

As to the outer product of pairwise disjoint sets, we have

THEOREM 5. For every set t there exists the set whose members are just those sets which contain a single member from each member of t.

If the set t is disjointed then the set whose existence was just claimed is called the *outer product* [1]) of the members of t and is denoted by Πt [2]). If t contains the member O we have $\Pi t = \mathrm{O}$.

Proof. Since the members of the desired set are certain subsets of the union $\bigcup t$ we start from the power-set of $\bigcup t$, i.e. from $\mathbf{P}\bigcup t = T$, which exists by the axioms of union and power-set. Let the condition $\mathfrak{P}(x)$ be "for *each* $s \in t$ the intersection $s \cap x$ is a singleton". Then the set $T_{\mathfrak{P}} \subseteq T$ exists by Axiom V; its members are those subsets of $\bigcup t$ which contain just one member from each member of t.

The remark regarding $\mathrm{O} \in t$ is self-explanatory; indeed, since O contains no member there is no set having a common member with O.

Thus, of the three operations on sets introduced in §6 of *Theory* – union, intersection and outer product – the performability of the first has been *postulated* by the axiom of union while the performability of the two others within ZF has been *proved* (Theorems 4 and 5) by means of the axioms of power-set and subsets, which were required anyway. (Theorem 5 also enables us to show the existence of the insertion set; see *T*, pp. 111–112.)

Once Axiom V is introduced, Axioms II–IV can be replaced, as easily seen, by the following weaker axioms, respectively: For any two elements a and b there exists a set y which contains a and b (and, possibly, additional members); for any set a there exists a set which contains all the members of the members of a; for any set a there exists a set which contains all the subsets of a.

Let us now explore some consequences of Axiom V which we may call *negative* consequences, at least as far as set existence goes.

THEOREM 6. There is no set which contains all sets. Furthermore, given any set a there is no set which contains all sets which are not members of a

[1]) The term "Cartesian product", used for the outer product in Fraenkel–Bar-Hillel 58 and in *T*, pp. 88–89, has been withdrawn because it is nowadays used for a somewhat different operation (see, e.g., § 3.5).

[2]) It was denoted in Fraenkel–Bar-Hillel 58 and in *T* by $\mathbf{P}a$; we now use $\mathbf{P}$ for the power-set operation.

(in particular, there is no set which is the *complement* of a).

Proof. Assume that v is a set which contains all sets. Let $\mathfrak{P}(x)$ be the condition $x \notin x$. By Axiom V there exists the set $v_{\mathfrak{P}}$, i.e., the set of all sets which are not members of themselves, which contradicts Theorem 2. Let a be any set, and assume that there is a set b which contains all sets which are not members of a. Then, as we saw above, by the axioms of pairing and union, $a \cup b$ is also a set. But $a \cup b$ is a set which contains every set, contradicting the first part of our theorem.

3.5. **Relations, Order, Functions.** Even though Axioms I–V do not suffice for the full development of set theory, as we shall see further on, they are all that is needed for the reduction of various notions of mathematics and set theory to the notion of set, and for establishing the elementary properties of these notions.

In §3.2 such a reduction has already been carried out for the notions of an ordered pair and an ordered n-tuple, in general. We now define:

DEFINITION. The *Cartesian product* $S \times T$ of the sets S and T is the set which consists of all ordered pairs $\langle x, y \rangle$, where $x \in S$ and $y \in T$ [1]).

The existence of $S \times T$ is proved as follows. If $x \in S$ and $y \in T$ then $x, y \in S \cup T$, $\{x\}, \{x, y\} \in \mathbf{P}(S \cup T)$, $\langle x, y \rangle = \{\{x\}, \{x, y\}\} \in \mathbf{PP}(S \cup T)$. Let $\mathfrak{P}(z)$ be the condition on z given by "z is an ordered pair $\langle x, y \rangle$ such that $x \in S$ and $y \in T$". $(\mathbf{PP}(S \cup T))_{\mathfrak{P}}$ is, by Axiom V, the required set $S \times T$.

Mathematicians are sometimes in need of "several different copies of the same set T". This seems to be forbidden by the axiom of extensionality, yet the notion of an ordered pair and the operation of Cartesian product give us a way of getting around the restriction imposed by this axiom. For every $s \in S$ the subset $\{s\} \times T$ of $S \times T$ (which consists of all ordered pairs $\langle s, y \rangle$, where $y \in T$) can be considered as a copy of T with the member $y \in T$ being represented by the pair $\langle s, y \rangle \in \{s\} \times T$. Obviously, the representatives $\langle s, y_1 \rangle$, $\langle s, y_2 \rangle$ of different members y_1, y_2 of T are different. Also if s_1, s_2 are two different members of S, then the "copies" $\{s_1\} \times T$ and $\{s_2\} \times T$ of T are disjoint.

The next notion with which we shall deal is the notion of a *binary relation* (henceforth just *relation*). A general relation cannot be viewed as an element for the same reason that a general collection cannot be viewed as an element, namely, the occurrence of antinomies [2]). Only those relations $\mathfrak{R}$ for which

[1]) This operation is similar to, but not identical with the operation of Cartesian product of T, which is called here 'outer product'.

[2]) Russell's antinomy can be trivially adapted to relations as follows. Let r be the

there is a set which consists of all x's and y's such that x stands in the relation $\mathfrak{R}$ to y (this set is called the *field* of the relation $\mathfrak{R}$) will be represented by sets, namely, by the set of all pairs $\langle x, y\rangle$ such that x and y are in the relation $\mathfrak{R}$[1]). Accordingly we define

DEFINITION. A set r is said to be a *relation* if all its members are ordered pairs.

We shall still speak also of relations which are not sets, or are not represented by sets. General relations are given in ZF by a condition $\mathfrak{P}$ on two variables, such as a condition $\mathfrak{P}(x, y)$ on x and y. E.g., the $\in$-relation is given by the condition '$x \in y$' on x and y. Whenever we speak of a relation which is a set, as in the definition above, we shall denote it with a lower-case letter, as in 'the relation r'.

Corresponding to our informal statement that if the field of a relation is a set then the relation can be represented by the set of all ordered pairs $\langle x, y\rangle$ such that x and y are in the relation, we have

THEOREM (-SCHEMA). For any condition $\mathfrak{P}(x, y)$ on x and y, if there is a set z which contains all elements x and all elements y such that $\mathfrak{P}(x, y)$ holds, then there exists a set u which consists of all pairs $\langle x, y\rangle$ such that $\mathfrak{P}(x, y)$ holds.

Proof. If z is a set as assumed, then every pair $\langle x, y\rangle$ for which $\mathfrak{P}(x, y)$ holds is a member of $z \times z$. Let $\mathfrak{Q}(t)$ be the condition on t given by 't is an ordered pair $\langle x, y\rangle$ such that $\mathfrak{P}(x, y)$ holds'. $(z \times z)_{\mathfrak{Q}}$ is the required set.

DEFINITION. The *domain* of the relation r is the set u of all elements x for which there is a y such that $\langle x, y\rangle \in r$[2]). The *range* of the relation r is the set v of all elements y for which there is an element x such that $\langle x, y\rangle \in r$. The *field* of the relation r is the union of its domain and its range. (The existence of such sets u and v follows easily from the axioms of union and subsets.)

DEFINITION. We say that a relation r is *on the set* a if r is a subset of $a \times a$, i.e., if the field of r is a subset of a.

We shall now see how the notions of order and an ordered set can be reduced to the notion of set.

DEFINITION. r is said to be an *order* (*ordering relation*) on the set a if r is a relation on a and the following (a)–(c) hold[3]).

relation which holds between x and y just in case x and y are relations and x does not stand in the relation x to y. As in Russell's antinomy we obtain that r stands in the relation r to itself just in case it does not stand in the relation r to itself (Whitehead–Russell 10–13 I, Ch. II, VIII).

[1]) Ternary relations, quaternary relations, etc. are treated similarly.

[2]) Several authors refer to our domain as range and to our range as domain.

[3]) This way of representing order by a set is due to Hausdorff 14, pp. 70 f. A differ-

(a). If $\langle x,y\rangle \in r$ and $\langle y,z\rangle \in r$ then $\langle x,z\rangle \in r$ (transitivity).
(b). For no x does $\langle x,x\rangle \in r$ (irreflexivity).
(c). For all different x and y in a, either $\langle x,y\rangle \in r$ or $\langle y,x\rangle \in r$ (comparability).
r is said to be a *well-ordering* of a if it is an order on a and
(d). Every non-void subset b of a (i.e., $0 \neq b \subseteq a$) has a first member x in the order r, i.e., for every member y of b other than x we have $\langle x,y\rangle \in r$. An *ordered set* is an ordered pair $\langle a,r\rangle$ in which r is an ordering relation on a [1]). A *well-ordered set* is an ordered set $\langle a,r\rangle$ in which r is a well-ordering of a.

One of the most fundamental notions in mathematics is that of a function [2]). A function F is, roughly, a rule which correlates with each element x out of some collection a single element [3]) denoted by $F(x)$. The function F can also be viewed as a binary relation $\mathfrak{R}$ which holds between x and y just in case y is $F(x)$. The characteristic property of such relations $\mathfrak{R}$ is that for every element x there is at most one element y which stands in the relation $\mathfrak{R}$ to x. (We said 'at most one element' rather than 'exactly one element' since the function is not necessarily supposed to be "defined" for all elements x.)

DEFINITION. A set f is said to be a *function* if f is a relation and for every x in the domain of f there is exactly one y such that $\langle x,y\rangle \in f$; this y is denoted by $f(x)$. A function f is said to be a *one-one* function if for any two different members x and z of its domain the values $f(x)$ and $f(z)$ are different too.

As in the case of the relations we shall still speak of functions which are not sets [4]). General functions are given in ZF by conditions $\mathfrak{P}(x,y)$ on two

ent way of representing an ordering relation by the set of all the "initials", was initiated by Hessenberg 06 and completed by Kuratowski 21 (for details see Fraenkel–Bar-Hillel 58, pp. 127–131). The present method is preferable because of its generality (it applies to all relations on a set, not just ordering relations) and its simplicity.

[1]) It is quite common to refer to this set as 'the ordered set a'; Notice that if c and s are two different orders on the set a then $\langle a,c\rangle$ and $\langle a,s\rangle$ are two different ordered sets and both are referred to as 'the ordered set a'. Yet there is nothing wrong with this way of speaking as long as it is clear which ordering relation r one has in mind.

[2]) von Neumann 25 and 28 even developed set theory with function, rather than set, as its basic notion, thus making it the basic notion of mathematics as a whole. (For a simpler development of the same idea, see R.M. Robinson 37.) For a newer attempt to base set theory on a generalized notion of function (abstract categories) see Lawvere 64 and 66, and other papers by the same author.

[3]) We shall not use the term 'function' for the so-called many-valued functions, which are just relations.

[4]) An example very similar to that of footnote 2 in p. 41 shows that not all functions can be regarded as elements.

variables such that for every element x there exists at most one element y for which $\mathfrak{P}(x,y)$ holds. Such conditions will be referred to as *functional conditions*. Whenever we speak of a function which is a set, as in the definition above, we shall denote it with a lower-case letter, as in 'the function f'.

The notion of equinumerosity of sets is now defined as follows.

DEFINITION. A set S is said to be *equinumerous* (equivalent) to a set T if there is a one-one function f whose domain is S and whose range is T. Such a function f is said also to be a *one-one mapping of S on T*.

We saw that various mathematical notions discussed in the present subsection were successfully reduced to the notion of set. That general relations and functions could not be reduced to sets can by no means be regarded as a failure; this is prevented by exactly the same reason that prevents the general collection from being a set, namely the antinomies. We can still claim success in the treatment given here to the notions of relation and function because in our axiomatic framework we can deal with relations and functions without introducing these notions as primitive notions. They are handled either as sets, according to the definitions above, or, in the general case, as conditions on two variables.

Looking back we cannot but wonder that the simple notion of the purely extensional set turned out to be so powerful as to encompass so many mathematical notions which are, at least at first sight, much more complicated than the notion of set. We shall also see later (in §5) how the notions of cardinal number, ordinal number and order type can be reduced to the notion of set.

3.6. **The Axiom of Infinity**. Axioms I–V (even with Axioms VII–IX added) *do not enable us to prove the existence of an infinite set*. Let us say that a set a is *hereditarily finite* if it is finite, its members are finite, the members of its members are finite, etc. (or, in other words, if the sets $a, \bigcup a, \bigcup\bigcup a$, ... are all finite). We notice that each one of Axioms II–V yields hereditarily finite sets when applied to such sets. Therefore, although we can prove by means of Axioms I–V the existence of infinitely many sets, e.g., $0, \{0\}, \{\{0\}\}$, ..., we cannot prove by means of these axioms the existence of a set which is not hereditarily finite [1]).

As long as we are interested only in elementary arithmetic and in finite sets, Axioms I–V are enough (or even just Axioms I, V and the axiom 'for

[1]) The arguments given here are an informal version of the proof of Bernays 37–54 VI that the axiom of infinity is independent of the other axioms of set theory.

every *a* and *b* there exists a set *c* which contains exactly all the members of *a* and *b* itself – i.e., $a \cup \{b\}$') [1]. Dealing with natural numbers without having the set of all natural numbers does not cause more inconvenience than, say, dealing with sets without having the set of all sets. Also the arithmetic of the rational numbers can be developed in this framework [2]. However, if one is already interested in analysis then infinite sets are indispensable since even the notion of a real number cannot be developed by means of finite sets only. Hence we have to add an existence axiom that guarantees the existence of an infinite set; most simply, of a denumerable set.

Before we proceed let us define the terms 'finite', 'infinite', 'denumerable' and 'reflexive', some of which we have already used. Since we shall not use these notions in the principal versions of the axioms, we shall not give a rigorous definition of these notion in ZF, but we shall remind the reader of those definitions of these notions in informal set theory which can easily be adapted to axiomatic set theory.

A set *a* is called *finite* if there exists a natural number *n* such that *a* is equinumerous with the set $\{0, 1, ..., n-1\}$ of all natural numbers which are smaller than *n*, i.e., the members of *a* can be put into a one-one correspondence with the natural numbers less than *n*. (As mentioned above, the notion of a natural number can be developed on the basis of Axioms I–V) [3]. *a* is called *infinite* if it is not finite. A set *a* is called *denumerable* if it is equinumerous with the set of all natural numbers. (We still do not know whether there is a set which is the set of all natural numbers.) A set is called *reflexive* if it is equinumerous with a proper subset of itself [4]. By means of Axioms I–V one can prove the following statements [5]. Assuming that the set of all natural numbers exists, a set is reflexive if and only if it has a denumerable subset

[1]) Cf. Zermelo 09, Bernays 37–54 II (or Suppes 60), Quine 63. However, as a consequence of the method of Ackermann 37, we shall not be able to prove without axiom VI the theorems of arithmetic which are proved in analytic number theory but cannot be proved by elementary arithmetical proofs – for the existence of many such theorems see Kreisel–Levy 68.

[2]) Quine 63, p. 119.

[3]) Several definitions of the notion of a finite set, equivalent to the present one but not using natural numbers, were proposed, among others, by Zermelo, Russell, Sierpiński, Kuratowski, and Tarski. A complete survey is given in Tarski 25. Suppes 60, § 4.2 develops the theory of finite sets within axiomatic set theory.

[4]) This was introduced as the definition of the notion of infinite set by Peirce 33 (pp. 210–249, 360 – cf. Keyser 41) and Dedekind 1888. Cf. also Bolzano 1851 (§ 20) and Cantor 1878.

[5]) *T*, § 2.5 and § 3.5. Cf. Suppes 60, § 5.3 (where the term 'Dedekind infinite' is used for 'reflexive').

(hence, trivially, every denumerable set is reflexive). Every reflexive set is infinite, in particular, all denumerable sets are infinite.

Dedekind, just like Bolzano [1]) four decades before, believed that he had proved the existence of infinite sets. However, not only are their methods incompatible with the restrictions of our axiomatic system but they are just those that lead to the logical antinomies (Chapter I). Dedekind, for instance, contemplated the set T whose members are all "objects of thinking" and proves its reflexiveness as follows. If t is any member of T, the thought 't is an object of thinking' is also a member of T. Hence the totality of all thoughts with the particular form 't is an object of thinking' defines a proper subset T_0 of T, and a one-to-one mapping between T and T_0 is formed by correlating with every $t \in T$ the member of T_0 that expresses 't is an object of thinking'; therefore T is a reflexive set. Yet the set T is of the type that involves antinomies. In fact, after the publication of Russell's antinomy Dedekind withdrew for some time his work – whose eminent importance, however, is independent of the mentioned argument.

From the axiomatic viewpoint there is no other way for securing infinite sets but *postulating* them [2]), and we shall express an appropriate axiom in several forms. While the first corresponds to Zermelo's original axiom of infinity, the second implicitly refers to von Neumann's method of introducing ordinal numbers [3]).

AXIOM OF INFINITY VIa [4]). **There exists at least one set Z with the following properties**

(*i*) $0 \in Z$

(*ii*) **if $x \in Z$, also $\{x\} \in Z$.**

AXIOM OF INFINITY VIb. **There exists at least one set Z_1 with the following properties**

(*iii*) $0 \in Z_1$

(*iv*) **if $x \in Z_1$, also $(x \cup \{x\}) \in Z_1$** [5]).

[1]) Dedekind 1888, §5; Bolzano 1851, §13; Russell 03, §339. Cf. the more elaborate examples given in O. Becker 27 (pp. 98 ff. of the book edition) and Scholz 28.

[2]) See, however, Bernays 61a, where the existence of infinite sets follows from an axiom schema which does not directly postulate their existence. A similar situation occurs within the set theory of Ackermann (§7.7).

[3]) Zermelo 08a, von Neumann 23 (for von Neumann's ordinals see §5.2).

[4]) Axioms VIa and VIb were called Axioms VII and VII*, respectively, in Fraenkel–Bar-Hillel 58.

[5]) A stronger axiom schema of infinity than VI is introduced in Fraenkel 27 (p. 114, Axiom VIIc). Fraenkel's axiom is equivalent, on the basis of axioms I–V, to the schema which asserts, roughly, that every "denumerable collection of elements" is a set (Bernays 37–54 III). Fraenkel's axiom is a direct generalization of VIa or VIb, in that instead of

AXIOM OF INFINITY VIc. **There exists at least one set Z_2 with the following properties**

(v) $O \in Z_2$

(vi) **if** $x \in Z_2$ **and** $y \in Z_2$, **also** $(x \cup \{y\}) \in Z_2$.

In symbols,

VIa. $\exists z[O \in z \wedge \forall x(x \in z \rightarrow \{x\} \in z)]$.

VIb. $\exists z[O \in z \wedge \forall x(x \in z \rightarrow x \cup \{x\} \in z)]$.

VIc. $\exists z[O \in z \wedge \forall x \forall y(x \in z \wedge y \in z \rightarrow x \cup \{y\} \in z)]$.

VIc implies VIa and VIb, since a set Z_2 as in VIc obviously satisfies the requirements for Z in VIa and for Z_1 in VIb. Axioms VIa and VIb have the advantage of assuming less than Axiom VIc but, on the other hand, the choice of the basic operation $\{x\}$ in VIa and $x \cup \{x\}$ in VIb is somewhat arbitrary, whereas, as it turns out, VIc asserts the existence of more "rounded off" sets [1]).

The "first" members of any Z satisfying VIa are evidently O, {O}, {{O}}, {{{O}}}, etc., and they are different from each other; for instance, $\{O\} \neq O$ because the former contains a member, viz. O, and the latter none. As to Z_1, $O \in Z_1$ implies $\{O\} \in Z_1$. Hence also $\{O\} \cup \{\{O\}\} = \{O, \{O\}\}$, and $\{O, \{O\}\} \cup \{\{O, \{O\}\}\} = \{O, \{O\}, \{O, \{O\}\}\}$, etc. are members of any Z_1 that satisfies VIb.

In drawing the next conclusions from the axiom of infinity we content ourselves with the form VIa; *mutatis mutandis* the results hold true for the form VIb as well.

By Axiom VIa there is a set Z which satisfies (i) and (ii). Let Z^* be the intersection of all sets Z' which satisfy (i) and (ii). To prove the existence of Z^* from our axioms we let $\mathfrak{P}(x)$ be the statement "x is a member of every set Z' which satisfies (i) and (ii)"; $Z_{\mathfrak{P}}$ is obviously the required set Z^*. Clearly, Z^* itself satisfies (i) and (ii); and since Z^* is obviously a subset of every set Z' which satisfies (i) and (ii), it is the least set with these properties.

Proceeding from an intuitive point of view, let us consider the set W which consists exactly of the sets O, {O}, {{O}}, {{{O}}}, ... W has indeed the properties (i) and (ii) of VIa and is included in any set Z which has these properties. The same observation was also made above concerning Z^*, therefore we have $W \subseteq Z^*$, $Z^* \subseteq W$ and, by extensionality, $Z^* = W$. Thus we can write

$$Z^* = \{O, \{O\}, \{\{O\}\}, \{\{\{O\}\}\}, \ldots\}.$$

the functions $\{x\}$ or $x \cup \{x\}$ of x, more general functions are admitted. Fraenkel's axiom follows from Axioms I–VII, but not from I–VI (since the system Π_2 of Bernays 37–54 VI satisfies Axioms I–VI, but does not satisfy Fraenkel's axiom).

[1]) See footnote 2 on the next page.

Z^* can be conceived as the set of all non-negative integers, for we may denote the null-set O by the numeral '0', {O} by '1', and generally $\{x\}$ by 'the successor of x' in the terminology of Peano's axioms for natural numbers [1]).

Starting from Axiom VIb instead of VIa we obtain instead of the set Z^* the set

$$Z_1^* = \{O, \{O\}, \{O, \{O\}\}, \{O, \{O\}, \{O, \{O\}\}\}, \ldots\}$$

which may as well be conceived as the set of all non-negative integers. In fact the members of Z_1^* are the finite ordinals of §5.2. Of any two different members of Z_1^*, one is both a member and a subset of the other, whereby a natural "order" is established in Z_1^* [2]).

Z^*, as well as any set Z which satisfies (*i*) and (*ii*) of VIa, is reflexive, and hence infinite, since by correlating each $x \in Z$ with $\{x\} \in Z$ we get a one-one mapping of Z onto a proper subset of itself which does not contain O. The same holds true for any set Z_1 as in VIb, in particular for Z_1^*, (and, a fortiori, for any set Z_2 as in VIc).

Hence versions VIa–VIc of the axiom of infinity only require an initial member O and a primitive function whereas the mapping needed is not postulated but constructed [3]).

One can produce a great many different versions of the axiom of infinity. Here, we shall mention only two more versions [4]):

VId.. *There exists a reflexive set.*

VIe. *There exists an infinite set.*

On the basis of Axioms I–V and the axiom of replacement, introduced below, all of Axioms VIa–VIe can be shown to be equivalent with each other [5]). Without the axiom of replacement one can show that Axioms VId

[1]) Peano's axioms consist of the following requirements concerning the set N of all natural numbers with the successor operation on it.

a) There is a particular number, called 0, which is not the successor of any number. b) Each number other than 0 is the successor of at most one number. c) (*Principle of mathematical induction*) Every subset of N which contains 0 and which contains with each number also its successor coincides with N.

[2]) If we go through the same construction, starting with Axiom VIc, we obtain a set Z_2^* which turns out to be the set $R(\omega)$ of §5.3. It follows easily from the axiom of foundation that this is exactly the set of all hereditarily finite sets.

[3]) This is essentially the method of Dedekind 1888.

[4]) Bernays 37–54 II, Bourbaki 56. Axiom V.1 of von Neumann 25 is equivalent to these on the basis of Axioms I–V.

[5]) Cf. any development of set theory which uses VIe, such as Bourbaki 56.

and VIe are equivalent [1]); and that neither of VIa and VIb does imply the other [2]). Since, as we saw above, VIc implies VIa and VIb, and each one of VIa, VIb implies VId and VIe, it follows that, on the basis of Axioms I–V alone, neither of VIa, VIb implies VIc, nor does either follow from VId or VIe. Axioms VId and VIe have the advantage of lack of arbitrariness and of assuming less than each of VIa–VIc. The disadvantages of VId and VIe are their reliance on the relatively complicated notions of finiteness and reflexivity, as opposed to the simple notions used in VIa–VIc, and the fact that the definition of the notion of a real number by means of Axioms I–V, VId (or VIe) is rather clumsy.

In our heuristic classification of the axioms, as to whether they are instances of the axiom schema of comprehension or not, the axiom of infinity can, to some extent, be viewed as such an instance. Each of VIa and VIb can be stated as "there exists a set which consists of all natural numbers", for the respective notion of natural number, and VIc can be stated similarly. Unlike the case of the axiom of infinity, the authors know no proof of Axioms VIII and IX from the axiom schema of comprehension which does not make an outright use of the idea behind Russell's antinomy or some similar antinomy.

3.7. **The Axiom Schema of Replacement.** The axiom of infinity, which in itself guarantees only the existence of denumerable sets, when added to Axioms I–V enables us to obtain more extensive sets, e.g., the sets $Z^{(2)}$, $Z^{(3)}$, ... where $Z^{(2)}$ is the power-set of the set Z^* above, and, for $k \geqslant 2$, $Z^{(k+1)} = \mathbf{P}Z^{(k)}$. Nevertheless, *Axioms I–VI are not sufficient to guarantee the existence of certain kinds of sets* whose counterparts in Cantor's theory have never been questioned. For example, as was mentioned above, one cannot prove from Axioms I–V, VIa the existence of Z_1^*. Presumably the simplest example of a set whose existence cannot be proved even by means of Axioms I–VIc is the denumerable set

$$A = \{Z^*, Z^{(2)}, Z^{(3)}, \ldots\}\ .$$

[1]) VId implies VIe since every reflexive set is infinite. To see that VIe implies VId we notice that if a is infinite then, even though one cannot prove, without using also the axiom of choice (VIII), that a is reflexive, one can still prove that $\mathbf{PP}a$ is (Tarski 25).

[2]) This is proved by methods similar to that used at the end of Bernays 37–54 VI. The result relies on the assumption that the axiomatic system which consists of Axioms I–V and any of Axioms VIa–VIe is consistent, i.e., free of contradiction. (If one such system is consistent any other such system is consistent too.)

The significance of the set A is emphasized by the fact that the union-set of A has a cardinal greater than the cardinal of any member of A (hence a cardinal $\geqslant \aleph_\omega$), while our previous axioms are not strong enough to yield a set of this cardinality [1]).

In addition to the insufficiency of our axioms with respect to particular (extensive) sets, the general method of definition by transfinite induction and, in particular, the proof that to every well-ordered set there is a corresponding ordinal number (§5.2) [2]) cannot be carried out on the basis of Axioms I–VI. The required supplement is the following *axiom schema of replacement.*

Before we present a formal version of the axiom of replacement let us give an intuitive account of it. As was asserted on p. 32, our guiding principle in admitting axioms of comprehension is to admit only axioms which assert the existence of sets which are not too "big" compared to sets already ascertained. If we are given a set a and a collection of sets which has no more members than a it seems to be within the scope of our guiding principle to admit that collection as a new set. We still did not say exactly what we mean by saying that the collection has "no more" members than the set a. It turns out that it is most convenient to assume that the collection has "no more" members than a when there is a "function" which correlates the members of a to all the sets of the collection in such a way that to each member of a corresponds a set in the collection and each set in the collection is correlated to one or more members of a. However, even though applications of the axiom of replacement do not seem to directly admit sets of larger cardinals than those already available, the combination of this axiom with Axioms I–VI is very powerful and enables us to prove the existence of sets with extremely large cardinals (whereas by means of Axioms I–VI alone we could not get even sets with the cardinal $\aleph_\omega$).

Let us use "$\mathfrak{P}(t, x)$ *is a functional condition on the set* a" as short for "for every set t which is a member of a there is *at most one* set x such that $\mathfrak{P}(t, x)$ holds".

AXIOM SCHEMA (VII) OF REPLACEMENT [3]) (or *Substitution*). **For any set a, if $\mathfrak{P}(t, x)$ is a functional condition on a then there exists a set**

[1]) Bernays 37–54 VI (take the system Π_2 with the generalized continuum hypothesis added).

[2]) von Neumann 23 (see *T*, pp. 181 ff, and Suppes 60, Ch. 7). In the system π_2 of Bernays 37–54 VI in which Axioms I–VI hold, but not Axiom VII, there are well-ordered sets of every denumerable order type, but the system does not contain even the ordinal $\omega \cdot 2$. (Hence, obviously, definition by transfinite induction fails in it.)

[3]) It was suggested first by Fraenkel 22 and, independently, by Skolem 23 (No. 4). Previous hints are to be found in Cantor 32 (p. 444) and Mirimanoff 17 (p. 49).

which contains exactly those elements x for which $\mathfrak{P}(t,x)$ holds for some $t \in a$.

In other words, if the domain of a function is a set, its range is also a set.

In symbols,

$$\forall z_1 \ldots \forall z_n \forall a [\forall u \forall v \forall w (u \in a \wedge \varphi(u,v) \wedge \varphi(u,w) \rightarrow v = w) \rightarrow$$

$$\exists y \forall x (x \in y \leftrightarrow \exists t (t \in a \wedge \varphi(t,x)))]$$

where u, v, w, y are not free in the formula $\varphi(t,x)$ and $z_1, \ldots, z_n$ are the free variables of $\varphi(t,x)$ other than t and x [1,2]).

In view of the explanation which preceded the formulation of Axiom VII it is clear why we made the requirement that $\mathfrak{P}(t,x)$ should be a *functional* condition on the set a. Indeed, if this requirement were to be dropped from Axiom VII contradiction would ensue; by applying Axiom VII to the condition $t \subseteq x$, which is obviously not a functional condition on any set a, and by taking for a the set $\{0\}$, one obtains the existence of the set of all sets x such that $0 \subseteq x$, which is just the set of all sets, contradicting Theorem 6 [3]).

In order to present Axiom VII as an axiom of comprehension we can also formulate it as

VII*. *For any set a there exists a set which contains exactly those sets x for which there is a $t \in a$ such that $\mathfrak{P}(t,x)$ holds, and for no $s \neq x$ does $\mathfrak{P}(t,s)$ hold,*

where $\mathfrak{P}(t,x)$ is any condition (not necessarily functional).

VII* obviously implies VII. To see that VII implies VII*, consider any condition $\mathfrak{P}(t,x)$. Let $\mathfrak{Q}(t,x)$ be the condition "$\mathfrak{P}(t,x)$ holds and for no $s \neq x$ does $\mathfrak{P}(t,s)$ hold".

[1]) $\varphi(u,v)$ denotes the formula obtained from $\varphi(t,x)$ by substituting u and v for all free occurrences of t and x, respectively, in $\varphi(t,x)$ (if u or v occurs in $\varphi(t,x)$ as a bound variable it should be replaced by another variable before substitution). $\varphi(t,w)$ is obtained similarly.

[2]) As in the case of the axiom of subsets, if we weaken Axiom VII by admitting only conditions $\mathfrak{P}(t,x)$ without parameters we obtain an axiom schema which still implies VII (by means of Axioms II–IV). Also, if we weaken Axiom VII by requiring $\mathfrak{P}(t,x)$ to be a *one-one* functional condition on a (i.e., $\mathfrak{P}(t,x)$ is a functional condition, and if, for $s, t \in a$, $\mathfrak{P}(s,x)$ and $\mathfrak{P}(t,x)$ hold then $s = t$), then the axiom schema thus obtained is equivalent to VII. To show that we proceed as follows. Let b be the set $\{a_t \mid t \in a\}$, where $a_t = \{s \mid s \in a \wedge \exists x(\mathfrak{P}(t,x) \wedge \mathfrak{P}(s,x))\}$, i.e., a_t is the set of all members s of a for which the functional condition $\mathfrak{P}$ yields the same value as for t. The existence of b follows from the axioms of power-set and subsets. VII is obtained by applying to the set b (in place of a) the functional condition $\mathfrak{Q}(u,x)$ given by "there is a $t \in u$ such that $\mathfrak{P}(t,x)$".

[3]) Suppes 60, §7.1.

$\mathfrak{Q}(t,x)$ is, obviously, a functional condition on any set a. Substituting $\mathfrak{Q}(t,x)$ for $\mathfrak{P}(t,x)$ in VII we get VII* (with $\mathfrak{P}(t,x)$).

Another form of the axiom schema of replacement which is very common in the literature is the following, in which the condition is required to be functional on the whole universe, not only on a:

$$\forall z_1 \dots \forall z_n [\forall u \forall v \forall w(\varphi(u,v) \wedge \varphi(u,w) \rightarrow v=w) \rightarrow$$

$$\forall a \exists y \forall x(x \in y \leftrightarrow \exists t(t \in a \wedge \varphi(t,x)))]$$

where u, v, w, y are not free in the formula $\varphi(t,x)$ and $z_1, \dots, z_n$ are the free variables of $\varphi(t,x)$ other than t and x. This version is easily shown to be equivalent to VII.

Axiom schema VII implies the axiom schema of subsets (directly) and the axiom of pairing (by means of the power-set axiom) [1]).

To prove the axiom schema of subsets we proceed as follows. Given any condition $\mathfrak{P}(x)$ we take for $\mathfrak{Q}(t,x)$ the condition "$t=x$ and $\mathfrak{P}(x)$", which is, obviously, a functional condition on any set a. Substituting $\mathfrak{Q}(t,x)$ for $\mathfrak{P}(t,x)$ in VII we get that there exists a set which contains exactly those sets x for which $\mathfrak{Q}(t,x)$ holds for some $t \in a$, i.e., there exists a set which contains exactly those sets x which are members of a and for which $\mathfrak{P}(x)$ holds.

To prove the axiom of pairing we proceed as follows. By Theorem 3 (p. 39) which uses only the axiom of subsets, which was just shown to follow from Axiom VII, the null-set O exists. By the power-set axiom there exists the set **PP**O$= \{\mathrm{O}, \{\mathrm{O}\}\}$. Let b and c be any sets. Let $\mathfrak{P}(t,x)$ be the condition "$t=\mathrm{O}$ and $x=b$ or $t=\{\mathrm{O}\}$ and $x=c$"; this is, obviously, a functional condition on any set a. Substituting $\{\mathrm{O},\{\mathrm{O}\}\}$ for a in VII and taking $\mathfrak{P}(t,x)$ as given here we get the existence of a set which contains just b and c.

Even though Axiom VII implies the axioms of pairing and subsets we shall not drop the latter axioms from our list. When one studies axiomatic set theory along the lines of the system ZF one encounters the axioms of pairing and subsets already in the beginning, where they are indispensable. Axiom VII is usually encountered much later, when more advanced topics are discussed. Also the axiomatic system which consists of Axioms I–VI (with, possibly, one or two of Axioms VIII and IX) is often encountered in the literature [2,3]).

A version of the axiom of replacement which is equivalent to VII on the basis of the axiom schema of subsets, but which does not imply the axiom schema of subsets, not even by means of Axioms I–IV, VI, VIII, and IX, is the following.

[1]) Zermelo 30.

[2]) Such a system is usually called *Zermelo's system* and is denoted by Z.

[3]) For a stronger version of Axiom VII which implies also some of the other axioms, see Bourbaki 54 (II, § 1, No. 6), Ono 57, A. Levy 60. For a weakening of Axiom VII which still retains much of the power of this axiom see Levy–Vaught 61.

For any set a, if $\mathfrak{P}(t, x)$ *is a functional condition on a then there exists a set which contains every set x for which there is a* $t \in a$ *such that* $\mathfrak{P}(t, x)$ *holds* [1]).
The only difference between this and VII is that the set whose existence is claimed is now allowed to contain members x other than those for which there exists a $t \in a$ such that $\mathfrak{P}(t, x)$ holds. (Compare the alternative formulations of Axioms II–IV on p. 40).

Axiom VII is in a certain sense also an axiom of infinity since together with Axioms I–VI it guarantees the existence of not just infinite but even extremely comprehensive sets [2]). However, once the axiom of infinity is left out, Axiom VII no longer implies the existence of any infinite set, not even by means of Axioms I–V, VIII and IX [3]).

One may ask here, as we did in the case of Axiom schema V, if, in the presence of Axioms I–VI (and possibly also VIII and IX), Axiom schema VII can be replaced by a finite number of single axioms. The answer is, again, negative [4]).

§4. THE AXIOM OF CHOICE

4.1. **Formulation of the Axiom. Its Introduction into Mathematics.** After having obtained, by means of the axiom of subsets, those subsets of a given set which are determined by a definite condition we raise the question whether possibly *other subsets*, not obtained in this way, may be conceived and admitted; and if so, how far such subsets are necessary for developing set theory. The present section deals with an axiom which yields such subsets.

We start from a disjointed set t. According to Theorem 5 on p. 40 the outer product Πt exists and its members, if any, are those subsets of $\mathsf{U}t$ whose intersections with each member of t are singletons. We shall call the members of Πt *selection sets of* t. If t does not contain the null-set among its members the question arises whether Πt might be the null-set O, i.e., whether

[1]) This is a result of A. Levy not yet published. The proof uses the method of forcing of Cohen 66.

[2]) From the viewpoint of cardinals one may say that Axiom IV permits us to advance by single steps, while Axiom VII, in conjunction with III, allows us to progress to the limit of an infinite progression of such single steps. See A. Levy 60 for a formulation of the conjunction of Axioms VI and VII which stresses its similarity to stronger axioms of infinity.

[3]) Bernays 37–54 VI, system Π_0.

[4]) This was proved by Montague 61a (relying on the assumption that ZF is free of contradiction – otherwise Axiom VII can be replaced by the axiom O≠O). For stronger results in this direction see A. Levy 65b and Kreisel–Levy 68.

there are no selection sets of t. The proof of Theorem 5, while showing that $O \in t$ implies $\Pi t = O$, does not answer our question; though one would expect that, in the present case where t does not contain O, $\Pi t \neq O$, no valid argument for it has been given so far.

The guess that there exists a selection set of t relies on the following argument. Since each member of t contains at least one member one might choose *one arbitrary member* in each $y \in t$. If there exists a set c which contains just all those arbitrary members, c is a subset of $\bigcup t$ and is indeed a selection set of t. In this case we therefore have $c \in \Pi t$, i.e. $\Pi t \neq O$, which is the desired result. If Cantor's "definition" of set (p. 15) is interpreted liberally enough, this introduction of the subset c of $\bigcup t$ can be considered as a valid argument which establishes the existence of the set c in naive set theory.

In our axiomatic theory, this way of introducing the subset c of $\bigcup t$ is not in accordance with the axiom of subsets [1]) – except for the trivial case that every member of t contains one member only, in which case $c = \bigcup t$ satisfies our condition. In the general case, the subset c of $\bigcup t$ has not been defined by a definite condition $\mathfrak{P}(x)$ that is characteristic, among all $x \in \bigcup t$, of the $x \in c$ and only of them. On the contrary, suppose $c \subseteq \bigcup t$ is of the desired kind and $y \in c$ belongs to a certain $y \in t$; then, replacing y by a different member y' of the same $y \in t$ will yield a new subset $c' \subseteq \bigcup t$ which differs from c, while c' is also a subset of $\bigcup t$ with the desired property. Thus, contrary to the subsets postulated by the axiom of subsets, the subsets of $\bigcup t$ needed for our purpose are not uniquely determined.

Of course, it is quite possible that some subset of $\bigcup t$ with the desired property may be obtained from the axiom of subsets or from other axioms and its existence will then guarantee that $\Pi t \neq O$. For example, let t be an infinite (say, a denumerable) disjointed set $\{t_1, t_2, ..., t_k, ...\}$ whose members t_k are non-empty sets of *natural numbers*. The existence of a selection set of t follows from the axiom of subsets applied to $\bigcup t$, where $\mathfrak{P}(x)$ is the condition given by "there is a member y of t such that x is the least number in the set y". The subset of $\bigcup t$ thus obtained contains from every member t_k of t just the least number in it. What enables us to get here a selection set by means of the axiom of subsets is the fact that in every non-empty set of natural numbers there is a unique least number.

The situation is entirely different when t is an infinite set whose members

[1]) The axiom schema of comprehension does not seem to be more useful here than the axiom of subsets since the required set c is anyway a subset of $\bigcup t$ (unless, of course, one uses the axiom schema of comprehension in a way which makes use of the idea behind some antinomy).

are arbitrary sets of *real numbers*. Then, in general, we do not know a rule which simultaneously assigns to each member of t one of its members (except for the case that the sets have a special quality which enables us to form such a rule; for instance, when each member of t contains algebraic numbers, in which case such a rule can be obtained via any enumeration of the set of all algebraic numbers – T, p. 42). Therefore, the axiom of subsets does not seem to help us to get a selection set of t in this case, and, a fortiori, in more general cases. If we want to always have a selection set for t we are in need of a special axiom, namely the

AXIOM (VIII) OF CHOICE [1]). **If t is a disjointed set which does not contain the null-set, its outer product Πt is different from the null-set.** In other words, among the subsets of $\mathsf{U}t$ there is at least one whose intersection with each member of t is a singleton.

Axiom VIII may be written in symbols in the form

$$\forall t[\forall x[x\in t\rightarrow \exists z(z\in x)\wedge\forall y(y\in t\wedge y\neq x\rightarrow\neg\exists z(z\in x\wedge z\in y))]\rightarrow$$

$$\exists u\forall x(x\in t\rightarrow \exists w\forall v[v=w\leftrightarrow(v\in u\wedge v\in x)])] .$$

In view of Theorem 5 on p. 40, Axiom VIII yields:

The outer product of the members of a disjointed set t equals O if and only if $O\in t$.

The terms 'choice' and 'selection set' originate from a psychological consideration which was formulated by Zermelo [2]) as follows: One may express the axiom (VIII) also by saying that it is always possible to choose from each member $M,N,R,$ ... of t a single member $m,n,r,$ ... and to collect all these into a set. (The disjointedness of t guarantees that the set thus obtained has no more than one member in common with each member of t.) The consequences of this psychologistic formulation, which is liable to misunderstanding, will be described in §4.4 and §4.6.

Using the notion of function one may express Axiom VIII as follows:

VIII*. *For any disjointed set t in which the null-set is not contained, there exists a function f (at least one) whose domain is t such that for each member s of t, f(s) is a member of s.* Such a function f is said to be a *choice function* on t.

The equivalence of VIII and VIII* is shown as follows. Given a selection

[1]) This name originates with Zermelo (see below). B. Russell called it the *Multiplicative Axiom*.

[2]) 08a, p. 266. Cf. Zermelo 04 and 08.

set c of t we can take as a choice function on t that subset of $t \times \bigcup t$ which consists of all the members which are of the form $\langle s,x \rangle$, where $\{x\} = c \cap s$, for some $s \in t$. On the other hand, given a choice function f on t its range c is a selection set of t.

A more useful version of the axiom of choice, which applies to arbitrary sets t rather than only to disjointed ones, is the following VIII**, which differs from VIII* only in that the word 'disjointed' is omitted.

VIII**. *For any set t in which the null-set is not contained there exists a choice function f, i.e., a function f whose domain is t such that for each member s of t, $f(s) \in s$* [1]).

The equivalence of VIII** and VIII is proved as follows. VIII** obviously implies VIII*, and hence also VIII. Let us now assume VIII and prove VIII**. Let u be a set which does not contain the null-set. Let t be the set which consists of the sets $\{s\} \times s$, for all $s \in u$. (The existence of such a set t follows easily from the axiom of subsets since for every $s \in u$, $\{s\} \times s \in \mathbf{P}(u \times \bigcup u)$.) We shall now see that t satisfies the requirements of VIII. Every $s \in u$ has a member x, since $s \neq 0$, hence the corresponding member $\{s\} \times s$ of t has a member $\langle s,x \rangle$ and therefore $\{s\} \times s \neq 0$. Given two different members s_1, s_2 of u, the corresponding members $\{s_1\} \times s_1$ and $\{s_2\} \times s_2$ of t are disjoint, since if $\langle x,y \rangle$ is a member of $(\{s_1\} \times s_1) \cap (\{s_2\} \times s_2)$ then $x = s_1$ and $x = s_2$, which contradicts $s_1 \neq s_2$; thus any two different members of t are disjoint and t is a disjointed set. By Axiom VIII, t has a selection set f. f is a subset of $\bigcup t$ and thus every member of f is a member of some member $\{s\} \times s$ of t, i.e., every member of f is an ordered pair $\langle s,x \rangle$ where $x \in s \in u$. This means that f is a relation and its domain is included in u. We shall now see that f is a function and its domain is exactly u. If $\langle s,x \rangle \in f$ and $\langle s,y \rangle \in f$ then, as we saw, x and y are members of s, hence $\langle s,x \rangle$ and $\langle s,y \rangle$ are both members of the member $\{s\} \times s$ of t. Since f is a selection set of t it has only one member in common with $\{s\} \times s$, hence $\langle s,x \rangle = \langle s,y \rangle$ and $x = y$. Thus for every s in the domain of f there is exactly one x such that $\langle s,x \rangle \in f$, i.e., f is a function. To see that the domain of f is exactly u, let s be any member of u; then $s \times s \in t$, and f, being a selection set of t, has a member $\langle s,x \rangle$ in common with $\{s\} \times s$, hence s is in the domain of f. For every s in the domain u of f, $f(s)$ is the set x such that $\langle s,x \rangle \in f$, but we saw that in this case $x \in s$, therefore $f(s) \in s$. Thus f satisfies all the requirements of VIII**.

The axiom of choice is probably the most interesting and, in spite of its late appearance, the most discussed [2]) axiom of mathematics, second only to Euclid's axiom of parallels which was introduced more than two thousand

[1]) K. Ono suggested (by hearsay) the following version of the axiom of choice which, like VIII, avoids the notion of function yet, like VIII**, applies to arbitrary sets t: For every set t, its union-set $\bigcup t$ has a subset c which has *at most* one member in common with each member s of t and which *is not included in any other subset c' of* $\bigcup t$ which *has the same property*. Ono's version obviously implies VIII. That VIII implies Ono's version is shown by means of Zorn's lemma – (p. 79).

[2]) See below (§4.6). Cf. the historico-critical exposition in Cassina 36; other expositions of a general and non-technical nature are Fraenkel 35 and Zlot 60.

years ago. Prior to a closer examination of its character, its purpose, and its history, we shall glance over its "pre-history".

Presumably the first explicit, if negative, allusion is contained in a paper of G. Peano of 1890 [1]), concerning an existence proof for a system of ordinary differential equations, where he writes: However, since one cannot apply infinitely many times an arbitrary law by which one assigns ("*on fait correspondre*") to a class an individual of that class, we have formed here a definite law by which, under suitable assumptions, one assigns to every class of a certain system an individual of that class. – In our axiomatic language this would mean: Since one cannot presuppose the existence of a selection set of t as defined in Axiom VIII, we have constructed a condition furnishing a suitable subset of $\bigcup t$ by means of the axiom of subsets.

In 1902 [2]), Beppo Levi, while dealing with the statement that the union of a disjointed set t of non-empty sets has a cardinal greater than, or equal to, the cardinal of t, remarked that its proof depended on the possibility of marking (selecting) a single member in each member of t.

To be sure, Cantor (and others) had applied the principle in question prior to Peano's and Levi's remarks. But he did so inadvertently, without being aware of using a procedure which previously had not been applied in classical mathematics or logic.

In 1904, following a suggestion of Erhard Schmidt, Zermelo explicitly formulated the principle of choice and used it as the basis for his first proof [3]) of the well-ordering theorem (*T*, pp. 222–227), and in 1908 for his second proof [4]). However, he could not then presuppose the set to be disjointed and therefore, since the notions of an ordered pair and of a function as in §3.5 were not known, he could not formulate the principle as we did in VIII** and he had to use looser formulations using a notion of "functional correspondence". In 1906, Bertrand Russell [5]) formulated the axiom in its proper "multiplicative" form, restricted to a disjointed set t. In 1908 [6]), Zermelo

[1]) Peano 1890, p. 210.

[2]) Levi 02. According to a communication by letter from F. Bernstein, about 1901 G. Cantor and F. Bernstein tried to construct a one-to-one correspondence between the continuum and the set of all denumerable order types (which has the cardinal of the continuum; *T*, p. 147). When they met with an insurmountable difficulty, B. Levi proposed to solve the difficulty by introducing the principle of choice which he formulated in a general form.

[3]) Zermelo 04.

[4]) Zermelo 08 (cf. the first edition of *T*, pp. 319–321, or Hausdorff 14, pp. 136–138).

[5]) Russell 06, pp. 47–52.

[6]) Zermelo 08, p. 110 and 08a, pp. 266, 273 ff.

showed how the general formulation can be obtained from the multiplicative form by means of the other axioms.

In the present exposition, only the fundamental lines regarding the axiom of choice are given; the literature references will enable the interested reader to obtain exhaustive information. The chief points to be discussed here are: the *consistency* and the *independence* of the axiom; *specialized forms* of the axiom; its *existential character*; its *applications* in set theory and in mathematics on the whole; and, finally, the *reaction of mathematicians* to the claim that it is one of the principles underlying mathematical research.

4.2. **The Consistency and the Independence of the Axiom.** Let us first recall that on p. 22 we denoted with ZF the system which contains all the axioms of set theory *except the axiom of choice*. The most fundamental metamathematical problems connected with the axiom of choice are the problems of its *consistency* (i.e., whether when added to the axioms of ZF it does not yield a contradiction) and its *independence* (i.e., whether the axiom of choice is or is not a theorem of the system ZF). In view of the new idea underlying the axiom and the controversy caused by it, any results concerning these problems are highly interesting.

A question which arises naturally in this connection is the question of the consistency of ZF itself (i.e., without the axiom of choice). If one can obtain a contradiction in the system ZF then both the axiom of choice and its negation are theorems of ZF (since every statement follows from the axioms of a theory in which a contradiction is provable). By Gödel's theorem on consistency proofs, if ZF is consistent then the consistency of ZF cannot be proved unless one uses in the proof some means which go beyond the means of this powerful theory (see Ch. V, §7) [1]). Therefore, in order to get results concerning the consistency and the independence of the axiom of choice it is best to assume that ZF is consistent. *All the results concerning the consistency and the independence of the axiom of choice or its weakened forms, and all other metamathematical results, which will be stated in the present section, will rely, tacitly, on the assumption that* ZF *is consistent (i.e., free from contradiction).*

In 1922, Fraenkel proved the independence of the axiom of choice [2]) for an axiomatic system of set theory which admits infinitely many objects which

[1]) If ZF is consistent then its consistency is unprovable even by the means of ZFC, as follows easily from Gödel's result that if ZF is consistent so is ZFC (see below) together with Gödel's theorem on consistency proofs applied to ZFC.

[2]) Fraenkel 22a.

are not sets, i.e., individuals (cf. §2) [1]). This did not solve the problem of the independence of the axiom of choice with respect to ZF since the existence of non-sets is not compatible with the axioms of ZF, as the axiom of extensionality permits the existence of just one element which contains no member, viz. the null-set (Theorem 3 on p.39). Fraenkel's proof was improved by Mostowski and Lindenbaum in 1938 [2]).

Fraenkel's proof uses a certain group-theoretic method, analogous to the method of Galois theory in algebra. Without going into the mathematical subtleties of that proof let us see what makes it possible. In ZF our tools for obtaining "new" sets are the axioms of comprehension (p. 32). Since *there is no characteristic which distinguishes one individual from another* [3]), when one uses an axiom of comprehension to obtain a set the only "asymmetry" which this set will have with respect to the individuals is that asymmetry which enter by means of the parameters $z_1, ..., z_n$ of the condition used in the axiom of comprehension. Thus one can assume, without running into any contradiction, that there is an infinite set I of individuals and, at the same time, that every set is related in the same way to all individuals, except for a finite number of them; in particular, the only subsets of I are the finite subsets and their complements [4]). Since it is an easy consequence of the axiom of choice that every infinite set is the union of two disjoint infinite sets [5]), the assumptions which we just mentioned, and asserted to be non-contradictory, are not compatible with the axiom of choice.

The major drawback of the Fraenkel–Mostowski method is not in the mere fact that a different axiom system is used, but in that this method shows only that one cannot prove the axiom of choice for sets t such that some of the members of $\bigcup t$ or of $\bigcup\bigcup t$, etc., are individuals. This method does not shed any light on whether the axiom of choice is needed to get a choice function for a set t of sets of real numbers, or of sets of sets of real numbers, etc.

In 1938, Gödel proved the consistency of the axiom of choice [6]) and

[1]) For details concerning such systems and the question of their consistency, see footnote 1 on p. 24 and footnote 1 on p. 25.

[2]) Mostowski 38 and 39, Lindenbaum–Mostowski 38.

[3]) In mathematical terms one would say that every permutation of the individuals can be extended to an automorphism of the universe of elements. In the proof of this statement the appropriate version of the axiom of foundation plays a central role.

[4]) Mostowski 38 (cf. A. Levy 58, p. 12).

[5]) In ZFC one can prove that every infinite set includes a denumerable subset (*T*, p. 43); the latter is obviously a union of two disjoint denumerable sets.

[6]) Gödel 38, 39, 40 (or see Shoenfield 67, Cohen 63, 66, Karp 67, Jensen 67 or Mostowski 69). The proof of Gödel 40 was carried out for a system of set theory very

thereby completely solved the consistency problem, together with the yet more difficult problem of the consistency of the generalized continuum hypothesis (see §6.1).

Gödel introduced a certain process which generates sets and called the sets generated by this process *constructible*. (This term has to be taken with a grain of salt, since some highly non-constructive notions are used in its definition.) The constructible sets are generated by the process in a sequential order one after the other, and this sequential order well-orders the constructible sets, i.e., every set s which contains at least one constructible set also contains a constructible set which is obtained by the process before any other constructible set in s. If t is a set of non-empty sets whose members are constructible sets then we can *define* a choice function f on t by means of the axiom of subsets. f is defined as that subset of $t \times \bigcup t$ which consists of all the members of the form $\langle s, x \rangle$ where x is the first member of s to be obtained by the process.

The next step is to consider the axiomatic system of ZFC$^+$ obtained from ZF by adding to it the *axiom of constructibility*; this axiom asserts that all sets are constructible. By our remark above it now becomes obvious that the axiom of choice is a theorem of the system ZFC$^+$ since the existence of a choice function on any set t now follows from the axiom of subsets. Finally, it is shown that the system ZFC$^+$ is consistent, which proves the consistency of ZFC (since any contradiction derivable in ZFC can, a fortiori, also be derived in ZFC$^+$).

In 1951–1955, Specker, Mendelson, and Shoenfield [1]), independently, proved that the axiom of choice does not follow from Axioms I–VII (which are all the axioms of ZF except Axiom IX of foundation). The system which consists of Axioms I–VII does, of course, admit no individuals. Their proofs closely follow the proofs of Fraenkel and Mostowski, the only essential difference being that the individuals are replaced by a special kind of sets (which are called *unfounded sets* – see §5). This transition from individuals to unfounded sets seems very natural when one considers Quine's approach to individuals (pp. 29–30), according to which individuals can be viewed as sets of some special kind. The proof of the present result, like that of Fraenkel and Mostowski, still does not shed any light on the case where t is a set of sets of real numbers, or of sets of sets of real numbers, etc.

similar to the system G of §7.4, but the same proof is valid also for the system B of §7.6, and can therefore be directly translated to a proof for ZF. A different proof, which has no bearing on the problem of the consistency of the continuum hypothesis, is indicated in Gödel 65 and carried out by Myhill–Scott 71.

[1]) Specker 57, Mendelson 56a, and Shoenfield 55.

The problem of the independence of the axiom of choice with respect to ZF (i.e., without individuals and in the presence of the axiom of foundation IX), which is a much more difficult problem than the related problems whose solutions we mentioned [1]), held out till 1963, when it was finally completely and affirmatively solved by Paul Cohen [2]). He showed, among other things, that in ZF one cannot prove Axiom VIII of choice, *not even for the case where t is a denumerable set of sets of real numbers*. The proof is highly technical; it makes use of the idea and techniques of Gödel's work on constructibility, as well as of new techniques specially developed for this proof. Cohen also proved that some other consequences of the axiom of choice, which are, likewise, much weaker than the full Axiom VIII, cannot be proved in ZF. Several such examples will be discussed in the next subsection [3]). Cohen's method also yields many results not connected with the independence of the axiom of choice (see §6.1 and 6.2) [4]).

4.3. **Special (weakened) Forms of the Axiom.** We shall now discuss various statements of set theory which are consequences of Axiom VIII that seem to be weaker than the axiom but still do not seem to be provable from the axioms of ZF. The questions which we shall ask are whether what seems to us to hold does indeed hold, i.e., we shall ask: Do these statements indeed not imply Axiom VIII (with the aid of the axioms of ZF)? Are they indeed unprovable in ZF? (In some cases only one of these questions will be discussed here.) Given two such different statements, we shall sometimes try to establish their relationship in ZF, i.e., to find out whether one of them

[1]) See Shepherdson 51–53 III.

[2]) Cohen 63/4, 65, 66. (See also Shoenfield 67, Jensen 67, Mostowski 69, and Rosser 69.)

[3]) In many such examples, the proof used to obtain a result by the method of P. Cohen borrows much from the proof used to obtain the corresponding weaker result by the method of Fraenkel and Mostowski. For a systematic study of some aspects of the relationship between the two methods, see Jech–Sochor 66, Jech 71 and Pincus ∞.

[4]) The methods of Cohen have been modified by Vopěnka 64, 65–67, Vopěnka–Hájek 65–67, Sacks 69, Shoenfield 67, Jensen 67, and Solovay 70. The most remarkable modification was achieved by Scott–Solovay ∞. They use a model of set theory with truth-values in a complete Boolean algebra. They dispense with the ideas connected with constructibility, and as a consequence, the proof becomes easily adaptable also to type theory (see Ch. III, §2) and weaker set theories – Scott 67. An exposition of their method is given in Rosser 69, and also in Jech 71. The same construction, but with forcing instead of Boolean truth-values, is carried out by Shoenfield 71.

implies the other in the system ZF [1]). (Notice that in ZFC their relationship is trivial since both are theorems of ZFC.)

Since all the questions of provability and independence which we shall discuss in the present subsection are concerned, unless mentioned otherwise, with the system ZF, we shall omit throughout this subsection all further reference to it and we shall say "... is provable" instead of "... is provable in ZF" and "... implies (does not imply) ..." instead of "... implies (does not imply) ... in ZF", etc.

Axiom VIII is equivalent (in ZF) to the statements which asserts that every set can be well-ordered (pp. 79–80); in particular, Axiom VIII implies the statement that the set of all real numbers can be well-ordered. If there is a relation r which well-orders the set of all real numbers then every disjointed set t of non-empty sets of real numbers has a selection set c; c can be taken as that subset of $\bigcup t$ which consists of all real numbers x which are the least members, according to the well-ordering r, of some member s of t. (This is very much like the proof on p. 54, where $\bigcup t$ consists of natural numbers.) However, as mentioned in §4.2, Cohen showed that one cannot prove that every disjointed set t of non-empty sets of real numbers has a selection set; therefore we know that one cannot prove that the set of all real numbers can be well-ordered [2]).

In Axiom VIII the set t, a selection set of which is claimed to exist, is supposed only to be disjointed and not to contain the null-set, while the cardinality of t and of the members of t remains arbitrary. The simplest way of specializing is, then, *to impose restrictions upon these cardinalities* [3]).

The most far-reaching specialization is obtained by assuming t to be finite. In this case, however, *the axiom becomes redundant* because it can be proved. It is sufficient to consider the case that t contains a single member, for the

[1]) The solutions to such problems, for systems of set theory with individuals or without the axiom of foundation, are surveyed in A. Levy 65, where further references are given (and will, therefore, be given here only in a few cases).

[2]) See, e.g. Rosser 69. Feferman and Levy showed that one cannot prove that there is any non-denumerable set of real numbers which can be well-ordered; see Cohen 66, Ch. IV, § 10. Moreover, they also showed that the statement that the set of all real numbers is the union of a denumerable set of denumerable sets cannot be refuted.

[3]) Other specializations of Axiom VIII are obtained by imposing on t restrictions of a different nature. The specialization of Axiom VIII obtained by requiring the members of t to be compact Hausdorff topological spaces is implied by the prime ideal theorem for Boolean algebras (p. 65) and implies the axiom of choice for sets t of finite sets (Łoś - Ryll-Nardzewski 54, Rubin - Scott 54). Another specialization, due to Knaster (see Kondô 37), is obtained by requiring the members of t in Axiom VIII** to be linear perfect sets of points.

transition to any finite set t can be achieved by means of ordinary mathematical induction and of the axioms of pairing and of union without involving essential difficulties [1]).

When $t = \{s\}$ contains a single member, the problem is of a logical rather than of a set-theoretical nature. According to the conditions of our axiom, s is a non-empty set; accordingly, the task is to "choose" a single member from a non-empty set. But for this purpose the axiom of choice is not required, contrary to an opinion expressed in various publications [2]).

In fact, for $t = \{s\}$, the existence of a selection-set follows, by the predicate calculus, from the assumption that s is not empty and from the existence of a singleton $\{x\}$ for any given element x.

Contrary to the case of a finite set t, *the finiteness of the members of t does not trivialize the choice problem*. Already Russell had, in an informal way, hinted at the gap between the use of a condition and the application of the axiom of choice by contrasting an infinite set t of pairs of shoes with a (say, equinumerous) infinite set of pairs of stockings. In the former case a subset of $\bigcup t$ may be constructively defined as containing all left shoes, and this set is evidently a selection-set of t, obtained without using our axiom. On the other hand, as long as manufacturers adhere to the regrettable custom of producing equal stockings for both feet there is no condition which simultaneously distinguishes one stocking in each of the infinitely many pairs. Hence a set containing just one stocking from each pair exists only by virtue of the axiom of choice. If the set of pairs were, for example, denumerable then we could not without our axiom form a one-one mapping between the set t of all pairs and the set $\bigcup t$ of all stockings, proving hereby that the latter set was also denumerable.

If we consider only the cardinalities of the members of t then the *weakest non-trivial form* of the axiom of choice is obtained by assuming the axiom of choice only for sets t all of whose members are finite sets, or even simpler, just pairs.

We now ask the question whether the weakest form of the axiom of choice can be proved. Let us first consider the case where the set $\bigcup t$ can be ordered, i.e., where there is a relation r which orders this set. Since every $s \in t$ is a finite subset of $\bigcup t$, s has a first member with respect to the order r. Therefore the subset of $\bigcup t$ defined by the condition "x is the first member of some

[1]) Cf. Littlewood 54, Prop. 17.

[2]) Notably Kamke 39 (§ 12), Denjoy 46–54 I, P. Levy 50. In these papers it is also erroneously maintained that the general axiom of choice can be inferred, without any further assumption, from the (trivial) case where t contains a single member.

$s \in t$ with respect to the order r'' on x is a selection set of t. We saw that in this case the existence of a selection set is provable; in particular, this is the case when $\mathsf{U}t$ consists of real numbers (which are always ordered by magnitude). On the other hand, one cannot prove the existence of a selection set of t even for every disjointed *denumerable set of pairs (or triples, or quadruples, etc.) of sets of real numbers* [1]).

By what was said above concerning sets t for which the set $\mathsf{U}t$ can be ordered, we get that in cases where the members of t are finite and the existence of a selection set of t is unprovable, also the existence of a relation which orders $\mathsf{U}t$ (or any set which includes $\mathsf{U}t$) is unprovable. Therefore the result mentioned at the end of the last paragraph implies that *one cannot prove that the set of all sets of real numbers can be ordered* [2]) and hence *one cannot prove that it is possible to order the set of all real functions* (i.e., the functions whose domain is the set of all real numbers and whose range consists of real numbers) [3]).

The statement that every set can be ordered is usually referred to as the *ordering principle* (or the *ordering theorem*) [4]). We have already mentioned that the axiom of choice is equivalent to the statement that every set can be *well*-ordered; therefore, the axiom of choice implies the ordering principle. It is now natural to ask whether the ordering principle is equivalent to the axiom of choice. It turns out that the ordering principle does not even imply

[1]) Cohen 63/4, 65, 66. Cohen has sets of natural numbers instead of our real numbers, but the transition from sets of natural numbers to real numbers is immediate. Also Cohen mentions only pairs, but trivial modifications give also the results for triples, quadruples, etc. Cf. Feferman 65 where the corresponding result is proved for a non-denumerable set of pairs of additive cosets of the real numbers over the rational numbers.

[2]) Cohen 63/4, 65, 66, Mostowski 69, Ch. XIV, §5. In the same way Feferman 65 derives from what was said in the last footnote the stronger result that one cannot prove that the set of all additive cosets of the real numbers over the rationals can be ordered.

[3]) Since the set of real functions whose range is included in $\{0, 1\}$ is obviously equinumerous to the set of all sets of real functions.

[4]) For stronger statements see Kinna–Wagner 55 (cf. Mostowski 58, Halpern–Levy 71 and Felgner 71a), the *order-extension principle* in Szpilrajn 30 or Sikorski 64 (p. 211, (c)) (cf. Mathias 61 and Felgner 69), and Tarski 54. Kurepa 53 obtains a statement equivalent to the well-ordering principle by taking the conjunction of the ordering principle and the statement that every partially ordered set has a maximal "anti-chain" (see Rubin–Rubin 63, M15($\overline{\mathrm{K}}$)). By what was said above, the ordering principle implies the axiom of choice for sets t of finite sets. The converse implication does not hold – see Läuchli 64, Marek 66a, and Pincus ∞.

the statement that the set of all real numbers can be well-ordered [1]); hence also the weakest form of the axiom of choice does not imply this statement.

A consequence of the axiom of choice which implies the ordering principle is the statement that every Boolean algebra has a non-principal prime ideal (henceforth the *Boolean prime ideal theorem – BPIT*) [2]). Since the ordering principle is not provable, the BPIT is unprovable too. Moreover, *one cannot prove that there is a non-principal prime ideal in the Boolean algebra of all sets of natural numbers* (with the usual union, intersection and complementation operations) [3]). In the other direction, it turns out that even the BPIT does not imply that the set of all real numbers can be well-ordered [4]).

A useful consequence of the axiom of choice is the following *axiom of dependent choices* [5]): If b is a non-empty set, r a binary relation and for every $x \in b$ there is a $y \in b$ such that $\langle x, y \rangle \in r$, then there exists a sequence $(x_1, x_2, ..., x_k, ...)$ of members of b such that $\langle x_k, x_{k+1} \rangle \in r$ for every integer $k \geqslant 1$. The axiom of dependent choices implies the *axiom of choice for denumerable sets* t [6]). As we mentioned above, on p. 61, the axiom of choice is unprovable for the case where t is a denumerable set of sets of real numbers; thus the axiom of dependent choices is unprovable, too. Moreover, even if one assumes the axiom of choice for denumerable sets t one cannot prove the axiom of dependent choices even for the case where b is the set of all real numbers [7]). On the other hand, the axiom of dependent choices does not even imply the existence of a well-ordering of the set of all real numbers [8]).

[1]) See Halpern–Levy 71, where it is proved that the existence of a well-ordering of the real numbers does not even follow from the statements that every set is equinumerous to a subset of the Cartesian product of the set of all real numbers and some well-ordered set (which is even stronger than the statement of Kinna–Wagner 55 and, a fortiori, than the ordering principle).

[2]) See Sikorski 64 for the Boolean-algebraic notions mentioned here and for a proof of the BPIT from the axiom of choice, and Łoś–Ryll-Nardzewski 51 and 54 for the proof of the ordering principle from the BPIT. For various statements equivalent to the BPIT see Henkin 54, Łoś–Ryll-Nardzewski 54, Rubin–Scott 54, Scott 54, Tarski 54, Luxemburg 64, Sikorski 64, §47, and Mendelson 64, §12.

[3]) Feferman 65, Sacks 69, Mostowski 69, Ch. XIV, §6.

[4]) Halpern–Levy 71 (which uses the combinatorial theorem of Halpern–Läuchli 66). Theorem 33.1 of Sikorski 64 (on extension of homomorphisms) is implied by the axiom of choice and implies the BPIT; it is not known whether any of these implications is an equivalence – see Luxemburg 64.

[5]) Bernays 37–54 III (Axiom IV* on p. 86), Tarski 48 p. 96. For a generalization see A. Levy 64.

[6]) Bernays 37–54 III, p. 86.

[7]) Jensen 66, Pincus ∞.

[8]) Feferman 64, Sacks 69.

Dedekind [1]) defined a set to be finite if it is not reflexive (see p. 45). Is Dedekind's definition of finiteness equivalent to the definition given here? We mentioned (on p. 46) that one can prove that no finite set is reflexive. By means of the axiom of choice one can also prove that every infinite set is reflexive, i.e., that every infinite set includes a denumerable subset [2]), and hence we get that in ZFC Dedekind's definition of finiteness is indeed equivalent to the one given here. On the other hand, in ZF *one cannot even prove that every infinite set of real numbers includes a denumerable subset* [3]). As to infinite sets of sets of real numbers (or infinite sets of real functions) one cannot even prove that such a set is always the union of two disjoint infinite sets [4]).

The axiom of choice implies the statement that the union of every disjointed set t which does not contain the null-set includes a subset equinumerous to t. (Indeed, any selection set of t is equinumerous to t by the function f of VIII*.) In ZF one cannot prove this statement even for the case where t is a set of real numbers [5]). It is not known whether this statement implies the axiom of choice [6]).

There are many statements of the arithmetic of cardinal numbers which are equivalent to the axiom of choice. In particular, such is the statement

[1]) Dedekind 1888.

[2]) This follows already from the axiom of choice for a denumerable set t (Whitehead–Russell 10–13 II, *124, Bernays 37–54 III, p. 85).

[3]) See, e.g., Halpern–Levy 71. For a generalization see Jech 66a.

[4]) This is shown by means of a "model" very similar to that constructed by Cohen in his proof of the unprovability of the weaker form of the axiom of choice (cf. Jech–Sochor 66). On the other hand, every infinite set of *real numbers* is the union of two disjoint infinite sets – see Tarski 25, p. 95, and A. Levy 58, Th. 2. For various definitions of finiteness whose equivalence is provable only by means of the axiom of choice, see Tarski 25, pp. 93–95. The proof that those definitions are not equivalent in ZF is given in Jech–Sochor 66 and in Pincus ∞. The properties of the cardinals which are finite according to Dedekind's definition (the so-called *Dedekind-finite* cardinals) are studied in ZF by Ellentuck 65. Tarski has proved in ZF that if there exists one Dedekind-finite infinite cardinal then there are at least $2^{\aleph_0}$ such cardinals. Tarski's question, whether the existence of such a cardinal implies in ZF the existence of a pair of incomparable Dedekind-finite cardinals, is still open.

[5]) Tarski proved that if there is an infinite set v which has no denumerable subsets, then there is a disjointed set t of non-empty sets such that $\bigcup t$ is equinumerous to a subset of t, while t is not equinumerous to any subset of $\bigcup t$ (i.e., the cardinality of t is strictly greater than that of $\bigcup t$); see A. Levy 65, §3. A slight modification of Tarski's construction allows one to make $\bigcup t$ a set of real numbers if v is such.

[6]) This is unknown even for set theory with individuals (or without Axiom IX of Foundation). See A. Levy 65, §3 for related statements.

that for every transfinite cardinal m (i.e., for every cardinal $\geqslant \aleph_0$), $m^2 = m$ [1]). The outstanding open problem in this direction is whether the statement that for every transfinite cardinal m, $2m = m$, which is a consequence of the axiom of choice [2]), is equivalent to it [3]). This statement can be phrased in terms of sets as follows: For every reflexive set a, $\{0,1\} \times a$ is equinumerous to a. One cannot prove the latter statement even for all sets a of real numbers [4]).

An important consequence of the axiom of choice in analysis is the existence of a set of real numbers which is not Lebesgue-measurable [5]). The existence of such a set is not provable in ZF, not even by means of the axiom of dependent choices (which is needed for the development of measure theory) [6]).

Returning to the weakest form of the axiom of choice, let us denote with Z_n the axiom of choice for sets t all of whose members contain exactly n members each, where n is a finite number. The problem of the interdependence of the Z_n's for different n's is an interesting problem of a combinatorial nature; it has been solved only recently [7]).

4.4. **The Existential Character of the Axiom. Effectivity. Selectors.** Save for the properly intuitionistic attitudes (Chapter IV) which are justified from their own point of view, the majority of the attacks on the axiom of choice [8])

[1]) Tarski 24, see Rubin–Rubin 63, I, §6. For the formal treatment of cardinals in ZF see §5.4. For results concerning finite powers of cardinals in ZF see Ellentuck 66.

[2]) *T*, Theorem 15 on p. 219.

[3]) This is unknown even for set theory with individuals (or without Axiom IX of Foundation).

[4]) If b is a set which is infinite but not reflexive and a is the union of b with a (disjoint) denumerable set then a is reflexive and $\{0,1\} \times a$ is not equinumerous to a – see A. Levy 58. Since one cannot prove that every infinite set b of real numbers is reflexive one cannot prove that every reflexive set a of real numbers is equinumerous to $\{0,1\} \times a$.

[5]) For elementary texts dealing with this notion see Halmos 50 and Munroe 53. In Van Vleck 08 and Sierpiński 27 it is shown that the existence of a non-measurable set follows already from the axiom of choice, applied to a set t of pairs.

[6]) Solovay 70 or see Jech 71; one has to assume the consistency of the statement asserting the existence of an inaccessible number; without this assumption one can still prove that the axiom of dependent choices is consistent with the existence of a translation-invariant measure defined on *all* sets of real numbers and extending the Lebesgue measure – Solovay 64, Sacks 69. It is yet unknown whether the *axiom of determinateness* of Mycielski–Steinhaus 62, which implies that every set of real numbers is measurable, is consistent with ZF (see Mycielski 64–66 and Mycielski–Swierczkowski 64).

[7]) Mostowski 45, Gauntt 70; see also Szmielew 47 or Sierpiński 58, VI, §5.

[8]) For instance, in addition to the literature quoted in §4.6, J. König 14 (pp. 170 f.), Dingler 31 (pp. 88 f. of the first ed.), Richard 29.

derived from not sufficiently appreciating its *purely existential character*. In fact, the axiom does not assert the possibility (with scientific resources available at present or in any future) of *constructing* a selection-set; that is to say, of providing a rule by which in each member s of t a certain member of s can be named. On the contrary, providing such a rule would mean obtaining the respective subset of $\bigcup t$ by the axiom of subsets, without involving the axiom of choice. The latter just maintains the *existence* of a selection-set, i.e. the non-emptiness of the outer product πt (whose existence is guaranteed without our axiom). In other words, the axiom maintains that, its assumptions fulfilled, among the subsets of $\bigcup t$ such subsets as contain a single common member with each member of t *will not be absent*, even if we fail to construct such a subset by means of the axiom of subsets. Too little attention was paid to this fundamental point during the first decades of the present century and thereby many sterile discussions were caused.

We shall now study the notion of effectivity [1]) both for its own sake and for the sake of comparing it with the axiom of choice.

To give proper weight to a definition, no matter whether within mathematics and logic or without, the existence of objects (at least one object) satisfying the definition should be shown. Normally this is done by providing a particular object that satisfies the definition, i.e., by giving an effective example. Not always need the example be given in a constructive way; its formation may make use of a non-predicative procedure or be based upon joining an existential proof which shows that *there are* objects satisfying the definition, to a demonstration that *no more than one* such object can exist. One may maintain that also in this way an effective example was given.

The term 'effective' has been used in mathematics in many different meanings [2]), sometimes even by one and the same author, which caused much confusion and many futile arguments. What is usually meant by 'effective', as used in the last paragraph, is 'definable' (or 'nameable'); from now on we shall use the term 'effective' only in this sense. A *definable set* is a set given by a condition $\mathfrak{P}(x)$ on x *without parameters* and such that in ZF, or in ZFC, one can prove that there exists just a single element x which

[1]) See Sierpiński 58, pp. 25, 35–36, 48–49, 105–107 and Kuratowski 58, pp. 142–143, where further references are given. Some other, rather limited notions of effectivity resulted in the theory of the analytical and projective hierarchies, which is dealt with by Lusin 30 and Lyapunow–Stschegolkow–Arsenin 50 (see also Kuratowski 58 and Kuratowski–Mostowski 68), where further references are given.

[2]) In addition to the meaning which this term has here it is also often used in the sense of 'decidable' or 'recursive' (Chapter V).

satisfies this condition. It is just in this case that we can speak of "*the set x* such that $\mathfrak{P}(x)$ holds" and give this set a proper name [1]). For example, the null-set O is given by the condition "x is memberless", and the set Z^* (of p. 47) is given by the condition "x is a subset of every set Z which contains O and which for each of its members y also contains $\{y\}$". The notion of a definable set is not a notion of the object language, it is a metamathematical notion. This is not the fault of the way in which this notion was introduced here; there is a profound reason behind this fact. If the notion of definability were a notion of the object language it would enable us to reproduce Richard's antinomy (Chapter I) in ZF [2]). Even though we can refer to arbitrary conditions $\mathfrak{P}(x)$ in the object language (since conditions are finite strings, or sequences, of symbols and the symbols can be assumed to be certain sets), the semantical relation between the arbitrary condition and the elements fulfilling it cannot be defined in ZF. Thus we cannot refer in our object language to the non-definable sets. Similarly, we cannot refer to all definable sets by a single statement of the object language, yet we can refer to all definable sets by a statement-schema, i.e., by a particularly simple infinite set of statements (of the object language). Suppose we want to assert that no definable set is a well-ordering of the set of all real numbers; this can be expressed by the schema "If there is exactly one x which fulfils $\mathfrak{P}(x)$ then this x is not a well-ordering of the set of all real numbers". (The schema is the set of all such sentences obtained by taking all different conditions $\mathfrak{P}(x)$.)

We now return to a question which has been considered earlier (on p. 63). Suppose that the set t, for which we want to get a selection set, consists of a single non-empty set s. As stated before, in this case the existence of the selection set can be established without using the axiom of choice, *but this does not mean that we can give an effective example of a selection set of t.* One can give an effective example of a selection set of t just in case s contains some definable element. Let us choose, for example, s to be the set of all well-orderings of the set of the real numbers; as a consequence of the axiom of choice, s is not empty, yet one cannot prove in ZFC that s contains any definable member as *one cannot prove in* ZFC *that there is a definable well-ordering of the set of all real numbers* [3]) *(i.e., in* ZFC *one does not get*

[1]) For the formal treatment of giving proper names to objects by means of the definite article, see, e.g., Rosser 53, Ch. VIII.

[2]) Gödel 65.

[3]) Feferman 65. Cf. also A. Levy 65a, Th. 6, Mostowski 69, Ch. XV, § 2, or Rosser 69.

any contradiction from the schema "no definable set x is a well-ordering of the set of the real numbers"). Other results along the same line are: *One cannot prove in* ZFC *that there is a definable ordering of the set of all sets of real numbers, or of the set of all real functions* [1]) (which implies the former result). *One cannot prove in* ZFC *that there is a definable non-measurable set of real numbers* [2]). The proofs of these results use the basic method which P. Cohen employed to prove the independence of the axiom of choice.

The proof that the above set s is not empty makes essential use of the axiom of choice, but this in itself cannot be said to be the cause of the strange behavior of the set s, if strange it is. Let us consider the set s' which is defined as follows: s' is the set of all well-orderings of the set of the real numbers, if there are such well-orderings, and is the set which contains O as its only member, otherwise. The non-emptiness of s' can already be proved in ZF. In ZFC s and s' are, obviously, proved equal, hence one cannot prove in ZFC that s' has definable members. Admittedly, the definition of s' seems artificial, yet there is no scientific criterion which draws a line between natural and artificial definitions. Whether a definition is natural or artificial depends to a large extent on its verbal version; by rewording one can sometimes make a natural definition out of an artificial one [3]). We shall also see later (in §6.1) other examples, totally unrelated to the axiom of choice, of definable sets which cannot be shown in ZFC to have definable members [4]).

We introduced definable sets by means of parameterless conditions $\mathfrak{P}(x)$. If we lift the ban on parameters we obtain the notion of a definable operation (or function). A *definable operation* on $z_1, ..., z_n$ is given by a condition $\mathfrak{P}(x)$, without parameters other than $z_1, ..., z_n$, such that one can prove in ZFC that for any given $z_1, ..., z_n$ there is exactly one x which fulfils the condition $\mathfrak{B}(x)$; this set x depends, in general, on $z_1, ..., z_n$ and is taken to be the value

[1]) This and several other results are, essentially, proved by Feferman 65 – see A. Levy 65a.

[2]) This and additional results are proved by Solovay 70; his results are based on the assumption of the consistency of the existence of an inaccessible number (§6.4).

[3]) E.g., we can say that an ordering r of a set a is *as good as possible* when it is a well-ordering, or, if a has no well-ordering, if it is any ordering. We can now define a set s'', similar in its properties to the set s', as the set of all orderings of the real numbers which are as good as possible. However, there is a more subtle relationship between usage of the axiom of choice and existence of non-empty sets with no definable members – see A. Levy 69.

[4]) Such an example is also given by the definable set of all non-constructible sets of natural numbers. Even if we assume that this set is non-void we cannot prove in ZFC that it contains a definable member – A. Levy 65a.

of the operation for $z_1, ..., z_n$. Inasmuch as the definable sets are the sets to which one gives proper names, the definable operations are the operations which are given proper names. For example, the condition on x "x consists exactly of the members of z_1 and the members of z_2" yields the binary operation of union $z_1 \cup z_2$, "x consists of exactly those elements which are members of both z_1 and z_2" yields the binary operation of intersection $z_1 \cap z_2$, "x has z as its only member" yields the unary operation $\{z\}$, "x consists exactly of the members of the members of z" yields the unary operation of union-set $\bigcup z$, etc.

A question which is strongly related to the axiom of choice and to considerations of effectivity is the question whether for a given axiomatic system Q of set theory there is a definable unary operation $\sigma(z)$ (on z) such that one can prove in Q that for every non-empty set z, $\sigma(z)$ is a member of z. Such a unary operation will be called a *selector* (in Q) [1]).

We shall now see that if Q contains the axioms of union and subsets and a selector $\sigma(z)$ is available in Q then the axiom of choice is provable in Q. Given a disjointed set t which does not contain O, a selection set of t is obtained by means of the axioms of union and subsets as a subset of $\bigcup t$ which consists of all the members which are $\sigma(s)$ for some member s of t. An example of a system of set theory with a selector is the set theory ZFC$^+$ obtained from ZF by adding to it the axiom of constructibility (see p. 60 and §6.2). In ZFC$^+$ a selector $\sigma(z)$ is obtained by means of the functional condition "if z is a non-empty set then x is the first member of z obtained by Gödel's process, and if z is O then x is O too".

In ZFC no selector is available. This follows immediately from the possibility of the existence of definable non-empty sets with no definable members, such as the non-empty set s of all well-orderings of the real numbers. If a selector $\sigma(z)$ were available in ZFC then one could prove in ZFC the existence of a definable member of s, namely $\sigma(s)$, but we know that such a proof is impossible (p. 69). It is a remarkable fact that there is a statement of the object language which asserts indirectly (in ZF or ZFC) the existence of a selector, and just that [2]). (The assertion "there exists a selector" cannot be

[1]) Cf. Montague–Vaught 59a.

[2]) This is an axiom which asserts that every set is the value of a definable function for some ordinal arguments (we say: every set is *ordinal-definable*). Cf. Gödel 65 and Myhill–Scott 71. This axiom also asserts, exactly, that every non-empty definable set has a definable member. For the relationship between the ordinal-definable sets and the constructible sets see. A. Levy 65a and McAloon 66. One can also consider the notion of *a relative selector*, i.e., a definable *binary* operation $\sigma(y,z)$ for which one can prove that there is a set y such that for every non-empty set z, $\sigma(y,z)$ is a member of z. As easily

directly expressed by a statement of the object language since such a statement would have to be something like "there is a condition $\mathfrak{P}(x)$ with a single parameter z such that for every z there is just one x which fulfils the condition, and if z is not O then this x is a member of z", but, as was pointed out on p. 69. where the notion of definability was discussed, the semantical relationship between x (or z) and $\mathfrak{P}(x)$ cannot be expressed by the object language.

One can also get from ZF a system of set theory with a selector by brute force. This is done as follows. First the object language is enriched by adding the operation σ as a new primitive notion, in addition to the membership relation. This enrichment of the object language causes our notion of condition (introduced on p. 21) to be richer too, since now we can express conditions which we could not express before. We denote with ZFC_σ the system of set theory formulated in our enriched language whose axioms are all the axioms of ZF, where *the notion of condition in the axiom schema of replacement (and subsets) is the wider notion just mentioned*, as well as the additional:

AXIOM (VIII_σ) OF GLOBAL CHOICE. **For every non-empty set z, $\sigma(z)$ is a member of z**[1]**).**

seen, the availability of a relative selector in a system Q of set theory, which contains the axioms of union and subsets, in enough to establish the axiom of choice in that system. In ZFC not even a relative selector is available as, essentially, proved by Easton 70 (where it is shown that Axiom $\mathrm{VIII}_\sigma^{\mathrm{C}}$ of §7.3 is not provable in the system VNBC of §7.3). A statement of the object-language which, for ZF or ZFC, asserts just the existence of a relative selector is "there exists a well ordering r of some set such that every set is ordinal-definable relative to r, i.e., every set is definable in terms of ordinals and the relation r". (The proof is completely analogous to that of Myhill–Scott 71.) A system of set theory in which a relative selector is available but no selector is available is the system obtained from ZF by adding the axiom of relative constructibility $\exists a(V=L^*)$ of Schoenfield 59, or $\exists k(V=L_k)$ of A. Levy 60a, formulated in the language of ZF. The existence of a relative selector in this system is trivial, the non-availability of a selector in this system is, essentially, shown by Feferman 65, §4, wherein the model obtained with $\delta = 1$ there is no selector (since there is no definable well-ordering of the real numbers), but there is a set k such that $V = L_k$, namely $k = s_0$.

[1]) Bourbaki 54 uses selectors for all properties rather than only for sets, i.e., for every condition $\mathfrak{P}(x)$ he introduces $\tau_x \mathfrak{P}(x)$ as a new constant (or function, if $\mathfrak{P}(x)$ has parameters). τ_x is the ϵ-operator of Hilbert (see Hilbert–Bernays 34–39 II). The axioms include, essentially, the axiom schema "if there is an x such that $\mathfrak{P}(x)$ then $\tau_x \mathfrak{P}(x)$ is such" and an appropriate extension of the notion of condition in the axiom schema of replacement. The system of Bourbaki has a selector $\sigma(y)$ namely $\tau_x(x \in y)$. On the other hand $\tau_x \mathfrak{P}(x)$ can be *defined* in ZFC_σ as "$\sigma(y)$ where y is the set of all x's of least rank such that $\mathfrak{P}(x)$" (see §5.3 for the notion of rank).

As in the case, mentioned earlier, of a set theory Q with a selector, the axiom of choice is provable in ZFC_σ [1]). Let us now compare ZFC_σ to ZFC. Every theorem of ZFC is, obviously, a theorem of ZFC_σ. There are statements which are theorems of ZFC_σ, but do not belong to the language of ZFC because they contain the symbol σ, such as Axiom VIII_σ itself. However, every statement which is formulated in the language of ZFC and which is a theorem of ZFC_σ is also a theorem of ZFC [2]). This also settles the question of the consistency of ZFC_σ. If one could derive a contradiction in ZFC_σ then $0 \neq 0$ would become a theorem of ZFC_σ, and hence also of ZFC, contradicting Gödel's result on the consistency of ZFC.

4.5. **Some Typical Applications of the Axiom.** A comparison of the axiom of choice with Axioms II–VII may cause the reader to wonder why we so strongly stress its significance. It might appear as if its statement, excluding the non-existence of a certain kind of subsets of U*t*, applied to special problems and methods only and meant but little for the general theory. This supposition seems to be supported by the fact that the axiom was introduced only at the beginning of the present century; that is to say, at a time when the bulk of both the theory of abstract sets and the theory of sets of points, including the nucleus of the modern theory of real functions, had already been developed.

Yet this supposition does not to conform to the actual situation. On the contrary, *fundamental and general theorems and methods* in the theory of sets as well as in analysis, algebra, and topology *are based on the axiom of choice*. In some cases those theorems are based on the axiom of choice only in the sense that we do not know a way of avoiding its use, but also a remarkable number of them turn out to be *equivalent* to our axiom [3]). True, the axiom was introduced only at the beginning of the 20th century, but it had been utilized long before while only much later was it observed that in the

[1]) On the other hand, if we denote with ZF_σ the set theory obtained from ZF by adding to it Axiom VIII_σ, while *the axiom schemas of subsets and replacement are not strengthened* (i.e., only conditions which do not contain σ are permitted in those axioms schemas), then by the second ϵ-theorem of Hilbert-Bernays 34–39 II, § 1 (where we take $\sigma(z)$ instead of $\epsilon_x(x\epsilon z)$) every statement of the language of ZF (i.e., which does not contain σ) which is a theorem of ZF_σ is also a theorem of ZF. Therefore, the axiom of choice, which is not provable in ZF, is also *unprovable in* ZF_σ.

[2]) Felgner 71.

[3]) For a very extensive list of mathematical statements equivalent to the axiom of choice see Rubin–Rubin 63. For newer results see Ward 62, Bleicher 65, Frascella 65 Kruse 63, Grätzer 67, Felgner 67 and 69, in which further references are found. The reader will find a comprehensive and detailed technical treatment of the whole area related to the axiom of choice in Jech ∞. The authors became aware of that book too late to mention it whenever it is a relevant.

respective proofs an argumentation not used and recognized in earlier mathematics was involved.

Therefore the axiom of choice must be admitted among the other acknowledged principles of mathematics. According to Hilbert [1]) it rests on "a general logical principle which is necessary and indispensable already for the first elements of mathematical inference".

To enable the reader to form his own opinion in this matter we shall now present a few characteristic applications of our axiom. Four examples will be given, selected not only in view of their fundamental character and of a minimum of technicality entering but also to cover a maximum variety of domains: two examples from the general theory of sets and one from each, analysis and algebra [2]).

The first example, taken from the elements of *abstract set theory*, concerns the operations on cardinals (addition, multiplication, exponentiation; see §§6 and 7 of *Theory*) and partly on order-types (*T*, §8) [3]). Since the point is the same in all these cases it will be sufficient to take the simplest case, viz. the *addition of cardinals* [4]). To obtain the sum of infinitely many [5]) (finite or infinite) cardinals we assign to each cardinal as its representative a set with that cardinal [6]) on condition that the representatives be pairwise disjoint; then the cardinal of the union of the representatives is the sum of the cardinals. Accordingly the sum would depend on the arbitrarily chosen representatives, yet the independence is guaranteed by a theorem (*T*, p. 82) stating that different ways of choosing the representatives necessarily yield equinumerous unions, hence the same sum-cardinal.

[1]) Hilbert 23, p. 152.

[2]) No example from topology is given here to avoid technicalities, see, e.g., Kuratowski 58. Rubin–Rubin 63 contains several topological statements equivalent to the axiom of choice (cf. also Ward 62). Läuchli 62 proves that Urysohn's Lemma cannot be proved without the axiom of choice (see Jech–Sochor 66).

[3]) Of course, the non-vanishing of a product of non-zero cardinals is also an example, but this can barely be distinguished from the axiom of choice itself.

[4]) For the role of the axiom in the arithmetic of cardinals in general cf., e.g., Sierpiński 58, Chapters VIII–X, and Bachmann 55, Chapters IV–V (see also Läuchli 61).

[5]) If the number of terms is finite the procedure is the same, but the axiom of choice is not required.

[6]) Apparently, here arises the question how to "obtain" such representatives. If we use the notion of cardinal of ZF as on p. 98, the axiom of choice is already required to obtain a set of representatives. Actually, we do not have to use here cardinals at all and we can assume that representatives are used from the beginning. Accordingly, our example refers to the theorem in *T*, p. 82 rather than to cardinals proper; in fact, this theorem and its analogues are the key theorems of the infinite arithmetic of cardinals.

Now the proof of this theorem is based on simultaneous one-one mappings between the representatives attached to the same cardinal by different choices. More precisely, if the cardinal $f(t)=\mathbf{c}_t$, where t runs over a certain set T, is represented once by a set a_t and again by b_t (hence $b_t \sim a_t$), let $\psi^{(t)}$ be a certain one-one mapping of a_t on b_t; by combining the mappings $\psi^{(t)}$ for all $t \in T$ we easily obtain a one-one mapping of the union of the sets a_t on the union of the sets b_t. However, $\psi^{(t)}$ is not uniquely determined by the equinumerous sets a_t and b_t; save for trivial cases, there are various mappings between these sets, and infinitely many when a_t (hence b_t) is infinite. The existence of the set $\Psi^{(t)}$ of all one-one mappings of a_t on b_t is proved by applying Axiom V to the set $\mathbf{P}(a_t \times b_t)$; similarly one proves the existence of the set Γ whose members are all sets $\Psi^{(t)}$ when t runs over T. But what we actually need, is a function ψ which assigns to each member t of T *a single member* $\psi^{(t)}$ of $\Psi^{(t)}$; to obtain such a function the axiom of choice is required, Γ taking the place of the set t in Axiom VIII**.

Hence, the addition of cardinals depends on our axiom [1]), provided the number of terms $\mathbf{c}_t$ is infinite (even if the terms themselves are finite cardinals >1). The same applies to the other operations with cardinals and with order-types.

The axiom of choice is widely utilized in analysis; in particular in the theories of point sets and of real functions. Most of these applications involve technical notions of the theories concerned. Here we shall give an example from the very first elements of analysis with which all readers are familiar.

One might expect the most common instance to be the following. After having proved that for each point x of a given set there exists at least one neighborhood of x – i.e. an open interval containing x – with a certain property, one chooses for each given x a definite such neighborhood. Apparently here our axiom is used inasmuch as for each x an arbitrary neighborhood is chosen simultaneously. However, in general the axiom can be dispensed with through a restriction to neighborhoods with rational ends;

[1]) Without the axiom of choice, when the cardinal number 2 is added to itself denumerably many times the result can be $\aleph_0$ (when one considers $\cup\{\{1,2\},\{3,4\},\{5,6\},\ldots\}$) but may also be different from $\aleph_0$ (as in the case of $\cup t$, for a denumerable disjointed set t of pairs which has no selection set – the existence of such a set is not refutable in ZF, as was mentioned in p. 64). Also when the cardinal number $\aleph_0$ is added to itself denumerably many times the result can be $\aleph_0$ (since $\aleph_0 \cdot \aleph_0 = \aleph_0$) but may also be the cardinal of the continuum $2^{\aleph_0}$, since one cannot refute in ZF the statement that the continuum is the union of a denumerable set of denumerable sets – cf. Cohen 66, Ch. IV, § 10.

then only an (effective) denumerable set of possibilities is left for every x and it is easy to mark a definite one among them by a general rule. (Cf. exercise 11 in *T*, p. 47.)

Yet with respect to concepts of an even more fundamental character we do depend on the axiom of choice. As usual (cf. *T*, p. 169) a point p shall be called an *accumulation point* of a subset K of the real line if in every neighborhood of p there is a point of K different from p. On the other hand one may base the elements of analysis upon the notion of *limit point*, defining p as a limit point of K if there exists a sequence (k_υ) of different points of K $(\upsilon=1, 2, ...)$ such that the sequence has the limit p.

Without the axiom of choice one easily proves that if p is a limit point of K, p is also an accumulation point of K. On the other hand, let us also suppose that, for any subset K of the real line, every accumulation point of K is a limit point of K. From this supposition one proves the following statement $\mathfrak{L}$ [1]): If $S=(S_1, S_2, S_3, ...)$ is a sequence of pairwise disjoint non-empty sets of real numbers, there exists a sequence of real numbers $(p_1, p_2, p_3, ...)$ such that p_k's with different indices k belong to different S_n's. Conversely, it is easy to infer from $\mathfrak{L}$ without using the axiom of choice that every accumulation point of a subset K of the real line is a limit point of K as well. Hence, in ZF the equivalence between the notions "accumulation point of K" and "limit point of K" is a necessary and sufficient condition for the validity of $\mathfrak{L}$.

It can be shown that $\mathfrak{L}$ is equivalent in ZF to the axiom of choice for a denumerable set t of sets of real numbers [2]), which we know is unprovable in ZF (p. 61). Thus the axiom of choice is needed to establish the equivalence between two fundamental and elementary notions of analysis which usually are identified without further ado.

This equivalence implies relations of equivalence between other fundamental concepts of analysis which can be defined by means either of accumulation point or of limit point: not only those of 'derived set' and of 'closed [3]),

[1]) Sierpiński 19, p. 120.

[2]) The non-trivial direction, namely that $\mathfrak{L}$ implies the axiom of choice for a denumerable disjointed set $t=\{s_1, s_2, s_3, ...\}$ of non-empty sets of real numbers is proved as follows. Given such a set t, let S_n be the set of all real numbers which "represent in some canonical way" the finite sequences $(u_1, ..., u_n)$ for which each u_i is a member of s_i, $1 \leqslant i \leqslant n$. It is easily seen that from a sequence $(p_1, p_2, p_3, ...)$ of real members such that p_k's with different indices belong to different S_n's one can obtain a selection set for t (by means of an effective operation).

[3]) One may define a set K of points as *closed* either (as in *T*) on condition that every accumulation point of K belongs to K, or on condition that every limit point of K belongs to K; analogically for the following concepts.

dense-in-itself, perfect set' (T, pp. 169 f) but also the concept of *continuous function*. In fact, in the elements of analysis one uses either of the following two definitions of continuity:

a) $f(x)$, defined for $a<x<b$, is called continuous at the point x_0 of that interval if to every positive ϵ there corresponds a positive δ such that

$$|x-x_0|<\delta \text{ implies } |f(x)-f(x_0)|<\epsilon.$$

b) $f(x)$ is called continuous at x_0 if

$$\lim_{k\to\infty} x_k = x_0 \text{ implies } \lim_{k\to\infty} f(x_k) = f(x_0)\,.$$

While one easily proves that a function which is continuous at x_0 in the sense a) is also continuous there in the sense b), the converse assertion cannot be proved without the axiom of choice [1]); more precisely, in ZF the validity of the axiom of choice for all denumerable sets t of sets of real numbers is necessary and sufficient for proving the equivalence of the definitions a) and b).

Naturally, the same alternative exists for the definition of the derivative of a real function, where the situation is analogous to that regarding continuity.

It was a complete surprise for the mathematical world when in 1910 Steinitz [2]) called attention to the important task performed by the axiom of choice in algebra, both for certain problems of classical algebra and still more for abstract algebra (which became an important branch of mathematics through the influence of that very essay of Steinitz).

To render a typical problem of this branch intelligible also to readers not familiar with modern algebra we proceed from a starting-point known to everybody, if not properly algebraic.

The so-called *fundamental theorem of algebra* [3]) can be expressed as follows. A polynomial

$$p(x) = a_0x^n + a_1x^{n-1} + \ldots + a_{n-1}x + a_n \qquad (a_0 \neq 0)$$

[1]) Yet the transition from b) to a) can be accomplished without the axiom if b) is presumed *for the entire interval*, i.e. for every convergent sequence of points from the interval. See Sierpiński 19, pp. 131 ff.

[2]) Steinitz 10, §§ 19–24 and van der Waerden 55 (ch. VIII–X).

[3]) This name has historical reasons only. From a modern point of view the name should be attributed to the theorem stated below regarding an algebraically-closed extension.

with integral rational coefficients a_k and the positive integral degree n has at least one zero $x=\gamma_1$ within the field of all complex numbers; hence it has n, not necessarily different, zeros.

However, the field of all complex numbers – which, as well as the field of all real numbers, is a concept of analysis and not of algebra – quite incidentally enters this theorem, viz. because it is a well-known and would-be "elementary" concept. Returning to algebra, by an *algebraic number* we mean any (complex) number that is a zero of $p(x)$, i.e. a root of an algebraic equation $p(x)=0$ with integral rational coefficients (cf. *T*, p. 8). The set of all algebraic numbers – which is denumerable, contrary to the set of all complex or real numbers – has the following two properties: 1) it constitutes a field with respect to addition and multiplication; 2) also if the coefficients a_k are any *algebraic* numbers the polynomial $p(x)$ has a zero, hence n zeros, in the field of all algebraic numbers.

A field F (not necessarily a field of numbers) is called *algebraically-closed* if it does not admit an algebraic extension; that is to say, if every polynomial $p(x)$ "in F" (i.e., with coefficients from F) has a decomposition into linear factors $x-\gamma_k$, γ_k belonging to F. In particular, F is called an *algebraic* algebraically-closed extension of a field F_0 if F_0 is a subfield of F, F is algebraically-closed, and every member of F is algebraic with respect to F_0, i.e. a zero of a polynomial in F_0. Therefore the field of all complex numbers is algebraically-closed, but not an algebraic extension of the field R of all rational numbers since the transcendental numbers are not algebraic with respect to R. On the other hand, the field A of all algebraic numbers is an algebraic algebraically-closed extension of R.

The construction of the field A from R does not involve the axiom of choice, due to the denumerability of the field. However, one needs the axiom of choice to prove that there is essentially *only one* such extension of R (where by "essentially only one" we mean "only one, up to isomorphic extensions") [1]).

Now we may generalize this train of ideas by starting not just with the field R of the rationals but with *any field* F_0 whose members need not even be numbers. In this case, as proved by Steinitz, the above theorem still holds true, i.e. there exists one, and essentially only one, algebraic algebraically-closed extension F of F_0. Then, however, both parts of the theorem rest upon the axiom of choice [2]).

[1]) Läuchli 62, Jech–Sochor 66.

[2]) Läuchli 62, Jech–Sochor 66. One can also prove that the existence part of the theorem for a general field F_0, together with the uniqueness part of the theorem for the field R, implies in ZF the weakest form of the axiom of choice, i.e., where t is a denumerable set of pairs, which is unprovable in ZF (p. 64).

This is a typical and important use of the axiom of choice in algebra; yet no single example can do full justice to the role of this axiom in algebra [1]).

The use of the axiom of choice in analysis and algebra has yielded an important by-product: a principle of an apparently different character which nevertheless turns out to be equivalent to the axiom of choice and which is, in many fields of mathematics, a welcome substitute for the axiom, since by using it one can often avoid not only the axiom of choice, but also transfinite induction as well.

As proved by Hausdorff [2]), every partially (partly) ordered set (T, p. 131) includes at least one *maximal* completely (or totally) ordered subset, i.e., an ordered subset which is not a proper subset of any ordered subset. The proof uses the axiom of choice and follows the method of Zermelo's first proof of the well-ordering theorem (T, pp. 224–227). Similar principles were introduced later, in several cases independently of Hausdorff, by Kuratowski, Zorn, and others [3]). These principles are now usually referred to as *Zorn's lemma or maximum principles*. We shall present here also the following maximum principle (Z) [4]).

Let us say that a relation $<$ which partially orders a set A is *inductive* (on A) if every (completely) ordered subset B of A has an upper bound (i.e., if there is a member y of A such that $x<y$ or $x=y$ for every member x of B).

(Z) If $<$ is an inductive partial ordering of a set A then A has at least one maximal member with respect to $<$ (i.e., there is a member z of A such that for no member x of A does $z<x$ hold).

Zorn's lemma, in its various versions, is equivalent to the axiom of choice [5]). One can also prove directly from it various theorems of abstract set theory which are usually proved by means of the axiom of choice or the well-ordering theorem [6]).

A last example is again taken from abstract set theory; it is the very case for which our axiom was originally introduced, viz. the well-ordering theorem and the comparability of cardinal numbers (T, pp. 224–228).

In both proofs of the well-ordering theorem given by Zermelo [7]) the

[1]) See Läuchli 62 and Jech–Sochor 66 for several other essential uses of the axiom of choice in algebra.

[2]) Hausdorff 14, pp. 140 f.

[3]) See Rubin–Rubin 63, I, § 4, for references and a more detailed discussion.

[4]) Bourbaki 56, § 1, No. 4.

[5]) See Rubin–Rubin 63 for the proofs and references; see also Felgner 67. Büchi 53 investigates this equivalence from the point of view of type theory.

[6]) See Halmos 60 (Sections 17 and 24) and Zorn 44.

[7]) Zermelo 04 and 08. Bernays 34–57 IV, pp. 143–145, compares the axiomatic simplicity of both proofs and gives an intermediate one.

fundament is a simultaneous choice of particular ("distinguished") members in every non-empty subset of the set s to be well-ordered; in other words, the supposition that a function f exists whose domain is the set of the non-empty subsets x of s, such that always $f(x) \in x$.

The comparability theorem may either be inferred from the well-ordering theorem in view of the comparability of well-ordered sets or be proved directly through the axiom of choice (*T*, pp. 231–233).

But the axiom of choice, the well-ordering theorem, and the comparability theorem are *equivalent* even in the following sense. It is evident that the well-ordering theorem implies Axiom VIII and is equivalent to it. For an arbitrary well-ordering r of the union $\bigcup t$ simultaneously well-orders every subset of $\bigcup t$, hence every member of t; therefore a selection-set of t exists by the axiom of subsets; for instance, the subset of $\bigcup t$ which contains out of each member s of t that member of s which is the first member of s in the well-ordering r.

Incidentally, any well-ordering of an arbitrary set s assigns to every non-empty subset of s a "distinguished" member of the subset; thus one directly obtains from it also Axiom VIII**, where the disjointedness is not assumed.

The transition to the comparability theorem from either the axiom of choice or the well-ordering theorem has just been mentioned. The converse direction is taken in Hartogs' proof [1]. Starting with an arbitrary set s we obtain, without using the axiom of choice, a certain well-ordered set C which proves to be neither equinumerous to s nor to any subset of s. (This, by the way, means that no set exists whose cardinal surpasses the cardinal of every well-ordered set.)

Hence, *if we accept the comparability of sets as a principle*, s is equinumerous to a subset of the well-ordered set C and has a cardinal less than C. s can, therefore, be well-ordered on account of a one-to-one mapping on a certain subset of C. Thus comparability implies the well-ordering theorem, hence the axiom of choice [2]. Accordingly, *these three statements are equivalent principles*; taking one of them as an axiom we obtain the others as provable theorems. That the axiom of choice is favored in the foundations of set theory and of mathematics in general has its reason in the general logical character of this axiom.

4.6. **Mathematicians' Attitude towards the Axiom.** We conclude our survey of

[1]) Hartogs 15 (or Rubin–Rubin 63, I §3, or Suppes 60, p. 247).

[2]) For the consistency with ZF of statements which strongly negate the comparability theorem see Marek–Onyszkiewicz 66, Jech 66, and Takahashi 67.

the axiom of choice by a glance over the attitudes taken by mathematicians towards the axiom since its explicit formulation in the beginning of the present century. (The "prehistory" of the axiom was described on pp. 59 f.)

The negative attitude of most intuitionists, because of the existential character of the axiom, will be stressed in Chapter IV. To be sure, there are a few exceptions, for the equivalence of the axiom to the well-ordering theorem (which is rejected by all intuitionists) depends, *inter alia*, on procedures of a supposedly impredicative character; hence the possibility exists of accepting the axiom but rejecting well-ordering as it involves impredicative procedures. This was the attitude of Poincaré.

Save for the intuitionistic point of view, both the positive and the negative attitude towards our axiom are *far more strongly influenced by emotional or practical reasons than by arguments of principle*.

For those accepting and using the axiom, the chief reason is its – or the well-ordering theorem's – indispensability for proving important theorems of analysis and set theory; this argument proved so strong that even scholars who in principle rejected existential procedures did not refrain from using the axiom to a certain extent in their analytic researches. As to algebra, the historical development can be hardly illustrated more strikingly than by the following quotation (translated) from Steinitz' pioneer work [1]). "As yet many mathematicians take a negative attitude towards the principle of choice. With the increasing recognition that there are mathematical questions which cannot be decided without the principle, the opposition against it will presumably fade more and more. On the other hand, for the sake of purity of method it seems suitable to avoid the principle as far as its application is not required by the nature of the problem concerned. I endeavored to make this boundary clearly visible."

Yet the opponents have also been considerably influenced by psychological rather than logical reasons – in spite of the opinion of as authoritative an observer as Lebesgue who maintained that no discussion between the two parties had been possible "because they had no common logic" so that they could do no better than insult each other. (Cf., however, the quotations below on p. 84.) As a matter of fact, as long as the (implicit and unconscious) use of the axiom by Cantor and others involved just arithmetical operations with cardinals and order-types or guaranteed the non-vanishing of a product whose factors differ from zero – in short, involved only generalized arithmetical concepts and properties well-known from finite numbers – nobody took offence. The same applies to its use in some elementary proofs of anal-

[1]) Steinitz 10, *Einleitung*.

ysis. Yet at the moment (1904) when the axiom, explicitly formulated, was used by Zermelo to prove and confirm one of the earliest assertions of Cantor, viz. the well-ordering theorem, mathematical journals [1]) were flooded with critical notes rejecting the proof, mostly arguing that our axiom was either illegitimate or meaningless.

True, various and widely different reasons for the rejection were given; but one cannot help feeling that the common denominator from which the various reasons derived was the unwillingness to accept its consequence, namely the theorem in which the opponents did not believe. This unwillingness deepened enormously when it became obvious that the main purpose connected with well-ordering since 1880, viz. to ascertain the place of the power of the continuum in the series of Alephs, had not become furthered a bit by the well-ordering theorem. The difficulties of this continuum problem, more clearly perceived today than 90 years ago (see § 6.1), had by the turn of the century induced many mathematicians to believe that the (linear) continuum and more complicated sets could not be well-ordered at all; in other words, that $2^{\aleph_0}$, $2^{2^{\aleph_0}}$, etc. were no Alephs. Cantor's contrary conviction, displayed dramatically at the Third International Congress of Mathematicians (1904; cf. *T*, pp. 222–3), carried little persuasion even for many set-theoreticians, let alone mathematicians in general. Now when Zermelo in his short notes 04 and 08, simple in technique in spite of their sagacity, proved just the contrary and confirmed Cantor's conviction yet without providing a method of ascertaining the Alephs concerned, one was inclined to believe that Zermelo's proofs yielded too much and were incorrect. On the other hand, the majority of critics did not contrive to find mistakes in the proofs and therefore distrusted their basis, the axiom of choice – not so much for itself as for its consequences. Of course, this scepticism is based upon not appreciating the existential character of the well-ordering theorem which a priori makes its usefulness for the decision of "constructive" questions like the continuum problem highly doubtful.

To be sure, not all opposition directed against Zermelo's proofs derived from antagonism to our axiom. There is the attitude of Poincaré mentioned above, rejecting an (actual or apparent) use of an impredicative procedure in

[1]) See in particular vol. 60 of *Mathematische Annalen* (Zermelo's first proof had appeared in vol. 59), Hadamard 05, and note IV in Borel 14. As becomes apparent especially in this note there was an early bifurcation among the opponents with regard to the case that the set *t* is denumerable; in this case Borel, as well as Denjoy 46–54, tended to accept the assumption that a selection set exists while Lebesgue saw at most psychological reasons for distinguishing between different transfinite cardinalities of *t*.

the proof. Poinçaré, in spite of his intuitionistic inclination (see Chapter IV), was ready to accept the axiom of choice, admitting the possible existence of rules which cannot be constructively, or anywise completely, formulated; in the present case the existence of subsets of U*t* which cannot be obtained by means of the axiom of subsets. This attitude is in accordance with his considering non-contradiction as the decisive criterion of mathematical existence.

Incidentally, the proofs of the well-ordering theorem have also been attacked with actually erroneous arguments [1]) while on the other hand insufficient proofs of the theorem have been proposed.

Apart from the well-ordering theorem some statements of a quite different character – in particular geometrical statements – have been proved by means of the axiom of choice, which because of their paradoxical character induced some mathematicians to reject the axiom [2]). Presumably the earliest statement (1914) of this kind is Hausdorff's discovery that half of a sphere's surface is congruent to a third of it [3]); other paradoxical consequences were found later [4]).

From the point of view of those mathematicians who rejected the axiom of choice the proof of a theorem by means of this axiom does not, of course, establish its truth, yet some of them were willing to grant that such a proof establishes the unprovability of the negation of the theorem by present mathematical methods. (By Gödel's result on the consistency of ZFC the negation of a theorem of ZFC is unprovable in ZF – the consistency of ZF being assumed [5]).)

It may surprise scholars working in the field of abstract or applied set theory that even after more than half a century of utilizing the axiom of choice and the well-ordering theorem, a number of first-rate mathematicians (especially French) have not essentially changed their distrustful attitude; not even such as have been working most successfully in the domain of point sets and of real functions. Some lectures and discussions delivered at an international conference on foundations of mathematics in Zurich 1938 [6]) are most

[1]) In a sarcastic form many objections against the first proof are refuted in Zermelo 08, where references to the literature (up to 1908) are given.

[2]) E.g., Borel 46 and 47 and Bouligand 47; cf. the reaction of P. Levy 50. Denjoy (46–54, V) justly remarks that certain complications in analysis caused by the rejection of the axiom are no less paradoxical than the Banach–Tarski theorem (Footnote 4).

[3]) Hausdorff 14, p. 469.

[4]) In particular, Banach–Tarski 24, von Neumann 29a, R.M. Robinson 47.

[5]) This opinion was expressed by Lusin (Sierpiński 58, p. 95) and also mentioned in Fraenkel 27, p. 86.

[6]) Notably Lebesgue 41, Sierpiński 41, and the respective discussions.

characteristic of what may be called *a stagnation of controversy during several decades*, in spite of an enormous actual development of research in the field under discussion. It is remarkable what Lebesgue says there of his attitude toward the axiom of choice: *Indicating thus my demands, it was not at all my intention to decide in favor of the purely negative attitude I have taken. I am far from regarding this attitude as a final one; it is just a provisional attitude, which I have been characterizing as a measure of caution, or even just of routine, for many years, since the very beginning of the controversy* (translated from French) [1]).

It has been mainatined that, after in 1904 "the powder keg had been exploded through the match lighted by Zermelo" in his first proof of the well-ordering theorem, the strange situation emerged that those who up to that moment had utilized set theory were opposed to the axiom of choice while other mathematicians were ready to accept the axiom [2]). Whether this is correct or not, at any rate the situation has since thoroughly changed due to various reasons: to the applications of the axiom to problems of almost every domain of mathematics; to the penetrating and many-sided investigations on statements equivalent to the axiom (from 1915 on) whose import is independent of the admission of the axiom; to the researches on the independence of the axiom (from 1922 on) in its weaker and stronger forms, confronted with our inability to prove many important results without the axiom; to the fact that none of the conclusions drawn from the axiom has led to a contradiction, a fact finally crowned by Gödel's proof of the consistency of the axiom, i.e. its compatibility with a reasonable system of axioms.

It is true that the character of our axiom does not conform to the realm of

[1]) *l.c.*, p. 118. It is worth while to reproduce (in translation) some of Lebesgue's subsequent remarks; he says with regard to our axiom: In the past, audacity and caution have been collaborating at each important progress. Why not do it once again? ... I desired to make it clear ... that the discussion has neither been nor is purely logical .... Each of us takes pains to understand, and to be certain of, a substratum underlying the words used. To this purpose we utilize comparisons, instances taken from History of Science, we proceed audaciously or timidly ... perhaps according to our age or to our race. Thus it is a question of trying to elaborate a new chapter of mathematics. What, then, is the use of logic? Certainly not to convince, to create confidence. ... At the historical moment of the paradoxes of set theory, when we emerged from discussions where none of us saw a way of repairing a logic that appeared ruined, nevertheless we continued applying that very logic to the problems we were studying; for an attitude of philosophical doubt taken in a discussion does not at all prevent the full certainty. ... Logic does not create confidence ... The researches on the foundations and the method of mathematics should give plenty of space to psychology and even to aesthetics.

[2]) See Lebesgue 41 and Sierpiński 41.

pure arithmetic or to geometrical rigor in the Greek sense. But this does not justify its being made a scapegoat, considering that modern "classical" analysis, and therefore geometry, in general – in contrast with intuitionistic mathematics (Chapter IV) – has not that character either [1]). Considering the fact that no possibility of construction is asserted there is no reason why the non-vanishing of the Cartesian product of non-vanishing sets or the idea of a well-ordering of the continuum should be regarded as hazardous hypotheses.

True, the axiom of choice was explicitly formulated only in the twentieth century and apparently was not implicitly used earlier than two decades before. But, at that, every mathematical principle was once expressed for the first time, mostly long after it had been used implicitly and unconsciously. The development of mathematics through the centuries has been achieved in two directions: by drawing *new conclusions* from previously admitted premises; as well as, in a less conspicuous way by adding *new premises* or principles to those admitted before, in accordance with the needs of science.

In fact one has arrived at the axiom of choice just as at other mathematical principles, viz. by *a posteriori* examining and logically analysing concepts, methods, and proofs actually found in mathematics whose original development in an intuitive manner rests on psychological rather than on logical foundations. This way of analysing then yielded the principle in question, and a reference to the intuitive or logical evidence of the principle was at best secondary [2]). Thus the Greek mathematicians were induced to include the axiom of parallels among the principles of geometry – an achievement whose ingenuity was fully appreciated only more than two thousand years later. When the independence of the axiom of parallels was ultimately proved by the middle of the 19th century nobody proposed to renounce those parts of geometry (metric and affine geometry) which depended on the axiom. In an analogous way we should admit the argumentations of algebra, analysis, and set theory that use the axiom of choice but at the same time examine what results can be obtained without the axiom and avoid it whenever possible. Hereby one learns to distinguish the domains of mathematics which are independent of the existential principles of choice and well-ordering from those where they are indispensable.

The analogy with geometry, where Absolute Geometry would correspond

[1]) Cf. Hadamard 05.

[2]) In the philosophical systems of Kant and Fries this idea plays an essential part. In particular in logic many principles are chiefly justified by the evidence of their consequences (cf. Schiller 28) and in geometry one may regard the existence of similar but non-congruent figures as much more evident than the axiom of parallels. Thus it seems hardly possible to accept the contrary attitude of Collingwood 33.

to the former domains, suggests the question: what shape will analysis and set theory assume by accepting a principle *contradicting* the axiom of choice? Such a "non-Zermelian" theory in some sense corresponds to non-Euclidean geometry [1]).

The last remarks seem to suggest the attitude taken in *Principia Mathematica*, namely not to ask whether the axiom of choice is "true", unavoidable, or admissible, but what additional parts of mathematics can be obtained by the admission of the axiom. This attitude appears even more legitimate today after the consistency and the independence of the axiom have been proved. This attitude is, however, opposed by Gödel's quasi-platonic conception of mathematical truth [2]), according to which the statement of the axiom has a categorical rather than a hypothetical character.

§5. THE AXIOM OF FOUNDATION

5.1. **Introducing the Axiom.** A very simple, but fundamental, question concerning the notion of set is left unsettled by Axioms I–VIII, namely: Does there exist a set *s* which is a member of itself [3])? If our set theory were based on the axiom schema of comprehension (p. 31) the answer would be trivially positive, since the set of all sets would be a member of itself (as is indeed the case in Quine's set theory in which there is a set which contains all sets – see Chapter III, §3). However, there is nothing in Axioms I–VIII to indicate the existence or the non-existence of such sets [4]). This observation makes

[1]) Alternatives to the axiom of choice, most of them concerning the theory of ordinal numbers, are discussed by Specker 57 and Bachman 55, §39 (some of these alternatives were proposed by Church 27). Out of the alternatives discussed by Specker 57, H (and C) was proved consistent by Feferman and Levy (see Cohen 66, Ch. IV, §10); the model used for this purpose also satisfies G_1. Alternative B is consistent (if and only if the existence of an inaccessible number is consistent) – Hájek 66. Other alternatives have not yet been proved consistent. An alternative of a different character is the axiom of determinateness of Mycielski and Steinhaus 62 (see also Mycielski 64–66 and Mycielski–Swierczkowski 64); for consequences of this axiom see also Addison–Moschovakis 68, Martin 68, and Y.N. Moschovakis 70.

[2]) Gödel 47 (in particular, footnote 2).

[3]) This question was first posed in the framework of axiomatic set theory, by Mirimanoff 17. (In type theory no set can be a member of itself, which is one aspect of Russell's vicious circle principle.)

[4]) The existence of such sets is compatible with Axioms I–VIII – see footnote 5 on p. 29 and footnote 1 on p. 101. Also their non-existence is compatible with the Axioms – see §5.5.

room for a new axiom which is called upon to decide this question; such is our next and last axiom of ZF, the axiom of foundation.

Each one of Axioms I–VIII was taken up because of its essential role in developing set theory and mathematics in general; if any single axiom were left out we would have to give up some important fields of set theory and mathematics. (Omitting Axiom I would, maybe, not jeopardize any field of mathematics, but would make the development of mathematics and set theory highly inconvenient.) The case of the axiom of foundation is, however, different; its omission will not incapacitate any field of mathematics. Yet, this axiom is of great interest in the study of the foundations of set theory.

In our search for an axiom which settles the question of the existence of a set which contains itself let us try and see to what extent we can „construct" sets without stumbling on such a set. Our present account of this "construction" will be informal; its rigorous development is postponed to §5.3. We shall temporarily widen the scope of our discussion by including in it also set theories which admit individuals but which are otherwise like ZF [1]).

We divide the universe of elements into *layers*. The bottom layer consists of all individuals – these are, in a certain sense, the simplest elements. (In ZF, which does not admit individuals, the bottom layer is empty.) The next layer consists of all sets of individuals – these can be considered to be the simplest sets. The third layer consists of all sets whose members are taken from the first two layers, i.e., sets which consist of individuals and sets of individuals, and so on. Whether there are individuals or not, in the second layer we get the null-set O, in the third layer we find the set $\{O\}$, the fourth layer contains the sets $\{\{O\}\}$ and $\{O, \{O\}\}$, and so on. On top of the first ω layers there is the ω-th layer, which consists of all the sets whose members belong to the first ω layers. In the ω-th layer we find the sets $Z^* = \{O, \{O\}, \{\{O\}\}, ...\}$ and $Z_1^* = \{O, \{O\}, \{O, \{O\}\}, ...\}$ (of pp. 47–48), one of which is taken to be the set of all natural numbers. In higher layers we find the set of all rational numbers (defined in one way or another), and then the set of all real numbers, etc.

As long as we proceed within the succession of the layers we will never reach a set which contains itself as a member, since all the members of a set in a given layer belong to lower layers. In the same way we can argue that within the layers there is no set which is a member of a member of itself, or a member of a member of a member of itself, etc.

Every "set" t of layers is immediately followed by a new layer which consists of all sets whose members belong to layers in t or to layers which are lower than some layer of t. Our construction of the layers is such that the

[1]) For such a set theory see footnote 1 on p. 25.

layers are stacked in a well-ordered fashion. Consequently, if one or more layers have members in common with a given collection of elements then there is a lowest layer which has members in common with that collection. In other words, if the collection of all elements x which fulfil a given condition $\mathfrak{P}(x)$ has members in common with some layer then there is a layer $\mathfrak{T}$ which is the lowest layer containing members of this collection, i.e., $\mathfrak{T}$ is the lowest layer containing elements x which fulfil $\mathfrak{P}(x)$. Let u be an element in the layer $\mathfrak{T}$ which fulfils $\mathfrak{P}(x)$; u is either an individual or a set. If u is a set then all the members of u, if any, belong to layers lower than $\mathfrak{T}$ and therefore none of them fulfils $\mathfrak{P}(x)$; if u is an individual it is still true that no member of u fulfils $\mathfrak{P}(x)$ since u has no members.

Let us refer to the sets which belong to the various layers as *well-founded sets*. It conforms to a large extent with the intuitive notion of set that the only objects which can rightfully be called sets are the well-founded sets. If we take up this point of view we can assert, according to the last paragraph, the following

AXIOM SCHEMA (IX) OF FOUNDATION (OR REGULARITY). **If there is some element x which fulfils $\mathfrak{P}(x)$ then there is a minimal element u which fulfils $\mathfrak{P}(x)$, i.e., u fulfils $\mathfrak{P}(x)$ but none of its members fulfils $\mathfrak{P}(x)$** [1]).

In symbols,

$$\forall z_1 \ldots \forall z_n [\exists x \varphi(x) \rightarrow \exists x(\varphi(x) \wedge \forall y(y \in x \rightarrow \neg \varphi(y)))],$$

where y is not free in the formula $\varphi(x)$ and $z_1, \ldots, z_n$ are the free variables of $\varphi(x)$ other than x.

We shall see later (in §5.3) that Axiom IX does indeed assert that all sets are well-founded. Let us now remark that even if one does not agree that only well-founded sets can be rightfully called sets, one can advance no argument for retaining sets which are not well-founded other than the desire for greater generality. This greater generality is not of much use since no field of set theory or mathematics is in any need of sets which are not well-founded. Opposing the desire for more generality there is always a desire for more restricted and definite notions, if by restricting the discussion no interesting

[1]) von Neumann 25, p. 239 and 29, p. 231; for different versions see already Mirimanoff 17 and Skolem 23, §6. Zermelo 30 introduced the name *Axiom der Fundierung*, because according to the axiom any "descending" sequence terminates (i.e., reaches its bottom or "foundation") after a finite number of steps. Bernays 37–54 II uses the name 'Restrictive Axiom'. A peculiar axiom which is essentially the conjunction of the axiom of extensionality and a weakened form of Axiom IX is given by Finsler 26 (Axiom II).

mathematical results are lost (see the discussion in §6.4). Thus one can accept Axiom IX not as an article of faith but as a convention for giving a more restricted meaning to the word 'set', to be discarded once it turns out that it impedes significant mathematical research.

Opposed to this way of looking at Axiom IX several authors [1]) put the intuitive reasoning which led to the notion of a well-founded set as the very basis for the notion of set, on a par with extensionality and comprehension. This point of view adopts one of the basic tenets of the logicistic attitude and in particular of type theory (see Ch. III, §2). The weakest part of this point of view is that the reasoning leading to the concept of a well-founded set uses the *well-ordering* of the layers, and the notion of well-ordering of, let us say, non-denumerable sets is far from being simple or conceptually fundamental. Axiom IX in itself does not involve the concept of a well-ordering, but it seems that this concept is unavoidable in any attempt to give an intuitive justification of Axiom IX. Also, it is traditional with the axiomatic attitude, as opposed to the logicistic attitude, that the attitude towards an axiom is mostly determined by the usefulness of the axiom in mathematics. Since, as we remarked above, Axiom IX is not essential for mathematics, it cannot be regarded as fundamental by the traditional axiomatic attitude [2]).

Axiom IX was formulated so as to be suitable also for set theories with individuals, but from now on we shall limit the discussion to the present axiomatic system which does not admit individuals.

By substituting $\sim\mathfrak{P}(x)$ for $\mathfrak{P}(x)$ in Axiom IX we get easily (by passing to its contrapositive) the following formulation of the axiom: *If* $\mathfrak{P}(x)$ *is a condition such that, whenever all the members of a set u fulfil* $\mathfrak{P}(x)$, *u itself also fulfils this condition, then every set fulfils* $\mathfrak{P}(x)$. In this formulation Axiom IX has the form of an axiom of induction [3]).

If we take for $\mathfrak{P}(x)$ in Axiom IX the condition $x\in y$ we get

IX*. *If y is a non-empty set then y has a member u such that* $u\cap y=0$.

One can also prove that IX* implies Axiom schema IX and is, hence,

[1]) Klaua 64, Kreisel 65, and Shoenfield 67.

[2]) Bernays 46 says "... the weaker form of the vicious circle principle, that no totality can contain members involving this totality ..., as Gödel says, ... is satisfied also by those systems of axiomatic set theory which have an axiom like Zermelo's Axiom der Fundierung (our Axiom IX), restricting sets to those called ordinary by Mirimanoff 17 (our well-founded sets). However, in axiomatic set theory the guiding idea for avoidance of the paradoxes is not that of the vicious circle principle but of "limitation of size" ...".

[3]) Tarski 55, where it is mentioned that, as a consequence of Axiom IX, one can also *define* by induction (recursion) on the membership relation.

equivalent to it [1]). Thus, in contrast to the axiom schema of replacement, Axiom schema IX can be replaced by a single axiom. We still chose the schema IX rather than IX* to serve as the principal version of the axiom because in some of the uses of the axiom Axiom schema IX is directly applicable whereas IX* is not [2]).

Let us consider now the case of a sequence $(s_1, s_2, s_3, \ldots)$ such that, for every $i \geqslant 1$, $s_{i+1} \in s_i$, i.e., $\ldots \in s_{i+1} \in s_i \in \ldots \in s_3 \in s_2 \in s_1$. Let y be the set $\{s_1, s_2, \ldots\}$. By IX*, y has a member u such that $u \cap y = 0$, but this cannot be the case since if u is s_k, for some $k \geqslant 1$, then $s_{k+1} \in u \cap y$. Thus Axiom IX contradicts the existence of such a sequence $(s_1, s_2, \ldots)$.

IX**. *There is no sequence* $(s_1, s_2, s_3, \ldots)$ *such that, for every* $i \geqslant 1$, $s_{i+1} \in s_i$ [3]).

One can also prove, using the axiom of choice, that IX** implies IX* and is, hence, equivalent to Axiom IX [4]).

The proof is as follows. Let r be a relation on y which consists of all ordered pairs $\langle u, v \rangle$ such that u and v are members of y and $v \in u$. If y is not as in IX* then y and r satisfy the hypothesis of the axiom of dependent choices (p. 65) and hence also its conclusion, which asserts the existence of a sequence $(s_1, s_2, s_3, \ldots)$ of members of y as in IX**.

In IX** the terms $s_1, s_2, s_3, \ldots$ of the sequence are not necessarily different from each other; thus Axiom IX rules out the existence of a set s which is a member of itself since in this case we get the sequence $\ldots \in s \in s \in s$, or of a set s which is a member of a member t of itself since in this case we get the sequence $\ldots s \in t \in s \in t \in s$, etc. In the cases where $s \in s$, $s \in t \in s$, etc. we can also apply IX* directly to the sets $\{s\}, \{s, t\}$, etc., respectively, to get a contradiction. Thus the axiom of foundation does indeed decide the question raised at the beginning of the present section.

For the system ZFC_σ, discussed in §4.4, IX* can be given a particularly neat formulation. In ZFC_σ, if y is a non-void set then $\sigma(y)$ is always a member of y. Since by IX* a non-void set y has always a member u such that $u \cap y = 0$ we can take $\sigma(x)$ to be such a u and thus take up the following version of IX*.

IX^*_σ $\quad \sigma(y) \cap y = 0$ [5]).

[1]) This was observed by Gödel – see Bernays 37–54 VI, p. 68. This equivalence is unprovable in various systems of set theory weaker than ZF – see Boffa 69 and Jensen–Schröder 69, where references to earlier work of Vopěnka, Hájek, and Hauschild is given.

[2]) Such as the proof of $\mathrm{IX}^{(3)}$ on p. 94.

[3]) Mirimanoff 17, Skolem 23 (§6).

[4]) The use of the axiom of choice here is essential – see Mendelson 58.

[5]) Bernays 58, p. 202.

IX$^*_\sigma$, in conjunction with VIII$_\sigma$, which asserts that $y \neq 0 \supset \sigma(y) \in y$, obviously implies IX*. On the other hand, IX* does not exactly imply IX$^*_\sigma$ in the presence of Axioms I–VIII$_\sigma$ since the only assertion we made in VIII$_\sigma$ concerning $\sigma(y)$, for $y \neq 0$, is that $\sigma(y) \in y$, and this, obviously, does not imply $\sigma(y) \cap y = 0$ even if y has always a member u such that $u \cap y = 0$. However, once we assume Axioms I–VIII$_\sigma$ and IX* we can *define* an operation σ' by $\sigma'(y) = \sigma(\{u \mid u \in y \,\&\, u \cap y = 0\})$, and easily prove Axioms VIII$_\sigma$ and IX$^*_\sigma$ with σ replaced by σ'.

Let us now return to the informal discussion which led us to the adoption of Axiom IX and try to make this reasoning precise. The major notion involved was the notion of the layers. If we want to give a correct definition of the layers it is convenient to have an indexing system for them. Since, as was mentioned above, the layers are stacked in a well-ordered fashion we shall use the ordinal numbers to enumerate the layers. This brings us to the topic of the ordinal numbers.

5.2. **Ordinal Numbers**. In the discussion of the ordinal numbers we shall not use the axioms of choice and foundation. Also throughout the rest of the present section we shall not use these axioms where we can do without them; whenever we shall use them this will be mentioned explicitly.

The ordinal numbers, as defined in T (p. 187), are the order types of well-ordered sets. In T (p. 138) the notion of an order type is not a *defined* notion of set theory; it is introduced by abstraction from the defined notion of similarity of ordered sets [1]). In a formal axiomatic theory this amounts to the introduction of a new primitive notion "the order type of $\langle a, r \rangle$", which we can write as $\overline{\langle a, r \rangle}$, together with the axiom: $\overline{\langle a, r \rangle} = \overline{\langle b, s \rangle}$ *if and only if the ordered sets* $\langle a, r \rangle$ *and* $\langle b, s \rangle$ *are similar*, and which the corresponding strengthening of the axiom schemas (where 'condition' now also stands for formulas which contain the new order type symbol). We shall see that in ZF the notion of an order type can be defined, so there is no need to introduce it in ZF as a new primitive notion. However, before we can deal with order types in general we have to deal first with the ordinal numbers.

We shall first discuss ordinal numbers informally, with the aim of later introducing a formal definition for this notion.

Let us denote with $W(\alpha)$ the set of all ordinals smaller than α; then we have

(1.1) If $\beta < \alpha$ then $W(\beta) \subseteq W(\alpha)$ (trivial).

[1]) See T, pp. 58/9 for the concept of abstraction. For the relation of *similarity* of ordered sets see T, pp. 134 f. and Suppes 60, p. 128.

(1.2) The relation $<$ well-orders $W(\alpha)$ and the order type of this ordered set is α (T, p. 197).

Zermelo and von Neumann [1]) were led by (1.2) to define the ordinals in such a way that α becomes *equal* to $W(\alpha)$, i.e., each ordinal is the set of all smaller ordinals. E.g., the least ordinal 0 is the set of all smaller ordinals, i.e., $0=O$; the next ordinal is $1=\{O\}$, then $2=\{O,\{O\}\}$, and so on; the least infinite ordinal is the set of all finite ordinals $\omega=\{0,1,2,\ldots\}=\{O,\{O\},\{O,\{O\}\},\ldots\}$ (this is the set Z_1^* of p. 48). On these ordinals the relation $<$ coincides with the $\in$-relation, i.e., $\alpha<\beta$ just in case $\alpha\in\beta$ $(=W(\beta))$.

DEFINITION. A set x is said to be *transitive* if every member of x is a subset of x, i.e., if every member of a member of x is a member of x $(\bigcup x\subseteq x)$.

Replacing $<$, $W(\alpha)$ and $W(\beta)$ by $\in$, α and β, respectively, in (1.1) we see that if we take each ordinal to be the set of all smaller ordinals then each ordinal is a transitive set. Carrying out the same replacements also in the first half of (1.2) we see that our new ordinals satisfy the requirements of the following definition.

DEFINITION. *The set x is said to be an ordinal if x is transitive and the $\in$-relation well-orders x*, i.e., if (a) x is transitive, (b) for all $u\in x$ we have $u\notin u$, (c) for all u, v, w in x, if $u\in v\in w$ then $u\in w$, and (d) every non-empty subset z of x has a member u such that $u\in v$ for every member v of z other than u. (There is no need to require that for all $u,v\in x$, $u\in v$ or $u=v$ or $v\in u$, since this follows from the requirement on the subsets z of x by taking $z=\{u,v\}$.) [2])

We shall now see that the ordinals as defined above do indeed behave as we expect them to. For this purpose we shall list here several theorems (numbered (2.1) to (2.8)) concerning the ordinals. Most of their proofs are easy and the reader can supply them himself with the aid of our hints or look them up in the literature [3]). The variables α, β, γ will vary over ordinals; we shall also write $\alpha<\beta$ for $\alpha\in\beta$ and $\alpha\leqslant\beta$ for $\alpha<\beta$ or $\alpha=\beta$.

[1]) von Neumann 23 and Zermelo (unpublished) about 1915 (quoted by Bernays 37–54 II, p. 6). These ordinals were discussed in detail by Mirimanoff 17 (who did not refer to them as ordinals).

[2]) Another definition, essentially due to Mirimanoff 17, defines x to be an ordinal if x is transitive and for all u, v in x, $u\in v$ or $u=v$ or $v\in u$ (Bernays 37–54 II, §5). The development of the theory of ordinal numbers from that definition, as well as the proof of the equivalence of the two definitions, makes an essential use of Axiom IX, since if x is a set such that $x=\{x\}$ then x is an ordinal according to the present definition, but is not a true ordinal (see Bernays 37–54 VII, p. 87).

[3]) E.g., Suppes 60, Sections 5.1 and 7.1.

(2.1) α is not a member of itself.

(2.2) If $\alpha<\beta$ and $\beta<\gamma$ then $\alpha<\gamma$.

(2.3) If $y\in\alpha$ then y is an ordinal. (Hence α is the set of all ordinals smaller than α.)

(2.4) If $y\subset\alpha$ and y is transitive then $y\in\alpha$. (Hint: Let u be the least member of α not in y, prove $u=y$.)

(2.5) For any two ordinals α,β $\alpha<\beta$ or $\alpha=\beta$ or $\beta<\alpha$. (Hint: Otherwise, (2.4) implies that $\alpha\cap\beta\in\alpha\cap\beta$.)

(2.6) If there exists an ordinal α which fulfils $\mathfrak{P}(x)$ then there is a least ordinal β which fulfils $\mathfrak{P}(x)$, i.e., β fulfils $\mathfrak{P}(x)$ and no $\gamma<\beta$ fulfils $\mathfrak{P}(x)$. (Hint: If α fulfils $\mathfrak{P}(x)$ but is not the least ordinal which fulfils $\mathfrak{P}(x)$ then take β to be the least ordinal in α which fulfils $\mathfrak{P}(x)$.)

(2.7) For every well-ordered set $\langle a,r\rangle$ there exists a unique ordinal α – the ordinal number of $\langle a,r\rangle$ – such that $\langle a,r\rangle$ is similar to $\langle\alpha,\in_\alpha\rangle$, where $\in_\alpha$ is the set of all ordered pairs $\langle\beta,\gamma\rangle$ with $\beta<\gamma<\alpha$.

(2.8) If all the members of a set x are ordinals then its union-set $\bigcup x$ is an ordinal too. ($\bigcup x$ is an upper bound of x, i.e., for every $\alpha\in x$, $\alpha\leqslant\bigcup x$.)

One can define functions on the ordinals by *transfinite induction* (which is sometimes also referred to as *transfinite recursion*). Prior to the formulation of this procedure let us give an auxiliary definition. Given a function F, by means of a functional condition (p. 50), and a set x such that $F(y)$ is defined for every $y\in x$, we denote with $F|x$ the set of all ordered pairs $\langle y,F(y)\rangle$, where $y\in x$. $F|x$ is a set by the axiom of replacement; it is called the *restriction of F to x*.

DEFINITION BY TRANSFINITE INDUCTION ON THE ORDINALS. For every function G (given by a functional condition) which is defined on all sets, one can formulate a functional condition which yields a function F defined on all ordinals such that for every ordinal α $F(\alpha)=G(F|\alpha)$. Moreover, this function F is unique in the sense that if F' is another function which is defined on all ordinals and such that $F'(\alpha)=G(F'|\alpha)$ for all ordinals α, then F' coincides with F on all ordinals, i.e., $F'(\alpha)=F(\alpha)$ for every ordinal α [1]).

5.3. **Well-founded Sets.** We shall now develop rigorous notions which correspond to the informal notions of the layers and the well-founded sets in §5.1. The set which we shall denote with $R(\alpha)$ will be, more or less, the union of the α first layers.

DEFINITION. We define the function R on all ordinals by transfinite

[1]) See proofs in Suppes 60, p. 207, Th. 8 and *T*, §10, Th. 7. For more general schemas of proof and definition by induction see Montague 55 and Tarski 55.

induction as follows: $R(\alpha)=\bigcup_{\beta<\alpha}\mathbf{P}(R(\beta))$ (where by the right-hand side we understand the union-set of the set of all $\mathbf{P}(R(\beta))$'s for $\beta<\alpha$) [1]).

This definition is the particular case of the general schema of definition by transfinite induction which is obtained by taking for G the function given by $G(x)=$ the union of all sets $\mathbf{P}z$, where z is a set such that for some $y, \langle y,z\rangle \in x$.

It is easily seen that $R(0)=\bigcup 0=0$, $R(1)=\bigcup\{\mathbf{P}(0)\}=\mathbf{P}(0)=\{0\}$, $R(2)=\{0,\{0\}\}$, $R(3)=\{0,\{0\},\{\{0\}\},\{0,\{0\}\}\}$, and so on. We shall now list a few theorems (numbered (3.1) to (3.7)) related to the function R, without full proofs [2]).

(3.1) For every ordinal α, $R(\alpha)$ is a transitive set, i.e., if $y\in x\in R(\alpha)$ then $y\in R(\alpha)$. (Hint: Prove by induction on α.)

(3.2) If $\beta<\alpha$ then $R(\beta)\in R(\alpha)$ (follows immediately from the definition of R) and $R(\beta)\subseteq R(\alpha)$ (by (3.1)).

(3.3) y is a member of some $R(\alpha)$ if and only if y is a subset of some $R(\alpha)$. (Hint: By (3.2) and $\mathbf{P}(R(\alpha))\subseteq R(\alpha+1)$.)

DEFINITION. A set x is said to be *well-founded* if it is a member (or a subset) of $R(\alpha)$ for some ordinal α. The rank $\rho(x)$ of a well-founded set x is the least ordinal β such that $x\subseteq R(\beta)$.

For a well-founded set x we have, as easily seen,

(3.4) $\rho(x)\leqslant\alpha$ if and only if $x\subseteq R(\alpha)$.

(3.5) $\rho(x)<\alpha$ if and only if $x\in R(\alpha)$.

(3.6) If $y\in x$ then y is well-founded and $\rho(y)<\rho(x)$.

We shall now prove:

(3.7) If all the members of a set z are well-founded then z, too, is well-founded.

Proof. By the axiom of replacement there is a set u which consists of all ordinals $\rho(y)$ for $y\in z$. By (2.8), the set u has a strict upper bound β; thus $\rho(y)<\beta$ for each $y\in z$. By (3.5) we have for each $y\in z$, $y\in R(\beta)$, i.e., $z\subseteq R(\beta)$, which establishes the well-foundedness of z.

Now we can formulate Axiom IX as

$\mathrm{IX}^{(3)}$. *All sets are well-founded.*

Let us now prove the equivalence of $\mathrm{IX}^{(3)}$ and IX and thereby clarify the meaning of Axiom IX. First we assume IX and prove $\mathrm{IX}^{(3)}$ by contradiction as follows. Assume that there is a set x which is not well-founded, then, by IX, there is a minimal set y which is not well-founded. Since y is a minimal non-well-founded set, every member of y is well-founded, hence, by (3.7), y is well-founded, which is a contradiction.

[1]) The function R is, essentially, the function ψ of von Neumann 29.

[2]) For proofs see Bernays 37–54 VI, pp. 66–67, and Shepherdson 51–53 II, § 3.2.

Let us now assume $\mathrm{IX}^{(3)}$ and prove IX. Given a condition $\mathfrak{P}(x)$ such that there is a set u which fulfils $\mathfrak{P}(x)$, let us consider the following condition $\mathfrak{Q}(y)$ on y – "y is the rank of some set x which fulfils $\mathfrak{P}(x)$". By $\mathrm{IX}^{(3)}$ u is well-founded, hence $\rho(u)$ is an ordinal which fulfils $\mathfrak{Q}(y)$. Therefore, by (2.6), there is a least ordinal α which fulfils $\mathfrak{Q}(y)$, i.e., there is a set x such that $\alpha = \rho(x)$ and x fulfils $\mathfrak{P}(x)$, whereas no set z with $\rho(z) < \alpha$ fulfils $\mathfrak{P}(x)$. Since, by (3.6), for each member z of x, $\rho(z) < \rho(x)$, no member of x fulfils $\mathfrak{P}(x)$; thus x is as required by IX.

In ZFC_σ (p. 72) one can define a one-one function which maps the universe of all sets on the collection of all ordinals or, what amounts to the same thing, one can define a relation which well-orders the universe of all sets in such a way that for every set y all the sets which precede y in this well-ordering constitute a set [1]). As a consequence, one can prove in ZFC_σ the following schema. *Given a collection of sets, determined by a condition $\mathfrak{P}(x)$, there is a one-one function F (given by a functional condition) whose range is this collection and whose domain is either the collection of all sets (the universe) or a set.* Informally, we can formulate this schema as follows. *Every collection of sets is either a set or equinumerous to the collection of all sets.* Since no set can be equinumerous to the collection of all sets (as follows immediately from the axiom of replacement and Th. 6 on p. 40) we can also formulate the schema as: *A collection of sets is a set if and only if it is not equinumerous to the collection of all sets.* This is the ultimate formulation of the "limitation of size" doctrine (p. 32) [2]).

5.4. **Cardinal Numbers. Order Types. Isomorphism Types.** We saw in §2.5

[1]) This is proved, essentially, in Bernays 37–54 VI, pp. 69–71 and in von Neumann 29 (see p. 137). That this cannot be proved in ZFC (for any function $\sigma(x)$) follows easily from the results of Easton 64 (see 70). That this cannot be proved in ZFC_σ without using the axiom of foundation is shown as hinted here. Within a system which contains Axioms I–VIII_σ and in which there is an infinite set u of indistinguishable sets y (such that, say, $y = \{y\}$), define a model (p. 292) that consists of the collection of all sets which are subsets of some transitive set that contains only finitely many members of u, and of the membership relation for such sets. In this model, Axioms I–VIII_σ (with the same σ) hold, whereas no relation well-orders (or even orders) the universe of sets. The statement in the text cannot be proved in a system of set theory which is like ZFC_σ but admits individuals. This is shown by a model as above, where u is now an infinite set of individuals. Notice that if u contains all the individuals then every set of individuals which is in the model is finite, hence there is no set of the model which contains all individuals.

[2]) A very similar statement was taken as an axiom by von Neumann 25 (Axiom IV2); cf. also von Neumann 29 (p. 227, *B*).

that many notions of set theory and mathematics can be reduced to the notion of membership, i.e., they can be defined within our theory. In the present subsection we shall see that this is also the case with respect to the notions of cardinal number and order type, as well as several other similar notions.

Whichever way is used to define the cardinality $|x|$ of the set x, the defined notion can be considered to correspond to the intuitive notion of cardinality only if we can prove

(4.1) $|x|=|y|$ if and only if the sets x and y are equinumerous.

The simplest way to define cardinal numbers would be to define the cardinality $|x|$ of the set x as the set of all sets which are equinumerous to x. (This is the Frege–Russell definition [1]), which is also used in Quine's set theory – Chapter III, §3.) However, even in ZFC such a set does not usually exist. E.g., if there were a set w which consists of all sets equinumerous to a given singleton, i.e., a set w which consists of all singletons, then $\bigcup w$ would be the set of all sets, contradicting Theorem 6 on p. 40.

Even though the Frege–Russell definition is unavailable in ZFC, there is another, still quite natural, way to define cardinal numbers in ZFC. It is by sheer luck that we can specify for each set x a particular set which is equinumerous to x, the choice of which is uneffected by replacing x with an equinumerous set. This is done as follows.

DEFINITION. a is said to be a *cardinal number* if a is an ordinal number *which is not equinumerous to any smaller ordinal*. (The cardinal numbers are called *initial numbers* in T, p. 216.) The *cardinality* $|x|$ of a set x is defined as the unique cardinal number a which is equinumerous to x. (The existence of such an a follows easily from the *well-ordering theorem*, (2.6), (2.7) and the transitivity of equinumerosity; the uniqueness of a follows from the transitivity of equinumerosity and (2.5).)

(4.1) is now easily shown.

We shall now proceed to define the notion of an order type in ZFC. Whichever way is used to define the order type $\overline{\langle a, r\rangle}$ of the ordered set $\langle a, r\rangle$, the definition must be such as to enable us to prove

(4.2) For any two ordered sets $\langle a, r\rangle$ and $\langle b, s\rangle$, $\overline{\langle a, r\rangle}=\overline{\langle b, s\rangle}$ if and only if $\langle a, r\rangle$ is similar to $\langle b, s\rangle$.

DEFINITION. The *order type* of the ordered set $\langle a, r\rangle$ is the set which consists of all ordered sets $\langle |a|, s\rangle$ which are similar to $\langle a, r\rangle$ (the order type of $\langle a, r\rangle$ is a subset of $\{|a|\}\times \mathbf{P}(|a|\times|a|)$ [2]). A set b is an *order type* if it is the order type of some ordered set.

[1]) Frege 1884, §68, and 1893–03 I, §42; Russell 03, §111.

[2]) This definition is due to A.P. Morse – see Scott 55.

(4.2) can easily be shown once one uses the axiom of choice to prove that for every set a, $|a|$ exists.

According to our definition of order types ordinal numbers are not order types. E.g., the ordinal number of the ordered set $\langle 0, 0 \rangle$ is 0 whereas its order type is $\{\langle 0, 0 \rangle\}$. If we want the ordinals to be order types too we can define the order types as follows: The order type of the ordered set $\langle a, r \rangle$ is the ordinal of $\langle a, r \rangle$ if $\langle a, r \rangle$ is a well-ordered set, and is the set of all ordered sets $\langle |a|, s \rangle$ which are similar to $\langle a, r \rangle$, otherwise.

In analogy to our definition of order types one can also define isomorphism types of groups, rings and Boolean algebras, homeomorphism types of topological spaces, etc.

In our definitions of the notions of a cardinal number, an order type, etc., in ZFC we made use of the full force of the axiom of choice (but we did not use the axiom of foundation); thus these definitions are not available in ZF. We shall now see that by means of the axiom of foundation one can give a general schema for definition by *abstraction*, in the sense of Cantor (see *T*, p. 59), by means of the notion of a $\simeq$-equivalence type [1]). This notion is a generalization of the notions of cardinal number, order type, etc.

Suppose we are given the collection of all sets which satisfy a condition $\mathfrak{P}(x)$. We say that a relation $\simeq$ is an *equivalence relation* on this collection if the following requirements (a)–(c) are satisfied.

(a) $x \simeq x$ for every set x in the collection. (*Reflexivity*)

(b) If $x \simeq y$ then $y \simeq x$. (*Symmetry*)

(c) If $x \simeq y$ and $y \simeq z$ then $x \simeq z$. (*Transitivity*)

We want to define the $\simeq$-equivalence type of every set x in the collection, i.e., we want to correlate to every set x in the collection an element $\tau(x)$ in such a way that we shall be able to prove

(4.3) For every x and y which fulfil $\mathfrak{P}(x)$, $\tau(x) = \tau(y)$ if and only if $x \simeq y$.

If all sets which fulfil the condition $\mathfrak{P}(x)$ constitute a set u then for each x in u we can define $\tau(x)$ to be that subset of u which consists of all members y of u such that $y \simeq x$. It is easily seen that (4.3) does indeed hold in this case. In naive set theory we can again follow the Frege–Russell method and define $\tau(x)$ to be the set of all sets y such that $y \simeq x$. However, in ZF we cannot in general be sure that there exists a set which contains all these y's. Therefore we define

DEFINITION. Given a relation $\simeq$ which is an equivalence relation on the collection of all sets which fulfil a given condition $\mathfrak{P}(x)$, we define, for the members x of this collection, the $\simeq$-equivalence type of x, $\tau(x)$, to be the set

[1]) This notion, as presented here, is due to Scott 55.

u of all sets y *of minimal rank* such that $y \simeq x$ (i.e., u is the set of all sets y such that $y \simeq x$ and for no set z with $\rho(z) < \rho(y)$ does $z \simeq x$ hold).

$\tau(x)$ is always a set, since, as easily seen, $\tau(x) \subseteq \mathbf{P}(R(\rho(x)))$. It is now easy to prove (4.3).

This general notion of the $\simeq$-equivalence types generalizes the notions of cardinal numbers, order types, isomorphism types of groups, etc., since the latter notions can be taken to be the types which correspond, respectively, to the equivalence relations of equinumerosity of sets, similarity of ordered sets, isomorphism of groups, etc. (4.1) and (4.2) become particular cases of (4.3). For example, according to our present definition, the cardinality $|x|$ of a set x is the set which consists of all sets y which are of minimal rank among the sets equinumerous to x.

Even though our present method is more general than that which we described on p. 96 above, in ZFC the method of p. 96 is sufficient, in practice, for all the cases where equivalence types are actually used. As far as ZFC is concerned the method of p. 96 is simpler since it does not appeal to the axiom of foundation or to the notion of rank, neither of which occurs in ordinary mathematical discussions [1]).

5.5. **Consistency and Independence of the Axiom of Foundations and of the other Axioms.** As mentioned on p. 58, where the question of the consistency and the independence of the axiom of choice was discussed, it seems impossible to obtain a convincing proof of the consistency of ZF. Therefore, what we can reasonably hope to do with respect to a given axiom of ZF is to assume that the axiomatic system which consists of all other axioms is consistent and, using this assumption, to prove that ZF is consistent.

From now on we shall refer to the axiomatic system consisting of Axioms I–VIII as 'the system I–VIII' and similarly to other systems.

The question of the consistency of the axiom of foundation relative to the other axioms has been answered positively by Gödel's proof of the consistency of the axiom of choice (pp. 59–60). Gödel, in essence, proved even the

[1]) In set theory based on Axioms I–VII, i.e., ZF without the axiom of foundation, one cannot define even a notion of cardinality for which (4.1) is provable. Therefore, if one wants the notion of cardinality to be available one has no choice but to add the operation $|x|$ as a new undefined notion to the language of set theory and add (4.2) as a new axiom (Tarski 24) and to strengthen the notion of condition in the axiom schemas appropriately. The fact that $|x|$ cannot be defined in the system I–VII follows from the fact that after $|x|$ is introduced in this system as a new primitive notion with axioms as above, one can prove in it new theorems which do not mention the notion of cardinality (see A. Levy 69a).

stronger result that if the system I–VII is consistent, so is ZFC [1]). We shall outline here an earlier proof, due to Skolem and von Neumann [2]), that if the system I–VIII is consistent so is ZFC [3]). We choose to present here von Neumann's proof, even though Gödel's proof yields a much stronger result, because the former is a very simple illustration of some of the methods used in relative consistency proofs in set theory and in other mathematical theories (see Chapter V, §3).

We shall now explain what we mean by an *interpretation* of (the language of) ZFC in the system I–VIII. Such an interpretation is given by means of a collection $\mathfrak{A}$ of elements together with a binary relation $\mathfrak{E}$ on this collection. Statements $\mathfrak{S}$ of ZFC are interpreted by letting the variables vary over the members of the collection $\mathfrak{A}$ and by taking $\in$ to be the symbol for the relation $\mathfrak{E}$. For every statement $\mathfrak{S}$ of ZFC we obtain a statement $\mathfrak{S}^*$ of the system I–VIII ($\mathfrak{S}^*$ happens to be in the same language as $\mathfrak{S}$, since the systems ZFC and I–VIII happen to use the same language) which asserts that the interpretation of $\mathfrak{S}$ holds. For example, if $\mathfrak{S}$ is $x \in y$ then $\mathfrak{S}^*$ is a statement which asserts that x and y are in the relation $\mathfrak{E}$; if $\mathfrak{S}$ is "For every set x there is a set y such that $x \in y$" then $\mathfrak{S}^*$ is "For every x in $\mathfrak{A}$ there is a y in $\mathfrak{A}$ such that x is in the relation $\mathfrak{E}$ to y". In all the interpretations with which we shall deal, 'equality' is interpreted as equality.

In order to prove the consistency of Axiom IX we shall produce a *model* of ZFC in the system I–VIII [4]), i.e., we shall produce an interpretation of ZFC in the system I–VIII such that the interpretation $\mathfrak{S}^*$ of every statement $\mathfrak{S}$ which is an axiom of ZFC is a theorem of the system I–VIII. In order to do this we must specify unambiguously the axioms of ZFC. For this purpose we shall use the following axioms as axioms for ZF: I, III, IV, VIa, VII, VIII and IX*. (Axioms II and V are left out because they are implied by the other axioms.)

In the system I–VIII, the function R and the notion of a well-founded set are available (since nowhere in the definition of these notions and in the proofs of their elementary properties did we utilize the axiom of foundation). Let us now interpret the statements of ZFC as follows: *The variables will*

[1]) See Shepherdson 51–53 I, pp. 164–5 or Rieger 57 (and footnote 6 on p. 59).

[2]) Skolem 23, §6, and von Neumann 29.

[3]) The same proof also shows that if the system I–VII is consistent so is ZF.

[4]) For a detailed discussion of the notion of an interpretation, and its relationship with the notion of a model, see Ch. V, §3. By the terminology used there, we have here just an interpretation of ZFC in the system I–VIII; the term "model" is used there for a somewhat different notion. The reason why we use here this term is explained on p. 292.

range only over the well-founded sets and the $\in$-symbol will retain the meaning of membership (i.e., $\mathfrak{A}$ will be the collection of all well-founded sets and $\mathfrak{E}$ will be the membership relation). Under this interpretation, if $\mathfrak{S}$ is the statement "There is a set which has no member", in symbols $\exists x \neg \exists y(y \in x)$, then $\mathfrak{S}^*$ is the statement "There is a *well-founded* set x which has no *well-founded* member".

Notice that originally only the range of the variables and the $\in$-relation are interpreted; the defined notions of ZFC become interpreted only by interpreting their definitions. For example, O is defined as the set which has no member; the interpretation O* of O is therefore defined as the well-founded set which has no well-founded member. Since, by (3.6), the members of a well-founded set are well-founded, the only well-founded set without well-founded members is O, thus O* = O. The equality of O and O* here is just an accident. Indeed, because the interpretation given here is a very simple and natural one, many defined notions coincide with their interpretations, but this is not necessarily the case with other interpretations.

We shall use asterisks also to denote the interpretations of various defined notions of set theory. E.g., $x \subseteq^* y$ means 'Every well-founded member of the set x is also a member of the set y' (see Definition I on p. 26). For well-founded sets x and $y, x \subseteq^* y$ holds if and only if $x \subseteq y$ (by (3.6)), thus we can say that $\subseteq^*$ and $\subseteq$ coincide for well-founded sets.

If $\mathfrak{S}$ is Axiom I then $\mathfrak{S}^*$ is "For all well-founded sets x, y if $x \subseteq^* y$ and $y \subseteq^* x$ then $x = y$". Since, for well-founded sets, $x \subseteq^* y$ means $x \subseteq y$ and $y \subseteq^* x$ means $y \subseteq x$, as we saw above, $\mathfrak{S}^*$ follows immediately from Axiom I.

If $\mathfrak{S}$ is Axiom III then $\mathfrak{S}^*$ is "For every well-founded set a there exists a well-founded set y which consists of the well-founded members of a". For a given well-founded set a we choose $y = \bigcup a$. If $a \subseteq R(\alpha)$ then, by (3.1), also $y \subseteq R(\alpha)$, hence y is well-founded. It is now easy to verify, using (3.1), that y is indeed as required in $\mathfrak{S}^*$.

It is also easy to prove that the interpretations of the following axioms hold: IV (since if $a \subseteq R(\alpha)$ then $\mathbf{P} a \subseteq R(\alpha+1)$), VIa (since the set $R(\omega)$ is well-founded, $0 \in R(\omega)$ and if $x \in R(\omega)$ then $\{x\} \in R(\omega)$), and VIII (since if t is well-founded so are $\bigcup t$ and all its subsets).

As to Axiom schema VII, we have to show that the interpretation $\mathfrak{S}^*$ of every statement $\mathfrak{S}$ of the form "For every set a, if for every $t \in a$ there is at most one set x such that $\mathfrak{P}(t, x)$ holds, then there exists a set y which contains exactly those sets x for which $\mathfrak{P}(t, x)$ holds for some $t \in a$" (where y is not free in $\mathfrak{P}(t, x)$) does indeed hold. $\mathfrak{S}^*$ is the statement "For every well-founded set a, if for every well-founded $t \in a$ there is at most one well-founded set x such that $\mathfrak{P}^*(t, x)$ holds, then there exists a well-founded set y which contains exactly those well-founded sets x for which $\mathfrak{P}^*(t,x)$ holds for some well-founded member t of a", where $\mathfrak{P}^*(t,x)$ is the interpretation of $\mathfrak{P}(t,x)$. Let $\mathfrak{Q}(t,x)$ be the condition given by 't and x are well-founded and $\mathfrak{P}^*(t,x)$". By the hypothesis of $\mathfrak{S}^*$, $\mathfrak{Q}(t,x)$ is a functional condition on a. Therefore, by Axiom VII (in the system I–VIII), there exists a set y which contains exactly those sets x for which $\mathfrak{Q}(t,x)$ holds for some member t of a. We have, $x \in y$ if and only if, for some member t of a, t and x are well-founded and $\mathfrak{P}^*(t,x)$ holds, i.e., $x \in y$ holds if and only if x is well-founded and for some well-founded member t of a $\mathfrak{S}^*(t,x)$ holds. Since all the members of y are well-founded, by (3.7), y itself is well-founded; thus $\mathfrak{S}^*$ is proved.

Let us now conclude the proof that our interpretation is indeed a model of ZFC by showing that if $\mathfrak{S}$ is Axiom IX* then $\mathfrak{S}^*$ is a theorem of the system I–VIII. $\mathfrak{S}^*$ is "Every well-founded set y which has at least one well-founded member has a well-founded member which has no well-founded member in common with y". Given such a set y, among the well-founded members of y there is at least one member u of least rank; by (3.6), u has no member in common with y, as claimed in $\mathfrak{S}^*$. Notice that we did not use Axiom IX in the proof of $\mathfrak{S}^*$; the proof was carried out in the system I–VIII.

To sum up what we did, we showed in the system I–VIII that there is a collection of objects, namely the collection of all well-founded sets, and a binary relation on this collection, namely the membership relation for well-founded sets, for which all the axioms of ZFC do indeed hold. This can be shown to guarantee that if there is no contradiction in the system I–VIII then there is no contradiction in ZFC. Moreover, this interpretation provides a very simple method by means of which if, God forbid, a proof of a contradiction were found in ZFC one would also obtain a proof of a contradiction in the system I–VIII. For the detailed arguments which establish these claims see Chapter V, §3. The idea is, roughly, as follows. If one is given a proof of a contradiction in ZFC one can reproduce in the system I–VIII, by means of the interpretation given above, the arguments which lead to the contradiction. Thus one will get in the system I–VIII a contradiction from the notion of well-foundedness, which is a correctly defined notion of this system.

Having dealt with the relative consistency of the axiom of foundation, we can also ask whether it is independent of the other axioms, assuming the consistency of the system I–VIII. As mentioned in passing at the beginning of the present section, one cannot even prove from Axioms I–VIII that no set contains itself [1]). Also, even if we add to Axioms I–VIII an axiom which

[1]) The neatest proofs of this and the other results mentioned here concerning the

asserts that for no finite $n>1$ does there exist a sequence $(s_1, ..., s_n)$ such that $s_1 \in s_n \in s_{n-1} \in ... \in s_2 \in s_1$, one still cannot prove the axiom of foundation.

We have considered till now the questions of the consistency and independence of the axioms of choice and foundation. We shall now ask the same questions concerning the other axioms. Let us mention first the problem of establishing the (relative) *consistency* of the various axioms, i.e., we shall ask whether one can show that if ZFC with a certain axiom omitted is consistent then ZFC is consistent too. For reasons which come out of Gödel's theorem on consistency proofs, and which will be explained in Chapter V, pp. 328–329, this cannot be done in the case of the axioms of union, power-set, infinity and replacement, not even if relatively strong means of proof are admitted [1]). The *independence* of each of those axioms (where one assumes for each of the axioms the consistency of the system consisting of all other axioms of ZFC) can be proved by appropriate models, or even by the same arguments which are used to show the impossibility of proving the consistency of those axioms [2]). The axioms of pairing and subsets follow from the other axioms, as we have seen. As to the axiom of extensionality, if ZFC is consistent then this axiom, in each one of its versions, is independent of the other axioms [3]). The answer to the question of whether it is possible to prove the relative consistency of the axiom of extensionality depends on the way in which equality is introduced (i.e., whether equality is taken to be a primitive notion of logic or set theory, or introduced by one of the three definitions we considered) and may depend also on the particular formulations of the other axioms [4]).

independence of the axiom of foundation are those of Rieger 57; for other proofs see Bernays 37–54 VII, Mendelson 56a, and Specker 57. The consistency with respect to the system I–VII of an axiom which is, in some respects, an extreme opposite of the axiom of foundation was proved by Scott (cf. A Levy 65b, Th. 47), Hájek 65, and Boffa 68.

[1]) In the terminology of p. 328, each one of these axioms is a strengthening axiom, since for each such axiom Φ we can prove in ZFC the existence of a set which is a model of the system which consists of all the axioms of ZFC except Φ.

[2]) See p. 329. Such proofs are given by Bernays 37–54 VI and Mendelson 56. The axiom schema of replacement does not even follow from all other axioms and finitely many of its own instances (see footnote 4 on p. 53).

[3]) A. Robinson 39.

[4]) Scott 61.

§6. QUESTIONS UNANSWERED BY THE AXIOMS

6.1. **The Generalized Continuum Hypothesis.** One of the earliest central problems in set theory, which could not be answered even by the means of naive set theory (as long as one did not use the idea behind some antinomy), is the *continuum problem*. In *Theory* pp. 69, 228–230 the history of Cantor's, and of the generalized continuum problem is sketched and references are given to the literature, where statements equivalent to the continuum hypothesis are introduced and where the hypothesis is used for proving various mathematical theorems [1]).

The generalized continuum hypothesis is the statement
H: $2^{\aleph_\alpha} = \aleph_{\alpha+1}$, for every ordinal α.
Cantor's continuum hypothesis is that particular case of *H* where $\alpha = 0$. Another version of the generalized continuum hypothesis is
H_1: If $\mathbf{c}$ is a transfinite cardinal then there is no cardinal $\mathbf{d}$ such that $\mathbf{c} < \mathbf{d} < 2^{\mathbf{c}}$ (in other words, for every reflexive set a if $b \subseteq \mathbf{P}a$ then $|b| \leqslant |a|$ or $|b| = 2^{|a|}$).

H_1 implies the axiom of choice in the system I–VII [2]). In particular, H_1 implies that for every α, $2^{\aleph_\alpha}$ is equal to some $\aleph_\beta$ with $\beta > \alpha$; if β were greater than $\alpha+1$ we would get $\aleph_\alpha < \aleph_{\alpha+1} < 2^{\aleph_\alpha}$, which contradicts H_1, thus H_1 implies also *H*. On the other hand, it is immediately seen, by means of the well-ordering theorem that the conjunction of *H* with the axiom of choice implies H_1 (in the system I–VII) and is, hence, equivalent to H_1 in that system.

In the stronger system I–VII, IX (which also contains the axiom of foundation) even H implies the axiom of choice (in fact, in that system the axiom of choice is equivalent to the statement that for every ordinal α, $2^{\aleph_\alpha} = \aleph_\beta$, for some β) [3]).

What we are now interested in is the question whether the continuum

[1]) Notably, Sierpiński 56 and Bachmann 55. Many of the consequences of the continuum hypothesis are also consequences of Martin's axiom, which does not yield any information on the ordinal γ for which $2^{\aleph_0} = \aleph_\gamma$ – see Martin–Solovay 70 and Solovay–Tennenbaum 71 (or Jech 71).

[2]) Specker 54 proves that if H_1 holds for $c=a$ and $c=2^a$ then 2^a (and hence also a) is an aleph, i.e., a cardinal of a well-ordered set. (The proof is also given in Bachmann 55, § 35, 1 and in Kuratowski–Mostowski 68, IX, § 6.) Earlier results in this direction are due to Lindenbaum, Tarski, and Sierpiński (see the proof in Cohen 66, Ch. IV, § 12). That *H* does not imply the axiom of choice in the system I–VII follows easily from the consistency of *H* combined with any Fraenkel–Mostowski–type "model" of I–VII in which the axiom of choice fails (such as given by Mendelson 56a or Specker 57).

[3]) Rubin–Rubin 63, p. 17, Kruse 63.

hypothesis, in its simple or general form, can be proved or refuted by means of the present, or of another, system of axioms of set theory. Cantor's efforts in the beginning of the 1880's were tragically unsuccessful, and during the next fifty years no actual progress was made, Hilbert's attempt notwithstanding [1]).

In 1938, Gödel proved, in his work which we have already discussed in §4.2, that the generalized continuum hypothesis cannot be refuted in ZFC (unless ZFC contains a contradiction). This was done as follows: In the system I–VII Gödel constructed a model, in the sense of §5.5 (p. 99), of the system ZFC$^+$ obtained from ZF by adding to it the *axiom of constructibility* (which asserts that all sets are constructible – see p. 60), thereby proving that if the system I–VII is consistent so is ZCF$^+$ [2]). Then Gödel proved in ZFC$^+$, in addition to the axiom of choice, also the generalized continuum hypothesis H [3]).

Since we shall now also discuss the independence of the continuum hypothesis and related statements, let us agree that *throughout this and the next subsection all the results concerning consistency and independence of various statements of set theory will rely, tacitly, on the assumption that* ZFC *is consistent* (or, what amounts to the same thing, that the system I–VII is consistent).

In 1963, P.Cohen established the independence of Cantor's continuum hypothesis, using the same methods he used in the proof of the independence of the axiom of choice, i.e., he showed that Cantor's continuum hypothesis (and, a fortiori, the generalized continuum hypothesis) cannot be proved in ZFC. Moreover, Cohen proved that one cannot refute in ZFC any statement of the form $2^{\aleph_0} = \aleph_\beta$, as long as '$\beta$' is the name of an ordinal which is defined in a "reasonable" way [4]), $\beta > 0$, and β is not the limit of a strictly increasing sequence γ_n, $n < \omega$, of ordinals. In particular, one cannot refute in ZFC any of the statements $2^{\aleph_0} = \aleph_n$ or $2^{\aleph_0} = \aleph_{\omega+n}$, where n is any fixed finite ordinal $\geqslant 1$ [5]).

Cohen's result, which we have just mentioned, is indeed the strongest independence result one could expect concerning $2^{\aleph_0}$; in ZF one can prove

[1]) Hilbert 25.

[2]) Gödel 38, 39, 40; also Cohen 66, Shoenfield 67, Karp 67, Jensen 67, Mostowski 69, Jech 71.

[3]) The proof of Gödel 40 was simplified by Doss 63 and Rieger 63. Still simpler proofs are given by the other authors mentioned in the previous footnote.

[4]) Such as the "absolutely definable ordinals (in the weaker sense)" of Hajnal 61.

[5]) Cohen 63/4, 65, 66; also Jensen 67 and 67b, Scott–Solovay ∞, Sacks 69, Mostowski 69, Rosser 69, Jech 71.

that if $2^{\aleph_0}=\aleph_\beta$ then $\beta>0$ (T, p. 112, Th. 2) and, as a consequence of the proof in T (pp. 118–119), that $c^{\aleph_0}$ (and, in particular, $2^{\aleph_0}$) cannot be the sum of a strictly increasing sequence of cardinal numbers (in which the König–Zermelo inequality – T, p. 98 – is used), β cannot be the limit of a strictly increasing sequence γ_n of ordinals. The requirement that β should be defined in a "reasonable" way must be made in order to avoid definitions of β such as "the successor of the ordinal γ such that $2^{\aleph_0}=\aleph_\gamma$". For such a β the statement $2^{\aleph_0}=\aleph_\beta$ is obviously refutable.

Proceeding along the same line of thought, let us now ask what ZFC does tell us concerning the value of $2^{\aleph_\alpha}$ for a general α. By the well-ordering theorem, for every ordinal α there is an ordinal β such that $2^{\aleph_\alpha}=\aleph_\beta$; therefore there is a function F defined on the ordinals whose values are ordinals such that for every α, $2^{\aleph_\alpha}=\aleph_{F(\alpha)}$. H asserts that for every α, $F(\alpha)=\alpha+1$. In ZFC we can prove the following statements (a)–(c) concerning the function F.

(a) $F(\alpha)>\alpha$ (Cantor's theorem, T, p. 112).

(b) If $\alpha<\beta$ then $F(\alpha)\leqslant F(\beta)$ (trivial).

(c) $F(\alpha)$ is not the limit (union) of a set of ordinals smaller than $F(\alpha)$ which is of cardinality $\leqslant\aleph_\alpha$. (This is again proved by means of the König–Zermelo theorem – T, p. 98.)

(d) If $\aleph_\alpha$ is a *singular* limit cardinal (i.e., α is a limit ordinal and is the limit (union) of a set of cardinality $<\aleph_\alpha$ of smaller ordinals) and for all ordinals δ such that $\gamma\leqslant\delta<\alpha$, where γ is some fixed ordinal $<\alpha$, $F(\delta)$ has the same value β, then also $F(\alpha)=\beta$ [1]).

It has been proved that, for every function F for which we can prove (a)–(c) in ZFC and which has a "reasonable" definition, one cannot refute in ZFC the statement which asserts that, for all the *regular* cardinals $\aleph_\alpha$, (i.e., where $\aleph_\alpha$ is not a singular limit cardinal – see also the definition on p. 110) [2]), $2^{\aleph_\alpha}=\aleph_{F(\alpha)}$ [3]). This completely settles the question as far as regular cardinals are concerned (and, in particular, for all cardinals of the form $\aleph_{\beta+1}$); the case of the singular cardinals is still mostly open. In particular, it is not yet known whether one can consistently assume that $2^{\aleph_\alpha}=\aleph_{\alpha+2}$ for every ordinal α (or even only for every ordinal $\alpha\leqslant\omega$) [4]).

[1]) Bukovský 65 (also discovered by S. Hechler).

[2]) For a detailed treatment of the notions of regular and singular cardinals, see Bachmann 55, §6.

[3]) Easton 70; see also Rosser 69.

[4]) There are versions of the generalized continuum hypothesis appropriate for set theory without the axiom of choice. For the formulation of those versions and for results related to their consistency and independence see Scarpellini 66, Derrick–Drake 67, Marek 66.

On account of the comparability of the cardinals one can prove in ZFC the existence of a set of real numbers of cardinality $\aleph_1$; yet all attempts to produce a definable set of real numbers such that its cardinality can be proved in ZFC to be $\aleph_1$ have failed [1]). If Cantor's continuum hypothesis is added to ZFC the answer to this problem becomes trivial since then the set of all real numbers, which is obviously a definable set, can be shown to be of cardinality $\aleph_1$. Thus, if we show that no definable set of real numbers can be proved in ZF to be of cardinality $\aleph_1$ then this also implies the independence of the continuum hypothesis. Indeed, one can prove the following stronger result: ZFC stays consistent when we add to it the statement $2^{\aleph_0} = \aleph_\beta$, where β is as above (p. 104), and the schema (see p. 69) "All definable sets of real numbers are finite, denumerable, or of cardinality $2^{\aleph_0}$" [2]).

Now that we know that neither the continuum hypothesis nor its negation is provable in ZFC, it is reasonable to ask whether set theory and analysis will bifurcate (or multifurcate) at the continuum hypothesis as plane geometry bifurcates (to Euclidean and hyperbolic geometries) at the axiom of parallels, on account of its independence of the other axioms of absolute geometry. According to Gödel [3]), "Only someone who (like the intuitionist) denies that the concepts and axioms of classical set theory have any meaning (or any well-defined meaning) could be satisfied with such a solution, not someone who believes them to describe some well-determined reality. For in this reality Cantor's conjecture must be either true or false, and its undecidability from the axioms as known today can only mean that these axioms do not contain a complete description of this reality". How are we going to know whether in the "well-determined reality" of the sets the continuum hypothesis is true or false? Gödel surmises that this question may be answered by "... other (hitherto unknown) axioms of set theory which a more profound understanding of the concepts underlying logic and mathematics would enable us to recognize as implied by these concepts". Gödel considers as natural candidates for this role axioms which say something on what is meant by the very notion of set, i.e., axioms which tell us whether the subsets of a given set are sets which are constructible or definable in some manner or are arbitrary multitudes of members of the given set; in Gödel's words: "... the continuum problem ... may be solvable by means of a new axiom which would state or at least imply something about the definability of sets".

However, one can hardly believe that many mathematicians will be inclined

[1]) See, e.g., Hardy 04 and Lusin 34.
[2]) A. Levy 70; it also follows from Solovay 70, Th. 3, part (4).
[3]) All the quotations from Gödel in the present subsection are taken from Gödel 47.

to accept some new axiom solely, or mainly, on the basis of a metaphysical belief in its power to reveal the true nature of the concept of set. Mathematicians usually tend to accept a new axiom if it serves as a cornerstone for significant mathematical theories (as was the case, e.g., with the axiom of choice) or, at least, if one can use it to prove a considerable number of interesting mathematical theorems. Such a position, though with somewhat stricter requirements, is also considered by Gödel: "... even disregarding the intrinsic necessity of some new axiom, and even in the case it had no intrinsic necessity at all, a decision about its truth is possible also in another way, namely, inductively by studying its "success", that is, its fruitfulness in consequences and in particular in "verifiable" consequences, i.e., consequences demonstrable without the new axiom, whose proofs by means of the new axiom, however, are considerably simpler and easier to discover, and make it possible to condense into one proof many different proofs. The axioms for the system of real numbers, rejected by the intuitionists, have in this sense been verified to some extent owing to the fact that analytical number theory frequently allows us to prove number theoretical theorems which can subsequently be verified by elementary methods. A much higher degree of verification than that, however, is conceivable. There might exist axioms so abundant in their verifiable consequences, shedding so much light upon a whole discipline, and furnishing such powerful methods for solving given problems (and even solving them, as far as that is possible, in a constructivistic way) that quite irrespective of their intrinsic necessity they would have to be assumed at least in the same sense as any well established physical theory" [1]).

Now, more than two decades after Gödel wrote the passages quoted above, his hope for a more profound understanding of the notion of set resulting in some new axioms which settle the continuum problem has not yet been fulfilled, in spite of the great advances obtained during this period in the metamathematics of set theory (in particular, Cohen's solution of the problem of the independence of the continuum hypothesis). On the other hand, the generalized continuum hypothesis turned out to be useful for proving many mathematical theorems [2]), some of them even "verifiable". Thus the general-

[1]) From this point of view one cannot rule out the possibility of a bifurcation of set theory, since, as in the case of plane geometry, it is conceivable that each one of two incompatible axioms may be extremely rich in verifiable consequences.

[2]) In addition to Sierpiński 56 and Bachmann 55, see, e.g., Erdös–Hajnal–Rado 65 and Erdös–Hajnal 66. In the case of Erdös 64, it is difficult to decide whether the consequences of the continuum hypothesis are more attractive than those of its negation. See J. Friedman 71 for a statement equivalent to the generalized continuum hypothesis.

ized continuum hypothesis is well on its way towards being accepted as one of the axioms of set theory, as used by most mathematicians. Nowadays most mathematicians would not doubt the truth of a mathematical theorem proved by means of the generalized continuum hypothesis, even though many of them would prefer a proof which does not use this hypothesis. The generalized continuum hypothesis, even more than the axiom of choice, thrives on the lack of competition; no alternative to the generalized continuum hypothesis is known to produce any interesting mathematical results. Also from the very formulation of the generalized continuum hypothesis it is clear that every alternative which specifies exactly the cardinalities of $2^{\aleph_\alpha}$ for all α's, i.e., which specifies the function F such that, for all α, $2^{\aleph_\alpha}=\aleph_{F(\alpha)}$, is almost bound to be somewhat artificial (unless some completely new ideas are applied) [1]).

6.2. **The Axiom of Constructibility**. This axiom has already been discussed earlier (pp. 64 and 104). In the system I–VII it implies the axioms of choice and foundation, as well as the generalized continuum hypothesis. We have also mentioned that Gödel proved, essentially, that the system I–VII stays consistent when the axiom of constructibility is added to it. P. Cohen proved that the axiom of constructibility cannot be proved in ZFC, not even by means of the generalized continuum hypothesis [2]). Moreover, he showed that even though the axiom of constructibility asserts that all sets are constructible, one cannot prove in ZFC (not even by means of the generalized continuum hypothesis) that there are more than denumerably many constructible real numbers [3]).

As an additional axiom for set theory the axiom of constructibility is somewhat attractive. It implies the axioms of choice and foundation as well as the generalized continuum hypothesis. It has also a good number of additional mathematically interesting consequences [4]). The most dramatical conse-

[1]) Maybe an axiom asserting that $2^{\aleph_0}$ is a very large cardinal will do; e.g. the hypothesis that there is a real valued measure on *all* subsets of the real line – see Solovay 71 and Kunen ∞.

[2]) Cohen 63/4, 65, 66, Shoenfield 67, Jensen 67, Scott–Solovay ∞, Sacks 69, Mostowski 69, Rosser 69. For results which establish the consistency of the existence of non-constructible sets which are very simple from the point of view of definability, i.e., Δ^1_3-sets, see Jensen–Solovay 70 and Jensen 70.

[3]) Cohen 66, Ch. IV, § 10. For the stronger result that it is consistent with ZF that all the constructible real numbers are Δ^1_3, see Jensen–Solovay 70.

[4]) Consequences in the theory of effective and classical hierarchies – Addison 59 and 62; consequences in infinite combinatorics – Jensen ∞a; consequences in algebra

quence of the axiom of constructibility is the negation of the famous *Souslin hypothesis* (*T*, p. 166) [1]). The axiom of constructibility also appeals to one's sense of economy – if we think of the ordinal numbers as a fixed given totality then the axiom of constructibility asserts that there are no sets other than those which can be proved to exist. However, it is by no means universally accepted that this economy is indeed a virtue; some mathematicians may consider it as a seclusion from the richness of set theory. Also, unlike the axiom of choice and the generalized continuum hypothesis, this axiom suffers from serious competition. There are alternatives to the axiom of constructibility which possess some attractions of their own. Such is *the hypothesis of the transcendence of the infinite cardinal numbers* [2]), which is implied by the still stronger *hypothesis of the existence of measurable cardinals* [3]).

6.3. **Axioms of Strong Infinity.** Now let us examine, informally, the methods available in ZFC for obtaining sets of larger and larger cardinality. Our starting point for obtaining infinite sets is the axiom of infinity which guarantees the existence of denumerable sets, i.e., sets of cardinality $\aleph_0$. By means of the power-set axiom we can now obtain sets of the cardinalities $2^{\aleph_0}, 2^{2^{\aleph_0}}, \ldots$. Let us consider the set $A = \{\omega, \mathbf{P}\omega, \mathbf{PP}\omega, \ldots\}$ (where ω is the least infinite ordinal). The cardinality $\mathbf{a}$ of its union-set $\bigcup A$ can be easily shown to be greater than those of all its members, i.e., $\mathbf{a} > \aleph_0, 2^{\aleph_0}, 2^{2^{\aleph_0}}, \ldots$. By means of the power-set operation we now get the sets $\mathbf{P}\bigcup A$, $\mathbf{PP}\bigcup A, \ldots$ of the respective cardinalities $2^{\mathbf{a}}, 2^{2^{\mathbf{a}}}, \ldots$. Then we consider the set $B = \{\bigcup A, \mathbf{P}\bigcup A, \mathbf{PP}\bigcup A, \ldots\}$; the cardinality $\mathbf{b}$ of $\bigcup B$ is still greater than all the cardinals considered till now, and so on. Thus we have seen how to obtain sets of larger cardinality by means of the operations of power-set and union (the latter applied to sets which are obtained by means of the axiom of replacement).

The way in which the operations of power-set and union effect the cardinalities of the sets considered, is as follows. The power-set of a set of cardinal-

and model theory – Keisler 68 and Silver 71a; consequences in the theory of recursive functions of ordinal numbers – Machover 61, Takeuti–Kino 62, and Tugué 64. The axiom of constructibility also contradicts the existence of measurable cardinals – see p. 113.

[1]) Jensen ∞ (see Jech 71). For an elementary exposition of Souslin's hypothesis and references see Rudin 69. The first proof of the independence of the Souslin hypothesis was found by Tennenbaum 68 (and independently by Jech 67). The consistency of the hypothesis with respect to ZFC was established by Solovay–Tennenbaum 71 (see Jech 71). The consistency of Souslin's hypothesis with ZFC together with the generalized continuum hypothesis is shown by Jensen ∞b. Jensen ∞c discusses generalizations of Souslin's hypothesis.

[2]) Takeuti 65 and 65a.

[3]) For references see footnotes 1 and 3 on p. 113.

ity $\mathbf{c}$ is a set of cardinality $2^{\mathbf{c}}$; the union set of a set D such that each $t \in D$ has the cardinality $\mathbf{d}_t$ is a set of cardinality at most $\sum_{t \in D} \mathbf{d}_t$ (where $\sum_{t \in D} \mathbf{d}_t$ stands for infinite addition – see T, p. 83). Therefore we can say that a set of cardinality $\mathbf{e}$ cannot be obtained from sets of smaller cardinality by means of the operations of power-set and union, if for every cardinal number $\mathbf{c}$ such that $\mathbf{c} < \mathbf{e}$ also $2^{\mathbf{c}} < \mathbf{e}$, and for every indexed sum $\sum_{t \in D} \mathbf{d}_t$ of cardinal numbers $\mathbf{d}_t$ such that for every $t \in D$, $\mathbf{d}_t < \mathbf{e}$ and also $|D| < \mathbf{e}$ we have $\sum_{t \in D} \mathbf{d}_t < \mathbf{e}$. Actually, since the operation with the constant outcome $\aleph_0$ is also available, we shall be interested only in such cardinals $\mathbf{e}$ as above which are greater than $\aleph_0$. Accordingly we define

DEFINITION. A cardinal number e is said to be *regular* if it is infinite and it is not the sum of $< \mathbf{e}$ cardinal numbers each of which is $< \mathbf{e}$ (i.e., whenever $\mathbf{e} = \sum_{t \in D} \mathbf{d}_t$ we have $|D| \geqslant \mathbf{e}$ or otherwise, for at least one $t \in D$, $\mathbf{d}_t = \mathbf{e}$). $\mathbf{e}$ is said to be (*strongly*) *inaccessible* if (1) $\mathbf{e} > \aleph_0$, (2) $\mathbf{e}$ is regular, and (3) for every $\mathbf{c} < \mathbf{e}$, we have $2^{\mathbf{c}} < \mathbf{e}$ [1]).

If there are inaccessible cardinals at all, then the least inaccessible cardinal $\mathbf{t}_0$ is "very big", indeed. $\mathbf{t}_0$ is greater than each of $\aleph_0, 2^{\aleph_0}, 2^{2^{\aleph_0}}, \ldots, \mathbf{a}, 2^{\mathbf{a}}, 2^{2^{\mathbf{a}}}, \ldots, \mathbf{b}, 2^{\mathbf{b}}, \ldots$ (where $\mathbf{a}$ and $\mathbf{b}$ are as on p. 109). Yet $\mathbf{t}_0$ is very small compared to the second inaccessible cardinal $\mathbf{t}_1$, which is greater than $\mathbf{t}_0, 2^{\mathbf{t}_0}, 2^{2^{\mathbf{t}_0}}, \ldots, \mathbf{t}_0 + 2^{\mathbf{t}_0} + 2^{2^{\mathbf{t}_0}} + \ldots, \ldots$ The third inaccessible cardinal is again much bigger, and so on. One asks, naturally, whether such enormous cardinals do indeed exist. With respect to the system ZFC one gets: *If* ZFC *is consistent then one cannot prove in* ZFC *the existence of an inaccessible number* $\mathbf{e}$ [2]). The proof of this proceeds as follows. Let us consider the cardinals as particular ordinal numbers, as on p. 96. Let $\mathfrak{A}$ be the collection of all sets if there is no inaccessible cardinal; otherwise, let $\mathfrak{A}$ be the set $R(\theta)$, where θ is the least inaccessible cardinal (R is the function defined on p. 94). $\mathfrak{E}$ is taken to be the usual membership relation for members of $\mathfrak{A}$. Now it can be shown,

[1]) This notion was first introduced by Sierpiński–Tarski 30 and Zermelo 30. The adverb 'strongly' is added in order to distinguish these inaccessible cardinals from the *weakly inaccessible cardinals* (which are defined to be the regular cardinals $\aleph_\alpha$ whose index α is a limit number) introduced by Hausdorff 08, p. 443, and 14. Every inaccessible cardinal is also weakly inaccessible. These two notions are easily seen to coincide if the generalized continuum hypothesis is assumed, but they do not coincide in ZFC (see Cohen 66 and Vopěnka 64). Definitions of the notion of an inaccessible number in the literature sometimes differ from the present one in that they admit $\aleph_0$, and sometimes even 2, as an inaccessible cardinal. The present definition of this notion is appropriate only in the presence of the axioms of choice; for another definition, appropriate even for ZF, see Levy 60 (Def. 1).

[2]) Firestone–Rosser 49 (cf. already Kuratowski 25 and Baer 29). Independence proofs are given in Mostowski 49, Shepherdson 51–53 II, and Mendelson 56.

without much difficulty, that the collection $\mathfrak{A}$ together with the binary relation $\mathfrak{E}$ is a model, in the sense of p. 99, of the system obtained from ZFC by adding to it the axiom "there is no inaccessible cardinal". This model establishes that the latter system is consistent if ZFC is, i.e., if ZFC is consistent then the existence of an inaccessible cardinal $\mathbf{e}$ cannot be proved in it.

The next question which comes up in this connection is the following: Can one prove in ZFC *that there is no inaccessible cardinal*? Mathematical experience so far seems to deny this possibility. Considerable mathematical research has been done concerning inaccessible cardinals yet nothing seems to indicate that the assumption of the existence of an inaccessible cardinal leads to a contradiction. Of course, such an answer is hardly satisfactory; but, unless one proves eventually in ZFC that there are no inaccessible numbers – which seems highly unlikely – this seems to be the strongest statement that one can ever make in this direction. For reasons which are closely related to Gödel's theorem on consistency proofs, and which are discussed on p. 328, it is impossible to give a convincing proof that if ZFC is consistent so is the system obtained from ZFC by adding to it the axiom "there exists an inaccessible cardinal", i.e., that if ZFC is consistent then the existence of inaccessible cardinals cannot be refuted in ZFC [1]).

Let us denote with ZFC# the axiomatic system which is obtained from ZFC by adding to it the axiom "there exists at least one inaccessible cardinal", and let us see what we can say in ZFC# concerning the existence of at least two inaccessible cardinals. The second inaccessible cardinal is related to ZFC# in the same way as the first inaccessible cardinal is related to ZFC. If ZFC# is consistent then the existence of a second inaccessible cardinal cannot be proved in ZFC#, and we cannot show that the existence of a second inaccessible cardinal is not refutable in ZFC#; yet, present mathematical experience indicates that this is indeed the case. (The proofs of these statements proceed along the same lines as the proofs of the corresponding statements concerning ZFC and the first inaccessible cardinal.)

In the literature one can find axioms which assert the existence of more and more inaccessible cardinals [2]). When one comes to deal with the question of their consistency and independence, the phenomena observed above are

[1]) Shoenfield 54a.

[2]) Mahlo 11 and 12/3 (concerns only weakly inaccessible cardinals), Tarski 38 and 39a (or Bachmann 55, §42), A. Levy 60 and 60b, Bernays 61a, Montague 62, Gaifman 67, and the references in footnotes 6 on p. 112 and 3 on p. 113. For axioms which involve also classes as well as sets (§ 7), see Bernays 61a, Takeuti 61 and 69. Many of these axioms are consistent with the axiom of constructibility – see Gödel 40 (note 10 in the second printing) and Tharp 66 (cf., however, p. 113).

repeated, i.e., one can show that a stronger axiom (that is, an axiom which asserts the existence of more inaccessible cardinals) is independent of a weaker one (if the weaker one is consistent), and that one cannot give a convincing proof of the consistency of the stronger axiom even if the consistency of the weaker one is assumed [1]).

In 1914, Hausdorff wrote [2]): "If there are regular initial numbers with a limit-index (and as yet one has not succeeded in discovering a contradiction in this hypothesis), the least among them has so exorbitant a magnitude that it will hardly ever come into consideration for the usual purposes of set theory." We shall call Hausdorff's regular initial numbers with a limit-index *weakly inaccessible cardinals;* among them one finds all the inaccessible cardinals [3]). Contrary to Hausdorff's prediction, the inaccessible cardinals proved to have significance not only for the foundations of set theory, but also for certain applications [4]). There is yet another aspect for which Hausdorff's prediction failed. By assuming the existence of inaccessible numbers we can prove many new theorems which have nothing to do with large cardinals, such as theorems about the arithmetic of the natural and real numbers [5]) (though these theorems do not seem to be of the kind that will be encountered by a mathematician who is not interested in metamathematics).

The attitude of the mathematicians towards the question of the existence of inaccessible cardinals, which has now become more acute than had been envisaged by Hausdorff, will be discussed in the next subsection. Let us only remark at this point that mathematicians would be much more inclined to accept the existence of inaccessible cardinals than to reject it.

The inaccessible cardinals turn up most naturally when one considers properties $\mathfrak{P}$ of cardinals which turn out to be such that (1) $\aleph_0$ has the property $\mathfrak{P}$; (2) if $\mathbf{a}$ has the property $\mathfrak{P}$ then $2^{\mathbf{a}}$ has the property $\mathfrak{P}$; and (3) if $\{\mathbf{b}_t\}$, $t \in A$, is an indeed set of cardinals such that for each $t \in A$, $\mathbf{b}_t$ has the property $\mathfrak{P}$, and $|A|$ has the property $\mathfrak{P}$ too, then $\sum_{t\in A} \mathbf{b}_t$ also has the property $\mathfrak{P}$ [6]). (1)–(3) imply that all the cardinals which are not greater than

[1]) The axiom of choice plays no significant role in the results mentioned so far in the present subsection; i.e., ZFC can be replaced everywhere by ZF, provided we use the appropriate definition of the notion of inaccessibility (footnote 1 on p. 110).

[2]) Hausdorff 14, p. 131.

[3]) See footnote 1 on p. 110.

[4]) See Sierpiński–Tarski 30, Koźniewski–Lindenbaum 30, Sierpiński 34 (with references to other papers), and all the papers referred to in footnotes 5 and 6 below.

[5]) Tarski 38 (p. 87), Mostowski 49, Kreisel–Levy 68.

[6]) For such properties $\mathfrak{P}$ and related research see Keisler–Tarski 64 (and Bukovský 65a), Erdös–Hajnal–Rado 65 (p. 109), and the many papers in their bibliographies (in particular, also the work of Hanf, Monk, and Scott).

an inaccessible cardinal have the property $\mathfrak{P}$ (i.e., if there is no inaccessible cardinal then all cardinals have the property $\mathfrak{P}$, and if there are inaccessible cardinals then all cardinals less than the least inaccessible cardinal have the property $\mathfrak{P}$). In several cases one can prove that many of the inaccessible cardinals, and in particular the first inaccessible cardinal, the second inaccessible cardinal, etc., have the property $\mathfrak{P}$, but one still does not seem to be able to prove that all cardinals have this property. In this case the assumption that there exists a cardinal which does not have the property $\mathfrak{P}$ is sometimes referred to as an *axiom of strong infinity*, since it entails the existence of very "large" cardinals. Among the axioms of strong infinity obtained in this way the hypothesis of the existence of a *measurable cardinal* [1]) deserves special attention. It is a very strong assumption in that it is stronger than almost all other known axioms of strong infinity [2]). It has other interesting consequences, in particular it contradicts the axiom of constructibility; moreover, it implies that there are only $\aleph_0$ constructible sets of natural numbers, and that some relatively simple sets of natural numbers are not constructible [3]). It has also consequences concerning classical mathematical problems not related to the metamathematics of set theory, such as the Lebesgue measurability of certain sets of real numbers [4]). The hypothesis of the existence of a measurable cardinal is compatible both with the generalized continuum hypothesis, and with the negation of the simple continuum hypothesis [5]).

6.4. **Axioms of Restriction.** In 1922 Fraenkel proposed to add to set theory an *axiom of restriction,* i.e., an axiom which, written after all other axioms, reads, roughly: "There are no sets other than those whose existence follows directly from the axioms written down so far" [6]). Such an axiom would be analogous to Peano's axiom of induction in arithmetic and inversely analo-

[1]) This notion is defined, e.g., in Scott 61a, where references to results of Banach and Ulam are given, and in Shoenfield 67, §9.10. This notion has uses in algebra (see Fuchs 58, §47) and in geometry (see Köthe 60, §28.8) – cf. also Keisler–Tarski 64 and Chang 65.

[2]) Keisler–Tarski 64, A. Levy 71, Solovay 66.

[3]) Scott 61a, Vopěnka 62, Gaifman 71, Solovay 67 (where earlier results of Rowbottom and Silver are referred to). The assumption of the existence of strongly compact (strongly measurable) cardinals is a stronger axiom of infinity; it contradicts even a weak version of the axiom of constructibility – Vopěnka–Hrbáček 66 and Kunen 70. Still stronger notions of large cardinals are those of the supercompact cardinals of Reinhardt–Solovay ∞ and the extendible cardinals – see also Magidor 71 and 71a.

[4]) Solovay 69, Martin–Solovay 69, Martin 70, and Mansfield 71.

[5]) Silver 71, Jensen 67a, Levy–Solovay 67.

[6]) Fraenkel 22, pp. 233–234 (*Axiom der Beschränktheit*).

gous to Hilbert's *Vollständigkeitsaxiom* (axiom of completeness) [1]) in geometry and analysis. While the latter, in order to gain categoricity, postulates that the domain is as *comprehensive* as compatible with the axioms, in the case of set theory, as in arithmetic, an axiom of restriction would demand the domain to be as *narrow* as compatible with the axioms [2]). This may mean that the domain is the "intersection" of all "models" of the system of axioms, provided that such an intersection can be consistently assumed to exist and to fulfil the axioms. Then, e.g., the non-existence of inaccessible numbers, as well as that of non-well-founded sets, could be proved (even without the axiom of foundation). Originally it was suspected that such an axiom of restriction cannot be formulated within our axiomatic theory [3]); but, as we shall see, it is possible to formulate axioms which can be very reasonably equated with Fraenkel's axiom of restriction; it is on the basis of the contents of these axioms, and on the desirability of restriction in general, that we have to accept or reject these axioms.

When we say that there are no more sets than required by the axioms we mean to assert the schema: "If $\mathfrak{Q}$ is a property such that every set whose existence follows from the axioms has the property $\mathfrak{Q}$ then every set has the property $\mathfrak{Q}$". The difficulty which we now encounter is to express accurately the statement: "Every set whose existence follows from the axioms has the property $\mathfrak{Q}$". We have infinitely many axioms which imply the existence of sets (since all the instances of the axiom of replacement are such), so a straightforward attempt to formulate that statement would lead us to an infinite formula; therefore we have to proceed somewhat indirectly. Thus, our first axiom of restriction will be the following schema:

FIRST AXIOM OF RESTRICTION. If $\mathfrak{Q}$ is a property for which (1)–(6) below hold, then every set has this property.

(1) If x and y have the property $\mathfrak{Q}$ so has $\{x, y\}$.

(2) If x has the property $\mathfrak{Q}$ so has $\bigcup x$.

(3) If x has the property $\mathfrak{Q}$ so has $\mathbf{P}x$.

(4′) If x has the property $\mathfrak{Q}$, and $\mathfrak{P}$ is any condition, then the subset $x_{\mathfrak{P}}$ of x determined by the condition $\mathfrak{P}$ also has the property $\mathfrak{Q}$.

(5) Some infinite set has the property $\mathfrak{Q}$.

(6′) If x has the property $\mathfrak{Q}$, $\mathfrak{P}(t, z)$ is a functional condition on x, and for every $t \in x$, $\mathfrak{P}(t, z)$ implies that z has the property $\mathfrak{Q}$, then the set y which consists of all elements z such that $\mathfrak{P}(t,z)$ holds for some $t \in x$ also has the property $\mathfrak{Q}$.

[1]) Hilbert 1899 (from the second ed. on, Axiom V2).
[2]) Axioms of this type are discussed in Carnap–Bachmann 36.
[3]) von Neumann 25, Fraenkel 27.

(4′) and (6′) are still not single statements, since they mention all conditions $\mathfrak{P}$, but they can, obviously, be replaced by

(4) If x has the property $\mathfrak{Q}$ and $y \subseteq x$ then y, too, has the property $\mathfrak{Q}$.

(6) If x has the property $\mathfrak{Q}$ and f is a function whose domain is included in x, and its range y consists of elements which have the property $\mathfrak{Q}$, then y, too, has the property $\mathfrak{Q}$.

One can prove, in the system I–VIII, that the first axiom of restriction is equivalent to the conjunction of Axiom IX (of foundation), and the statement which asserts that there are no inaccessible cardinals [1]). One can also prove that to the rather limited extent permitted by the general theorem on the non-categoricity of theories formulated in first-order logic (p. 300), an axiom like the First Axiom of Restriction, or its equivalent version mentioned above, will indeed guarantee the categoricity of set theory [2]).

In addition to the general arguments against the restriction axioms, which we shall present below, let us examine in particular what has been achieved by the First Axiom of Restriction. Since we have adopted the axiom of foundation anyway, all that we can prove by means of the First Axiom of Restriction is that there are no inaccessible cardinals and the rather direct consequences of that (see the preceding subsection). This is too little for the idea of restriction, which should be a very powerful tool (as are the axiom of induction in arithmetic and the axiom of completeness in geometry). In particular, we cannot conclude anything from this axiom concerning the continuum hypothesis [3]). All this points towards searching for a stronger axiom of restriction.

When we now consider again the First Axiom of Restriction, trying to strengthen it, we see that its relative weakness is due to the excessive strength of assumptions (4′) and (6′). What we wanted to state in (1)–(6) is that if the collection determined by $\mathfrak{Q}$ (to which we shall refer loosely as "the collection $\mathfrak{Q}$") can serve as a model for ZF then this collection contains all sets. When we consider, for example, condition (4′) we notice that in order for the collection to serve as a model for ZF it is not necessary that it should contain, together with a set x, every subset $x_{\mathfrak{P}}$ of x determined by an arbitrary condi-

[1]) The axiom of restriction of Carnap 54, §43, is also equivalent to the same statement (in the formal system given there).

[2]) To obtain this categoricity result one has to formulate set theory in second-order logic, or with classes (see §7), as is done to obtain categoricity in arithmetic and geometry. The proof of categoricity is, essentially, contained in Zermelo 30.

[3]) This easily follows from the proof of the consistency of the non-existence of inaccessible cardinals (see p. 110), knowing that each of the continuum hypothesis and its negation is consistent with ZFC.

tion $\mathfrak{P}$, as demanded by (4′); it is enough if this holds only for those conditions $\mathfrak{P}$ which are defined by referring only to the members of the collection $\mathfrak{Q}$. A corresponding weakening of (6′) is now also called for. However, the schemas (4′) and (6′) were luckily equivalent to the single statements (4) and (6), but we are not that lucky with the new versions of (4′) and (6′). Nevertheless, by a considerable amount of effort and technical ingenuity, one can actually produce a single statement which can be proved to assert almost exactly what we are looking for. This is the

SECOND AXIOM OF RESTRICTION (i) All the sets are constructible, and (ii) there are no transitive sets which are models of ZF [1]).

Part (i) of this axiom implies that all sets are well-founded; moreover, we know that it also implies the generalized continuum hypothesis. Part (ii) of this axiom implies, in particular, the non-existence of inaccessible cardinals. Thus we see that the Second Axiom of Restriction implies the First Axiom of Restriction, and is indeed quite a powerful axiom. Let us now also notice a feature that both axioms of restriction have in common. Each is equivalent to the conjunction of two statements, one of which states that certain "large" ordinals or sets of large rank do not exist (inaccessible ordinals in the first axiom, transitive models of ZF in the second axiom), and can therefore be called a limitation on the "size" of the ordinals; the other statement declares that certain complicated sets which do not necessarily have a large rank, or a rank at all, do not exist (non-well-founded sets in the first axiom, non-constructible sets in the second axiom) [2]).

Another axiom which was suggested [3]) as a formalized version of the axiom of restriction is an axiom which asserts that every set "has a name", i.e., for every set x there is a parameterless condition $\mathfrak{P}$ such that x is the only set which fulfils $\mathfrak{P}$ (in this case we say that $\mathfrak{P}$ is a *description* of x). This condition cannot be stated in our present set theory, but can be stated in the language of the von Neumann–Bernays set theory (§7) [4]). One may be erroneously led to regard this as an axiom of restriction if one takes the vague statement, "There are no sets other than those whose existence is provable from the axioms", too literally. The fact that a set x has a description $\mathfrak{P}$ does

1) Shepherdson 51–53 III. See also Mostowski 49, Cohen 63 and 66.

2) See, e.g. Takeuti 69, for the distinction between these two kinds of restriction.

3) Suszko 51.

4) The consistency of this axiom with the axioms of VNB (§7.1) is proved by Myhill 52 (in order to obtain this result one has to assume that one can add consistently to VNB the full induction schema of set theory with classes (p. 139); cf. Montague–Vaught 59a). Moreover, as shown by Cohen 66, every transitive model of ZF in which the Second Axiom of Restriction holds also satisfies this axiom.

not mean that x must occur in every interpretation of set theory. The description $\mathfrak{P}$ of x is, in general, formulated by referring to *all* sets, and therefore, under a narrower concept of set, $\mathfrak{P}$ may describe some other set x' or cease to be a description at all. Such an axiom does not refute the existence of inaccessible numbers or of non-well-founded sets (the latter holds, of course, only if we do not assume the axiom of foundation) [1]).

Having discussed the question of how to formulate the axiom of restriction, let us consider now the question of whether such an axiom is at all desirable. In the case of the axiom of induction in arithmetic and the axiom of completeness in geometry, we adopt these axioms not because they make the axiom systems categorical or because of some metamathematical properties of these axioms, but because, once these axioms are added, we obtain axiomatic systems which perfectly fit our intuitive ideas about arithmetic and geometry. In analogy, we shall have to judge the axioms of restriction in set theory on the basis of how the set theory obtained after adding these axioms fits our intuitive ideas about sets [2]). To restrict our notion of set to the narrowest notion which is compatible with the axioms of ZFC just for the sake of economy is appropriate only if we have absolute faith that the axioms of ZFC (and the statements which they imply) are the only mathematically interesting statements about sets. It is difficult to conceive of such absolute faith in the sufficiency of the axioms of ZFC [3]) (as one would have in, say, the full axiom of comprehension if it were not inconsistent). Even if one had such a faith in the axioms of ZFC, it is likely that he would settle rather for something like an axiom of completeness, if there were some reasonable way of formulating it.

When we discussed, in §6.2, the desirability of the axiom of constructibility it was mentioned that it is highly dubious whether it should indeed be taken up as one of the axioms of set theory. Let us now see what the arguments are against that part of an axiom of restriction which limits the "size" of the sets. The usual arguments for and against the adoption of some statements as axioms involve some interplay between consideration of mathematical elegance, on the one hand, and Platonistic attitudes on the other. For example, from a Platonistic point of view a strong case can be made for the existence of individuals, but from the point of view of mathematical elegance there is little that speaks for this assumption (i.e., the individuals do not con-

[1]) This can easily be shown by the methods of Myhill 52 (or Cohen 66).

[2]) See Gödel's opinion on this comparison in Benacerraf–Putnam 64, p. 270, and those of Mostowski 67 and Kreisel 67.

[3]) Similar arguments are used in Zermelo 30.

tribute considerably to the existence of interesting mathematical structures), and so our sense of economy may well have the upper hand over our Platonistic tendencies.

As to the question of the mathematical elegance of the axioms of restriction, we do not say that an axiom of restriction contributes to mathematical elegance just because we can prove stronger theorems by means of it. The way in which an axiom of restriction helps to prove a stronger theorem is often simply by denying the existence of sets which do not conform to the desired theorem. This would be regarded as dishonesty by the conscientious mathematician and is exactly the case with our axioms of restriction; it is, however, not the case with the axiom of foundation or the assumption that there are no individuals. There are no deep mathematical theorems which we are able to prove after assuming these axioms just because we have banished all opposition (unless this happens in a too obvious way, as in the theorem that no set is a member of itself, and then such a transparent dishonesty cannot really be called dishonesty). After adopting these axioms we have, essentially, the same theory, only that the treatment and view are much more streamlined.

If one takes the Platonistic point of view, there is a consideration, in addition to mathematical elegance, which opposes size restrictions. As a result of the antinomies we know that there is no set which contains all sets; a reasonable way to make this conform to a Platonistic point of view is to look at the universe of all sets not as a fixed entity but as an entity capable of "growing", i.e. we are able to "produce" bigger and bigger sets. The axiom of restriction points to the existence of some fixed natural universe of sets, but if the collection of all sets in this universe is again a Platonistic entity, then why should it not be admitted as a new set by allowing a wider universe than that allowed by the axiom of restriction? [1]) When we try to reconcile the image of the ever-growing universe with our desire to talk about the truth or falsity of statements that refer to *all* sets we are led to assume that some "temporary" universes are as close an "approximation" to the ultimate unreached universe as we wish. In other words, there is no property expressible in the language of set theory which distinguishes the universe from some "temporary universes". These ideas are embodied in the *principles of reflection*, which are, mostly, strong axioms of strong infinity [2]).

As a last remark let us mention that one can prove that if ZFC is consis-

[1]) See Zermelo 30. Related arguments led von Neumann (in 25) to believe that no axiom schema of restriction can be formulated.

[2]) A. Levy 60 and 60b, Montague 61a and 62, Bernays 61 and 61a, Takeuti 61 and 69, Tharp 67.

tent it remains so even upon addition of the Second Axiom of Restriction [1]). On the other hand we know that we cannot prove that if ZFC is consistent then it remains consistent after adding to it an axiom which asserts the existence of an inaccessible number. This, however, will usually not be taken as an argument for the axiom of restriction as opposed to an axiom of strong infinity; we know that the reason for the impossibility of proving the relative consistency of strong axioms of infinity is due to Gödel's theorem on consistency proofs, and so this fact does not give rise to suspicions concerning the consistency of strong axioms of infinity (see pp. 328–329).

§7. THE ROLE OF CLASSES IN SET THEORY

7.1. **The Axiom System VNB of von Neumann and Bernays.** In the present section we shall discuss the various systems of set theory which admit, beside sets, also classes. Classes are like sets, except that they can be very comprehensive; an extreme example of a class is the class which contains all sets. We shall analyze in detail the relationship of these systems to ZF. The main point which will, in our opinion, emerge from this analysis, is that set theory with classes and set theory with sets only are not two separate theories; they are, essentially, different formulations of the same underlying theory.

We shall carry out a detailed discussion of an axiom system VNB due to von Neumann and Bernays which exhibits the typical features of set theory with classes. Later we shall consider, in somewhat less detail, the other main variants of set theory with classes, ignoring systems which differ only technically from the systems which we shall discuss.

In our exposition of set theory we had to use the metamathematical notion of a *condition* many times. We used it in formulating the axioms of subsets, replacement and foundation; we used it also in the least ordinal principle (2.6) on p. 93), in the (meta)theorem on definition by transfinite induction, and in definition by abstraction (in §6.4). When we look at other axiomatic mathematical theories to see whether they also make such a frequent use of the notion of a condition, we see that in some theories, e.g., the elementary theories of groups and fields in algebra, such a metamathematical notion does not occur at all, but in most theories this metamathematical notion is avoided only at the expense of developing the axiomatic theory within set theory, and using the mathematical notion of set instead of the metamathematical notion of condition. Thus, instead of formulating the

[1]) Shepherdson 51–53 III.

axiom of induction of number theory as "For every condition $\mathfrak{P}(x)$, if 0 fulfils the condition and if, for every x which fulfils the condition, $x+1$ fulfils it too, then every natural number x fulfils the condition", we can say "For every set P, if $0 \in P$ and for every x, if $x \in P$ then $x+1 \in P$ too, then P contains all natural numbers"; and instead of saying "For every condition $\mathfrak{P}(x)$, if some natural number fulfils the condition then there is a least natural number which fulfils it", we say "Every non-empty set P of natural numbers has a least member". Here, where we axiomatize the notion of set, it may seem at first sight that our task is easier than ever, since we deal with sets anyway, but this is not the case. When one considers a mathematical theory A given within set theory, the most basic fact of set theory used in developing A is the following principle of comprehenson:

(*) For every condition $\mathfrak{P}(x)$ of the theory A there exists a set Q which consists exactly of those objects (of the theory A) which fulfil the condition.

If we apply (*) to set theory we get the axiom schema of comprehension (of §3.1), which we already know to be contradictory. To avoid this contradiction we seem to have no choice but to assume that not all the sets guaranteed by the principle (*) are sets of set theory. To distinguish between the sets of set theory, i.e., the objects which were discussed in §§1–6 above, and the sets guaranteed by the principle (*), we shall refer to the latter as *classes*. Thus we have two kinds of sets; the sets of the first kind are still referred to as sets or, synonymously, as elements, and their behavior is expected to be as determined by the theory ZF; the sets of the second kind are called classes and their theory will depend to a large extent on the class axiom of comprehension (*). Actually, the notion of a class is not entirely new; we have already talked about classes in §§1–6, but there we used for them the term 'collection'.

Now we set out to develop an axiom system for set theory with classes. As we shall see, there are several ways of doing that, some of which differ only in technical detail, while others show more fundamental differences. At this point let us make sure that we know what we intend by the notion of class. Having done that, we can write down axioms which, we believe, are true statements about our intended classes. We mean by a 'class', as we meant earlier by the informal term 'collection', the extension of some condition $\mathfrak{P}(x)$, i.e., the class, collection, set, or aggregate, whatever you wish, of all sets x which fulfil the condition $\mathfrak{P}(x)$. Let us recall here what we mean by 'condition'. A condition is any statement of the language introduced in §1. That language mentions only sets, membership and equality of sets and other notions defined in terms of these. Since we shall now deal with languages which mention also classes, we modify the term 'condition' by the adjective 'pure'; i.e., a *pure condition* is a condition which mentions only sets, membership and equality of sets and other notions defined in terms of these notions. In particular, a pure condition is *not* supposed to mention classes. In §§1–6 all the conditions we dealt with were pure since these were the only conditions we could express in the language of ZF. Using our present terminology we say that the classes are intended to be the extensions of the pure conditions.

Let us note, at this point, that our present way of looking at classes as extensions of pure conditions is by no means the only accepted one. In §7.4 we shall see a different point of view and study its implications concerning the axiomatic theory.

We denote the system of set theory which we are now developing by VNB (after von Neumann and Bernays). In most of its details it follows the system proposed by Bernays in 1937 [1]).

First, let us describe the language we shall use for VNB. Lower case letters will always stand for sets and capital letters for classes (other than 0 which still stands for the null-set). Now we have two kinds of membership, the membership of a set in a set (as in $0 \in \{0\}$), and the membership of a set in a class (as in $0 \in B$ where B is, say, the class of all sets which are not members of themselves). Both kinds of membership will be denoted by the same symbol $\in$ [2]). Expressions of the form $A \in B$ or $A \in x$ are considered, for the time being, to be meaningless and are not allowed in our language, i.e., our grammatical rules allow the use of 'is a member of' only after an expression which denotes a set. (We do not claim that the expressions $A \in B$ and $A \in x$ are *necessarily* meaningless. On the contrary, we shall see later that we can attribute a very natural meaning to these expressions.) We also admit here statements of the form $x = y$ and $A = B$ as basic statements of our language. The possibility of admitting these statements as defined, rather than primitive, notions will be discussed in a short while. Expressions of the form $x = A$ are, for the time being, not allowed in our language. In addition, we have available in our language all the sentential connectives, i.e., *...or...*, *...and...* , *if... then...* ... , *...if and only if...* , *it is not the case that...* and so on, and the quantifiers *for every set x ...* , *there exists a set x such that ...* , *for every class X ...*, *there exists a class X such that ...* (in symbols, $\forall x$, $\exists x$, $\forall X$, $\exists X$, respectively).

[1]) Bernays 37–54.

[2]) Bernays 37–54 denotes the latter kind of membership with η, i.e., he writes $x \, \eta \, B$ where we write $x \in B$. Even though we use here just one symbol $\in$ we can differentiate between the two kinds of membership according to whether we have on the right side of the $\in$ symbol a symbol for a set, such as $\{0\}$, or for a class, such as B.

When we go on developing VNB we shall use a much richer language since we shall introduce many new notions by means of definitions. We shall refer to the language described here, without any defined symbols, as the *primitive language* of VNB, and its symbols will be called the *primitive symbols*.

In §2 we had the choice of taking equality as a primitive notion (of logic, or set theory) or else as a defined notion. There we chose to adopt equality as a primitive notion of logic, and accordingly we now make the same choice with respect to classes. According to our intended notion of class, namely as the extension of a pure condition, two classes are identical if and only if they have the same members (this is what we mean by 'extension'). Therefore we adopt the following axiom.

AXIOM (X) OF EXTENSIONALITY FOR CLASSES. **If for every element x, x is a member of A if and only if x is a member of B, then $A = B$.**

In symbols: $\forall x(x \in A \leftrightarrow x \in B) \rightarrow A = B$.

Had we chosen in §2 to take equality as a primitive notion of set theory (attitude (b) on p. 26), or as a defined notion (attitude (c) there), then we would now still be faced with the same choice with respect to the classes. If we choose to *define* equality of classes, then we would have to define it so that the requirements of reflexivity, symmetry, transitivity, and substitutivity ((*i*)–(*iv*) on p. 25) would be satisfied. In the case of sets we saw that the most direct way of defining equality is to use requirement (*iv*) for atomic statements (which was denoted there with (*iv*′)) as the defining property of equality, since every other definition of equality must anyway be equivalent to that one. The atomic statements in which a class variable occurs are all of the form $x \in A$ and hence we would now define that $A = B$ if for every element x, $x \in A$ if and only if $x \in B$. As in the case of sets, any other definition of equality of classes would necessarily have to be equivalent to this one. This definition conforms also to our intuitive notion of class, as mentioned above. If we define equality of classes this way, and it seems to be the only reasonable way of doing it, then Axiom X of extensionality for classes does not have to be assumed as an axiom since it is now an immediate consequence of the definition of equality.

Considering the intended meaning of the term 'class', it seems that the following axiom schema is the most natural axiom of comprehension for classes.

(*) *There exists a class which consists exactly of those elements x which fulfil the condition* $\mathfrak{P}(x)$, where $\mathfrak{P}(x)$ is any pure condition on x.

The trouble is that (*) is a bit too weak. There are simple facts which are true about our intended classes which cannot be established by means of (*). E.g., let us consider the statement

(**) *For every class A there exists a class B which consists of all elements which are not members of A.*

This is certainly true for the intended classes since, if A is given by the pure condition $\mathfrak{P}(x)$, then B is given by the pure condition "it is not the case that $\mathfrak{P}(x)$". (*) is, therefore, sufficient to prove every instance of (**) for any particular class A determined by a pure condition, but is not sufficient to prove (**) in general [1]). Thus we see that if we adopt (*) as our only axiom of comprehension for classes, our handling of classes will be rather clumsy, which is a great disadvantage, as we introduced classes, to begin with, in order to get a more streamlined treatment. Therefore we prefer to choose the following axiom.

AXIOM (XI) OF PREDICATIVE COMPREHENSION FOR CLASSES[2]). **There exists a class A which consists exactly of those elements x which satisfy the condition $\mathfrak{P}(x)$**, where $\mathfrak{P}(x)$ is a condition which does not contain quantifiers over classes, i.e., $\mathfrak{P}(x)$ does not contain expressions of the form "*for every class X ...*" or "*there exists a class X such that ...*" (but $\mathfrak{P}(x)$ may contain quantifiers over elements).

It immediately follows from the axiom of extensionality for classes that the condition $\mathfrak{P}(x)$ in Axiom XI determines a *unique* class A; therefore we can speak of *the* class of all elements x which fulfil $\mathfrak{P}(x)$. Accordingly we lay down

DEFINITION VI. $\{x|\mathfrak{P}(x)\}$, where $\mathfrak{P}(x)$ is a condition which does not use class quantifiers, is defined to be *the* class of all sets x which fulfil the condition $\mathfrak{P}(x)$. The expressions $\{x|\mathfrak{P}(x)\}$ will be called *class abstracts*.

Now, let us make clear exactly in which way classes and sets are supposed to be mentioned in $\mathfrak{P}(x)$ in Axiom XI and in Definition VI. Expressions of the form $x \in A$, where A is a class variable, are of course permitted; quantifiers over class variables are forbidden. Which *defined* notions are to be allowed in $\mathfrak{P}(x)$? The criterion is simple – defined notions are allowed only if, when we replace these notions in $\mathfrak{P}(x)$ by their definitions, we obtain a condition expressed in the primitive language which contains no class quantifiers. E.g., if $\mathfrak{Q}(x)$ is a condition which does not use any class quantifier, then the expression $\{x|\mathfrak{Q}(x)\}$ is allowed in $\mathfrak{P}(x)$, since $y \in \{x|\mathfrak{Q}(x)\}$ can be replaced by

[1]) This can be shown rigorously by a method like that used below in §7.2 to prove that every statement of the language of ZF which is provable in VNB is also provable in ZF.

[2]) This axiom is called predicative since it asserts only the existence of those classes which are given by a definition which does not presuppose the totality of all classes, unlike Axiom XII in §7.5 below.

$\mathfrak{Q}(y)$ which does not involve any class quantifiers.

To get an idea of what we can do by means of Axiom XI, we shall now define a few constant classes and operations on classes. Notice that these defined notions satisfy the requirement mentioned above, i.e., they can be replaced in expressions containing them by their definitions, without introducing any quantifiers on class variables.

DEFINITION. $\Lambda = \{x \mid x \neq x\}$ (the *null* class).
$\mathrm{V} = \{x \mid x = x\}$ (the *universal* class).
$A \cup B = \{x \mid x \in A \text{ or } x \in B\}$ (the *union* of A and B).
$A \cap B = \{x \mid x \in A \text{ and } x \in B\}$ (the *intersection* of A and B).
$-A = \{x \mid x \notin A\}$ (the *complement* of A).
The basic properties of these classes and operations are easily proved from their definitions.

Now we shall see why we admitted in Axiom XI conditions $\mathfrak{P}(x)$ exactly as stated above. Let us first become convinced that Axiom XI is not too wide for our purposes. To see this let us observe that Axiom XI is indeed true for our intended classes. Let $\mathfrak{P}(x)$ be a condition as in Axiom XI, with the class parameters $A_1, \ldots, A_k$. By our assumption on $\mathfrak{P}(x)$, if it contains symbols other than the primitive symbols, we can replace them by their definitions without introducing class quantifiers, and thus get a condition in the primitive language equivalent to $\mathfrak{P}(x)$. Since in the following there will be no need to distinguish between two equivalent conditions, we can assume that it is already $\mathfrak{P}(x)$ which contains only symbols of the primitive language, without containing any quantifiers over class variables. If $A_1, \ldots, A_k$ are classes as intended, then for some pure conditions $\mathfrak{Q}_1(x), \ldots, \mathfrak{Q}_k(x)$ we have $A_i = \{x \mid \mathfrak{Q}_i(x)\}$, for $1 \leqslant i \leqslant k$. Since there are no class quantifiers in $\mathfrak{P}(x)$, the only classes mentioned in it are $A_1, \ldots, A_k$, and the only places in which they are mentioned are expressions of the form $y \in A_i$. Since $y \in A_i$, for $A_i = \{x \mid \mathfrak{Q}_i(x)\}$, can be replaced by $\mathfrak{Q}_i(y)$, $\mathfrak{P}(x)$ is equivalent to a pure condition $\mathfrak{Q}(x)$, and therefore the class $\{x \mid \mathfrak{Q}(x)\}$, which is a class as intended since $\mathfrak{Q}(x)$ is a pure condition, consists exactly of the elements x which fulfil the condition $\mathfrak{Q}(x)$. To sum up, the conditions $\mathfrak{P}(x)$ that we allow in Axiom XI differ essentially from pure conditions only in that they may also contain class parameters; once these parameters are given values which are classes determined by pure conditions, the condition $\mathfrak{P}(x)$ itself becomes equivalent to a pure condition.

We shall also see in §7.2 that while Axiom XI is a schema it can be replaced, equivalently, by a finite number of its instances, all of which are simple enough to want to have around if one hopes to deal with classes in a neat way. This is another reason why Axiom XI cannot be said to contain too much.

Having hopefully convinced ourselves that we did not admit in Axiom XI too many conditions, we now have to convince ourselves that Axiom XI is not too narrow for our purposes. If we were to admit more conditions, and have a natural and simple criterion for which conditions to admit, it seems that we would have to admit at least all the conditions that involve no more than one existential (or universal) class quantifier. However, in this case we would get statements which are not true of our intended classes, since there are conditions $\mathfrak{Q}(x)$ of VNB, involving one class quantifier, such that for every pure condition $\mathfrak{P}(x)$ the statement "for all x, $\mathfrak{Q}(x)$ if and only if $\mathfrak{P}(x)$" is refutable in VNB [1]), and hence there is no intended class A such that, for every element $x, x \in A$ if and only if x fulfils $\mathfrak{Q}(x)$.

One can formulate a statement in VNB which asserts that there are no classes other than those determined by pure conditions [2]). This is, of course, true of our intended notion of class, but it is not implied by the axioms of VNB [3]). Shall we add it as an axiom to VNB? Such a statement is relatively complicated and does not seem to be useful for proving theorems which one may ordinarily consider in set theory [4]). Since we are interested in the intended classes only to the extent that using them streamlines set theory, and this has been achieved by Axiom XI, there is no need to add an axiom which rules out all classes other than those determined by pure conditions.

The way in which we introduced the classes is not confined only to set theory; on the contrary, Axiom XI can be used to introduce classes in any mathematical theory. Suppose we start with some mathematical theory T formulated in the first-order predicate calculus (with or without equality), and we add to it the following items: (a) a new kind of variables which we call class variables, (b) all the statements of the form $x \in A$, where x is a

[1]) Such a condition $\mathfrak{Q}(x)$, with one existential class quantifier in front, is given by the formula *Stsf* of Mostowski 51, p. 115, or can be obtained by diagonalization over the pure conditions as in Kruse 63a.

[2]) This statement is formalized by Kruse 63a (it is also evident from Mostowski 51 how to do it). An even stronger statement is considered in Myhill 52.

[3]) Provided that the theory QM of § 7.5 is consistent. In QM we can prove that there is a class which is not determined by a pure condition, since the existence of $\{x \mid \mathfrak{Q}(x)\}$, where $\mathfrak{Q}(x)$ is as above, is provable in QM. This independence result can also be shown to follow from the weaker assumption of the consistency of VNB, by combining the methods of Easton 64 and Feferman 65.

[4]) In particular any statement which mentions only sets and which is provable in VNB using the assumption that all classes are determined by pure conditions is also provable in VNB without that assumption. This can be shown by the method used in § 7.2 below to show that every statement which mentions only sets and which is provable in VNB is already provable in ZF.

variable of T and A is a class variable, as new atomic statements, (c) all the statements which can be obtained from the old and new atomic statements by means of the logical connectives and the quantifiers, and (d) Axioms X and XI of extensionality and predicative comprehension for classes as additional axioms. The theory obtained from T by these additions is, up to possible trivial differences in notation, the predicative second order theory of T [1]).

Until now we set down the axioms we needed for the classes. As for the sets, we take the axioms of ZF, with some changes which are, essentially, just technical.

First we take up all the single axioms of ZF, i.e., the axioms of **Extensionality** (I), **Pairing** (II), **Union** (III), **Power set** (IV), and **Infinity** (VI) [2]). Since the main reason for introducing classes is to avoid metamathematical notions in the formulation of most of the axioms and theorems, and since the intended classes are just the extensions of pure conditions we can now replace the conditions by classes in the axiom schemas of ZF.

AXIOM (V^c) OF SUBSETS. **For every class P and for every set a there exists a set which contains just those members x of a which are also members of P.** In symbols: $\forall P \forall a \exists y \forall x(x \in y \leftrightarrow x \in a \wedge x \in P)$.

Before we put down the axiom of replacement we have to deal first with the notions of relation and function. In § 3.5 we saw that these notions are, in many respect, similar to that of condition. Having all but replaced the mathematical notion of condition by the mathematical notion of class, we can now apply similar methods to the metamathematical notions of relation and functional condition. In the case of a general mathematical theory we cannot handle relations by means of classes, and we need new metamathematical notion of condition by the mathematical notion of class, we can now apply similar methods to the metamathematical notions of relation and functional condition. In the case of a general mathematical there is a set which contains all x's and all y's for which $\mathfrak{P}(x,y)$ holds, whereas here we can deal with all relations).

DEFINITION. A class A is said to be a *relation* if it consists only of ordered pairs. If A is a class then $\mathfrak{D}(A)$ (the *domain* of A) is the class of all elements x for which there is a y such that $\langle x,y\rangle \in A$, and $\mathfrak{R}(A)$ (the *range* of A) is the class of all elements y for which there is an x such that $\langle x,y\rangle \in A$. A class A is said to be a *function* if it is a relation and, for all x, y, z, if $\langle x,y\rangle \in A$ and

[1]) See, e.g., Church 56, §58.

[2]) The latter axiom refers only to sets; therefore we have to replace now the capital letters used in its formulation by lower case letters.

$\langle x, z\rangle \in A$ then $y=z$. If F is a function, and $x \in \mathfrak{D}(F)$, then we denote by $F(x)$ the element y such that $\langle x, y\rangle \in F$.
In symbols:

$$\mathfrak{Rel}(A) =_{\mathrm{Df}} \forall x(x \in A \rightarrow \exists y \exists z(x = \langle y, z\rangle)).$$

$$\mathfrak{D}(A) =_{\mathrm{Df}} \{x \mid \exists y(\langle x, y\rangle \in A)\},\ \mathfrak{R}(A) =_{\mathrm{Df}} \{y \mid \exists x(\langle x, y\rangle \in A)\}.$$

$$\mathfrak{Fnc}(A) =_{\mathrm{Df}} \mathfrak{Rel}(A) \& \forall x \forall y(\langle x, y\rangle \in A \wedge \langle x, z\rangle \in A \rightarrow y = z).$$

THEOREM (-SCHEMA). Let $\mathfrak{Q}(x,y)$ be a condition on x and y (i.e., a relation in the old, metamathematical, sense), then there is a unique relation A such that, for all x and y, $\langle x, y\rangle \in A$ if and only if $\mathfrak{Q}(x, y)$ holds. If $\mathfrak{Q}(x, y)$ is a functional condition, then there is a unique function F such that: $x \in \mathfrak{D}(F)$ if and only if there is a y such that $\mathfrak{Q}(x, y)$ holds, and for every x in $\mathfrak{D}(F)$, $\mathfrak{Q}(x, y)$ holds if and only if $y = F(x)$.

Proof. Given the condition $\mathfrak{Q}(x,y)$, let A be the class which consists of all sets z which are ordered pairs $\langle x, y\rangle$ such that $\mathfrak{Q}(x, y)$ holds. Obviously, $\mathfrak{Q}(x, y)$ holds if and only if $\langle x, y\rangle \in A$. This immediately implies that $x \in \mathfrak{D}(A)$ if and only if, for some y, $\mathfrak{Q}(x, y)$ holds. If, in addition, $\mathfrak{Q}(x, y)$ is a functional condition, then A is obviously a function. For $x \in \mathfrak{D}(A)$ we have, by definition of $A(x)$, $\langle x, A(x)\rangle \in A$; hence $\mathfrak{Q}(x, A(x))$ holds. Since $\mathfrak{Q}(x, y)$ is a functional condition we have that, for all y, $\mathfrak{Q}(x,y)$ holds if and only if $y = A(x)$.

AXIOM ($\mathrm{VII^c}$) OF REPLACEMENT. **If F is a function, and a is a set, then there is a set which contains exactly the values $F(x)$ for all members x of a which are in $\mathfrak{D}(F)$.**
In symbols:

$$\forall F(\mathfrak{Fnc}(F) \rightarrow \forall a \exists b \forall y(y \in b \leftrightarrow \exists x(x \in a \wedge x \in \mathfrak{D}(F) \wedge y = F(x)))).$$

In formulating Axiom $\mathrm{VII^c}$ we were guided, for the sake of convenience, by the symbolic version of VII on p. 52, rather than by the main version of Axiom VII. Unlike Axiom VII, Axiom $\mathrm{VII^c}$ does not seem to imply the axiom of pairing, since ordered pairs are used in an essential way in the notion of function on which Axiom VII relies. On the other hand, in the presence of the axiom of pairing Axiom $\mathrm{VII^c}$ implies Axiom $\mathrm{V^c}$ (of subsets – the proof is exactly the same as the proof in §3.7 that Axiom VII implies Axiom V).

AXIOM ($\mathrm{IX^c}$) OF FOUNDATION. **Every class P which has at least one member has a minimal member u, i.e., u is a member of P, but no member x of u is a member of P.**
In symbols: $\forall P(\exists u(u \in P) \rightarrow \exists u(u \in P \wedge \forall x(x \in u \rightarrow x \notin P)))$.

Axiom $\mathrm{IX^c}$ follows the version IX of the axiom of foundation which is a schema. Alternatively, we can use here versions IX* or IX** which are single statements. The proof that $\mathrm{IX^c}$ is equivalent to IX* and IX** is essentially the same as the corresponding proof for the schema IX in §5.1 and §5.3.

When we compare the classes with the sets, we see that there is a certain amount of overlap; we introduced the classes as the extensions of the pure conditions, but the sets are also the extensions of some of the pure conditions. For every set y we have the class $\{x \mid x \in y\}$ which has exactly the same members (but there is no set which has exactly the same members as the class $\{x \mid x \notin x\}$). It actually turns out that the distinction between the set y and the class $\{x \mid x \in y\}$ serves no purpose. Therefore, to avoid the looks, if not the substance, of such a distinction, let us define as follows.

DEFINITION VII. $z = A$ (and $A = z$) if z and A have exactly the same members.
$A \in B$ if, for some member z of B, $z = A$.
$A \in y$ if, for some member z of y, $z = A$.
A is a *set* if, for some z, $z = A$.
A is a *proper class* if A is not a set.

Now we have arrived at the convenient situation where equality and membership are defined for any two objects. It is easily seen that we now have full substitutivity of equality (i.e., that $x = A$ implies that $\mathfrak{P}(x)$ holds if and only if $\mathfrak{P}(A)$ holds). Since we have, as immediately seen, $\Lambda = \mathrm{O}$ we can, by the full substitutivity of equality, use O and Λ synonymously. The operation which we define on classes are also defined for sets since we can replace the class variables in their definitions by set variables. The outcome of such an operation is a class, which may also be a set (in the sense of Definition VII).

Using our present terminology we can rewrite the axioms of VNB as follows.
I (Extensionality of sets). As in ZF.
II (Pairing). $\{x \mid x \textit{ is } a \textit{ or } x \textit{ is } b\}$ is a set.
III (Union). $\{x \mid x \textit{ is a member of a member of } a\}$ is a set.
IV (Power set). $\{x \mid x \subseteq a\}$ is a set.
$\mathrm{V^c}$ (Subsets). $P \cap y$ is a set.
VIb (Infinity). $\{x \mid x \textit{ is a finite ordinal}\}$ is a set [1]).
$\mathrm{VII^c}$ (Replacement). If F is a function then
$\{x \mid x = F(y) \textit{ for some } y \in a \cap \mathfrak{D}(F)\}$ is a set.
$\mathrm{IX^c}$ (Foundation). $A \neq 0 \rightarrow \exists u (u \in A \wedge u \cap A = \mathrm{O})$.
X (Extensionality of classes). As on p. 122.
XI (Predicative comprehension). As on p. 123.

7.2. **Metamathematical Features of** VNB. In the transition from ZF to VNB

[1]) It is easily seen that the set Z_1^* of p. 48 consists exactly of all finite ordinals in the sense of § 5.2.

the axiom schemas of ZF became single axioms of VNB; however, VNB has the extra axiom schema of comprehension – Axiom XI. The conditions $\mathfrak{P}(x)$ permitted in Axiom XI can be assumed, as we have already mentioned, to be formulated in the primitive language. In this language there is only a finite number of ways in which atomic statements can be made up, and only a finite number of ways in which more complicated statements can be formed from simpler ones. If we can make appropriate classes correspond to the various statements, and we find out how our construction of statements affects the corresponding classes, we can translate the finitely many rules for the construction of statements into finitely many rules for the construction of classes, and thus replace Axiom XI by a finite number of its instances [1]). Since a general statement involves an arbitrary finite number of free variables $x_1, ..., x_n$ we have to deal here also with ordered n-tuples; actually, it will suffice to consider only ordered pairs and triples.

The axioms which correspond to the propositional connectives are:

AXIOM XI1. **For every class A there is a class C which consists of all elements x which are not in A.**
This axiom corresponds to negation : $C = \{x \mid \textit{it is not the case that } x \in A\} = -A$ (C is called the *complement* of A).

AXIOM XI2. **For all classes A and B there is a class C which consists of all common members of A and B.**
This axiom corresponds to conjunction: $C = \{x \mid x \in A \textit{ and } x \in B\} = A \cap B$. The axiom which corresponds to the atomic statement $x \in y$ is:

AXIOM XI3. **There is a relation E such that $\langle x, y \rangle \in E$ just in case that $x \in y$** [2]).
The axiom which corresponds to the existential quantifier "there exists a y such that" is:

AXIOM XI4. **For every relation A there exists a class C which consists exactly of the first members of the ordered pairs which are members of A.**
$C = \{x \mid \textit{there exists a } y \textit{ such that } \langle x, y \rangle \in A\} = \mathfrak{D}(A)$. ($C$ is the domain of A.)
The next axiom is needed because statements can contain parameters, and therefore we have to reckon with classes of the form $\{x \mid x \in y\}$ or $\{x \mid x = y\}$.

AXIOM XI5. **For every set y there exists a class C which consists exactly of all the members of y.**

[1]) This is due to von Neumann 25 (with some of the ideas originating with Fraenkel 22a). For set theories like our VNB, where this is more surprising, it was shown by Bernays 37–54 I. We shall see in §7.5 that the fact that in Axiom XI the statements are not supposed to contain class quantifiers is used here in an essential way.

[2]) There is no need for an axiom which corresponds to the atomic statement $x = y$ since it is equivalent to the formula $\forall z(z \in x \leftrightarrow z \in y)$ which does not involve equality.

$C=\{x \mid x \in y\}$. Using Definition VII this becomes: For every set there is a class equal to it. An alternative to this axiom is

AXIOM XI5*. **For every set y there exists a class C which contains y as its only member.**
$C=\{x \mid x=y\}$. Using Definition VII, this becomes: For every singleton set $\{y\}$ there is a class equal to it.
Now we have to add three axioms of a rather technical nature, which are needed in order to handle statements with more than one variable.

AXIOM XI6. **For every class A there exists a relation C which consists of all ordered pairs $\langle x, y \rangle$ such that $y \in A$.**
$C=\{\langle x,y\rangle \mid y \in A\}$ – read: C is the class of all ordered pairs $\langle x,y\rangle$ such that $y \in A$. Notice that $\{\langle x,y\rangle \mid y \in A\}$ is a new notation not previously used.

AXIOM XI7. **For every relation A there exists a relation C which consists of all ordered pairs $\langle x, y\rangle$ whose *inverses* $\langle y, x\rangle$ are in A.**
$C=\{\langle x,y\rangle \mid \langle y,x\rangle \in A\}$.

AXIOM XI8. **For every relation A there exists a relation C which consists exactly of all ordered pairs $\langle\langle x,y\rangle, z\rangle$ such that $\langle x,y,z\rangle$ $(=\langle x,\langle y,z\rangle\rangle)$ is in A.**
$C=\{\langle\langle x,y\rangle,z\rangle \mid \langle x,y,z\rangle \in A\}$.
Using the ideas outlined above, preceding the list of Axioms XI1–XI8, one can prove that Axiom XI follows from these axioms [1]).

Let us compare the systems VNB and ZF. First we notice that the language of VNB is richer, i.e., every statement of ZF is also a statement of VNB, yet no statement of VNB which involves class variables is a statement of ZF. Moreover, some of the statements of VNB express things which cannot be expressed in ZF at all. To make the latter assertion clearer we point out that this has a twofold meaning; first, there is a closed statement of VNB which is not equivalent in VNB to any statement of ZF [2]); and, second, as we mentioned above (p. 125), there is a condition $\mathfrak{Q}(x)$ of VNB such that for every pure condition $\mathfrak{P}(x)$ (i.e., for every condition $\mathfrak{P}(x)$ of ZF) the statement "$\mathfrak{Q}(x)$ if and only if $\mathfrak{P}(x)$" is refutable in VNB.

Every statement of ZF which is provable in ZF is also provable in VNB. This is easily seen, since all the single axioms of ZF are also axioms of VNB, and all the schemas of ZF immediately follow from the corresponding axioms of VNB, by means of Axiom XI. Let us prove, for example, Axiom V (of

[1]) See the proof in Bernays 37–54 I, § 3, Gödel 40, pp. 8–11, or Mendelson 64, Ch. 4, § 1. The original proof is due to von Neumann 28, Ch. II, § 1.

[2]) Axiom $\mathrm{VIII}_{\sigma}^{\mathrm{c}}$ of § 7.3 is such a statement. A stronger and more general example can be obtained by combining the method of A. Levy 65b, § 7 with the truth definition of ZF of Mostowski 51.

subsets) in VNB. Let a be a set and $\mathfrak{P}(x)$ a condition of ZF, i.e., a pure condition. By the axiom of comprehension there is a class A which consists of the elements x which fulfil $\mathfrak{P}(x)$; by Axiom V^c there is a set y which consists of the common members of a and A, i.e., of those members of a which fulfil $\mathfrak{P}(x)$.

So far we have seen that whatever we can express and prove in ZF we can also express and prove, respectively, in VNB. We have also seen that some statements can be expressed in VNB but not in ZF. The next natural question which comes up is whether whatever can be expressed in ZF and proved in VNB can already be proved in ZF; more precisely, if $\mathfrak{S}$ is a statement of ZF which is provable in VNB, is $\mathfrak{S}$ necessarily provable in ZF? We shall give a positive answer to this question, and therefore we can say that even though VNB is a theory richer than ZF in its means of expression, VNB is not richer than ZF as far as proving statements which mention only sets is concerned. Thus, if one is interested only in sets and regards classes as a mere technical device, one should regard ZF and VNB as essentially the same theory, and the differences between those theories as mere technical matters.

Our present task is not as easy as our earlier task of showing that every theorem $\mathfrak{S}$ of ZF is also a theorem of VNB. There we used the fact that the proof of $\mathfrak{S}$ in ZF can be trivially reproduced in VNB. Now, if we are given a proof in VNB of a statement $\mathfrak{S}$ of ZF, there is not always a way in which this proof can be reproduced in ZF, since the proof of $\mathfrak{S}$ in VNB may involve statements which cannot be expressed in ZF. Attacking the problem from a different angle, we observe that if the statement $\mathfrak{S}$ is not provable in ZF, there is no reason why $\mathfrak{S}$ should be provable in VNB; after all, the new axioms of VNB do not give us any information about sets – they just assert the basic facts about the pure conditions, or the classes determined by these conditions, and those facts hold irrespective of whether $\mathfrak{S}$ is true or false. This informal argument can be turned into a precise argument, as we do below.

We shall now show that if the statement $\mathfrak{S}$ of ZF is not provable in ZF, it is also not provable in VNB. Suppose $\mathfrak{S}$ is not provable in ZF; then, by the completeness theorem of the first-order predicate calculus [1]), there is a set m and a binary relation $\in'$ on m (i.e., $\in' \subseteq m \times m$) such that $\langle m, \in' \rangle$ is a model of ZF in which $\mathfrak{S}$ does not hold. We now define "classes" for this model. Let us say that u is a *model-class* if for some condition $\mathfrak{P}(x)$ (of ZF) with n parameters and for some $y_1, ..., y_n \in m$, u is the subset of m which consists exactly of all the members x of m which satisfy the condition $\mathfrak{P}(x)$ in the model, when the values of the parameters are taken to be $y_1, ..., y_n$. By the very arguments which we used in § 7.1 to justify the predicative axiom of comprehension and Axioms V^c, VII^c, and

[1]) p. 296.

IX^c, one shows that if one interprets the notion of set as "member of m", the notion of class as "model-class", that of membership of a set in a set as $\in'$, and that of membership of a set in a class as membership (in a model-class), then all the axioms of VNB become true, while $\mathfrak{S}$ stays false. Thus $\mathfrak{S}$ is not provable in VNB [1]).

We have now shown that if a statement $\mathfrak{S}$ of ZF is provable in VNB it is also provable in ZF, but we did it by means of an indirect argument. Thus, even if we know the proof of $\mathfrak{S}$ in VNB our method above does not show us how to get a proof of $\mathfrak{S}$ in ZF by any method other than the ungainly way of scanning all proofs of theorems in ZF till we find a proof of $\mathfrak{S}$, which we know must be among them. To get another method which will actually show how to obtain a proof of $\mathfrak{S}$ in ZF once a proof of $\mathfrak{S}$ in VNB is known, we have to return to the idea mentioned above of reproducing in ZF the proof of $\mathfrak{S}$ in VNB. As we have mentioned there, this is not always possible; however, it turns out that if a statement $\mathfrak{S}$ of ZF has a proof in VNB, we can always change this proof to another proof of $\mathfrak{S}$ in VNB which is particularly simple and can therefore be reproduced in ZF [2]).

A particularly important consequence of what we have discussed just now is that if ZF is consistent so is VNB. Suppose VNB were not consistent then every contradictory statement of ZF, e.g., "some set is a member of itself, and no set is a member of itself", would be provable in VNB. By the metamathematical theorem discussed above every such contradictory statement is also provable in ZF, hence ZF is inconsistent too. Moreover, if Q is a theory obtained from ZF by adding to ZF a set T of axioms which involve only sets and Q′ is the theory obtained from VNB by adding to it the same set T of axioms, then Q′ is consistent if and only if Q is consistent [3]). Thus all the results mentioned in §4 and §6 concerning the consistency and the independence of the generalized continuum hypothesis and the axiom of constructibility with respect to ZF go over to corresponding results with respect to VNB. In order to be able to transfer more consistency and independence

[1]) This proof is due to Novak 51, Rosser–Wang 50, Mostowski 51.

[2]) The idea, due to Paul J. Cohen, is to change first the proof of $\mathfrak{S}$ in VNB to a cut free proof (as, e.g., in Schütte 50); the latter can be easily reproduced in ZF. An earlier method is given by Shoenfield 54. These proofs of Shoenfield and Cohen show that the Gödel-number (see p. 306) of a proof of $\mathfrak{S}$ in ZF depends on the Gödel-number of a proof of $\mathfrak{S}$ in VNB via a primitive recursive function, whereas the proof given above suffices to establish this dependence only via a general recursive function.

[3]) This is shown as follows. Q is consistent if and only if no statement $\mathfrak{S}$ which is a negation of a conjunction of finitely many statements out of T is provable in ZF. This, we know, holds if and only if no such statement $\mathfrak{S}$ is provable in VNB, which, in turn, holds if and only if Q′ is consistent.

results from ZF to VNB, let us notice the following. When we showed that a statement $\mathfrak{S}$ of ZF is provable in ZF if and only if it is provable in VNB, it mattered only that VNB contains the axioms of predicative comprehension (XI) and extensionality (X) and that, other than that, VNB consists exactly of the single axioms of ZF and of axioms corresponding to the axiom schemas of ZF. Therefore, the same result applies also to other pairs of corresponding theories. For example, we proved in §5.5 that the axiom of foundation is not refutable in the system which consists of all axioms of ZF other than the axiom of foundation. From what we said just now it follows that the axiom of foundation is also not refutable in the system which consists of all the axioms of VNB other than the axiom of foundation. Therefore, *all* the consistency and independence results of §§4–6 apply, literally, also to VNB. In fact, in each particular case it is evident anyway, without using the present general principle, that the proof of the relative consistency (or independence) applies equally well to VNB as it does to ZF.

7.3. **The Axiom of Choice in** VNB. We can add to VNB the local axiom of choice of ZF, i.e., Axiom VIII, and obtain thereby a system which we denote with VNBC. By the results mentioned above, the statements $\mathfrak{S}$ of ZF which are theorems of VNBC are exactly the theorems of ZFC. However, when one wants to have an axiom of choice in VNB one usually chooses a very natural *global* axiom of choice which is strongly related to the global axiom of choice VIII_σ of ZFC_σ and which is presented below.

Suppose that we start with a ZF-type set theory Q which has a selector $\sigma(x)$ in the sense of §4.4. It does not matter whether Q is obtained from ZF by addition of an axiom which allows us to define the selector (such as the axiom of constructibility), or if Q is obtained from ZF by adding σ as a new operation symbol and adding Axiom VIII_σ which asserts that $\sigma(x)$ is indeed a selector; but, in the latter case, we have to widen our intended notion of class to also include extensions of conditions which involve selection, in addition to sets and membership. Let us consider, informally, the class F of all sets x which are ordered pairs $\langle y, \sigma(y)\rangle$, where y is a non-void set. This class F is obviously a function, and for every non-void set y we have $F(y) = \sigma(y)$, and since $\sigma(y) \in y$ we have $F(y) \in y$. This leads us to

AXIOM ($\text{VIII}_\sigma^\text{c}$) OF GLOBAL CHOICE. **There exists a function F whose domain contains all non-void sets, and such that for every non-void set y, $F(y)$ is a member of y** [1]).

[1]) For many equivalent versions of Axiom $\text{VIII}_\sigma^\text{c}$ see Rubin–Rubin 63, Part II; cf. also Isbell–Wright 66.

We shall denote with VNBC_σ the system of set theory obtained from VNB by adding to it Axiom $\mathrm{VIII}^{\mathrm{c}}_\sigma$. It is easily seen that Axiom VIII of local choice is a theorem of VNBC_σ (as we observed for ZFC_σ in §4.4), and hence all the theorems of VNBC are also theorems of VNBC_σ. Let us point out that the language of VNB_σ does not contain any selector symbol; Axiom $\mathrm{VIII}^{\mathrm{c}}_\sigma$ offers the advantages of a selector without the disadvantage of having to extend the language. The subscript σ is used in the designation of Axiom $\mathrm{VIII}^{\mathrm{c}}_\sigma$ and VNBC_σ, even though they do not involve the symbol σ at all, for the purpose of stressing their close relationship with Axiom VIII_σ and ZFC_σ, respectively.

In the same way in which one proves the metatheorem that the theorems of VNB which mention only sets are exactly the theorems of ZF one can also prove that the theorems of VNBC_σ which mention only sets are exactly the theorems of ZFC_σ which do not mention σ [1]). Since, as we said in §4.4, the theorems of ZFC_σ which do not involve σ are exactly the theorems of ZFC, we know that the theorems of VNBC_σ which involve only sets are exactly the theorems of ZFC and are, hence, also exactly the theorems of VNBC which involve only sets. Thus, as far as sets are concerned, VNBC and VNBC_σ are the same theory. On the other hand (assuming the consistency of VNB), the two theories do not completely coincide, since Axiom $\mathrm{VIII}^{\mathrm{c}}_\sigma$ is not a theorem of VNBC [2]).

It follows from what we said above that if ZFC is consistent so is VNBC_σ. Moreover, if any set T of statements of ZFC which can be added to ZFC as new axioms without causing a contradiction, the same addition will not introduce a contradiction in VNBC_σ either. Thus the generalized continuum hypothesis and the axiom of constructibility are consistent with VNBC_σ. Moreover, the axiom of constructibility (or any other axiom which implies in ZF the existence of a selector [3]) implies Axiom $\mathrm{VIII}^{\mathrm{c}}_\sigma$ in VNB, since a class F as required by Axiom $\mathrm{VIII}^{\mathrm{c}}_\sigma$ is given by the class of all ordered pairs $\langle x, \sigma(x)\rangle$, where x is a non-void set and σ is some fixed selector.

If we formulate in terms of classes the result we mentioned at the end of §5.3, we get the following theorem, which is proved by means of the global axiom of choice and the axiom of foundation. *A class A is proper if and only*

[1]) Actually, one can also give a natural translation of all the statements $\mathfrak{S}$ of ZFC_σ, including those which mention σ, into statements of VNB such that $\mathfrak{S}$ is a theorem of ZFC_σ if and only if its translation is a theorem of VNBC_σ.

[2]) Easton 64. If Q' is a theory obtained from VNB by adding to it axioms which involve only sets and Axiom $\mathrm{VIII}^{\mathrm{c}}_\sigma$ is a theorem of Q' and if Q is the theory obtained from ZF by the addition of the same axioms, then there is in Q a relative selector (see footnote 2 on p. 71). The proof is, again, as at the end of § 7.2.

[3]) Or even a relative selector.

if it is equinumerous to the class V *of all sets* (where by A being equinumerous to B we mean that there is a one-one function F whose domain is A and whose range is B). Indeed, von Neumann chose to introduce a very closely related statement as an axiom which replaces the axioms of (subsets,) replacement and global choice [1]).

In analogy to Axioms VIII_σ and IX_σ of ZFC_σ, we can also adopt the following strong versions of the axioms of global choice and foundation

(*) $A \neq \mathrm{O} \supset \sigma(A) \in A$

(**) $\sigma(A) \cap A = \mathrm{O}$.

In the presence of Axioms $\mathrm{VIII}_\sigma^{\mathrm{C}}$ and IX, $\sigma(A)$ can be defined in terms of a class F as in Axiom $\mathrm{VIII}_\sigma^{\mathrm{C}}$ [2]).

7.4. **The Approach of von Neumann.** The way we introduced and motivated VNB in §7.1 is not the way this was done historically for set theory with classes. The first axiom system for set theory with classes was put forth by von Neumann in 1925 [3]). The main technical difference between his system and VNB is that he used the notion of function as the basic notion rather than those of set and class. Von Neumann recognized that the notions of function and class are interchangeable as basic notions for set theory. He gives, as a reason for his choice of the notion of function, the fact that every axiomatization of set theory uses the notion of function anyway, and hence it is simpler to use the notion of function as the basic notion. Now we are using functions (or functional conditions) mostly in our formulation of the axiom of substitution, but Fraenkel's axiom system (mentioned on p. 37), which apparently influenced von Neumann, was also using functions in the axiom schema of subsets. Taking the axiom system of Fraenkel as a starting point, it really seems very reasonable to take the notion of function, rather than that of set, as the basic notion. However, it turned out, in spite of a later simplification of von Neumann's system [4]), that this approach is rather clumsy, and that it is after all simpler to take the notions of set and class, or only that of class alone, as the basic notions of set theory.

What von Neumann regards as the first main feature of his theory is the following. In ZF the guiding principle in writing down the axioms is the *doctrine of limitation of size* (see p. 32), i.e., we do not admit very comprehensive sets in order to avoid the antinomies. Von Neumann regards as the

[1]) See p. 137 below.

[2]) Bernays 58, Ch. VIII. $\sigma(A)$ is defined as $F(t)$, where t is the set of the members of A of minimal rank (in the sense of §5.3).

[3]) von Neumann 25 and 28.

[4]) R.M. Robinson 37.

main idea of his set theory the discovery that the antinomies do not arise from the mere existence of very comprehensive sets, but from their elementhood, i.e., from their being able to be members of other sets. He uses the name 'set' not only for what we call sets, but also for what we call classes. To illustrate his ideas better we shall present a system G of axioms of set theory with classes which is very close to one given by Gödel [1]).

In VNB we defined a set a and a class A to be equal when they have the same members. For every set a there is a class $\{x \mid x \in a\}$ equal to it. As far as our theory is concerned, the set a and the class $\{x \mid x \in a\}$ serve exactly the same purpose. Therefore we can actually identify them and base the system G on classes only.

In the theory G we have just one kind of variables for which we use capital letters. These variables are understood to refer to classes. The only primitive relation symbol, in addition to equality, is the membership symbol $\in$ [2]). We start with a definition.

DEFINITION. X is a *set* if and only if there is a class Y such that $X \in Y$.

Now, having *defined* the notion of set we can use lower case variables for sets, in the same way that we used, in § 5.2, Greek-letter variables for ordinals. In other words, whenever we say "there exists a y such that ..." we mean "there exists a Y such that Y is a set and ..." and whenever we say "for every y ..." we mean "for every Y which is a set ...".

From this point on we just take up Axioms II, III, IV, $\mathrm{V^c}$, VI, $\mathrm{VII^c}$, $\mathrm{IX^c}$, X and XI of VNB. The only difference is that, while the lower case variables in these axioms were part of the primitive language of VNB, they are now defined restricted variables in G. It is easy to prove that in VNB and G we can prove exactly the same theorems [3]) (we have, of course, to make sure that we translate correctly, since VNB and G use different languages).

Von Neumann's system is like G in the sense that all sets are also classes [4]). To get the flavor of his approach let us, for a while, call the classes sets and the sets elements. The elements are the sets which are also members of sets. There are some sets which are not elements, such as the set of all elements which are not members of themselves. Thus the doctrine of limitation of size

[1]) Gödel 40; see also Mostowski 39.

[2]) The idea of using in a von Neumann–Bernays set theory just one kind of variables and a single binary relation symbol is due to Tarski; see Mostowski 49, p. 144.

[3]) See, e.g., Kruse 63a.

[4]) Von Neumann's system also admits individuals (in the sense of § 2). We shall not discuss individuals in the framework of set theory with classes, since admitting individuals here does not give rise to any new interesting discussions. For a system which is like G but admits individuals, see Mostowski 39.

is retained by von Neumann, but only in the sense that very comprehensive sets cannot be members of sets and not in the sense that such sets have to be avoided altogether.

Von Neumann regards, as another major innovation of his system, the fact that, whereas in ZF the limitation of size doctrine serves only as a guide to the introduction of axioms, in his system it is actually incorporated as the following axiom:

(∗) *A set A is an element if and only if there is no function F which maps it on the set* V *of all elements* (i.e., the domain of F is A and its range is V) [1]).

This axiom is equivalent to the conjunction of the axioms of union (III), (subsets – V^c), replacement (VII^c), and global choice (VIII^c_σ) [2]). (∗) is attractive from an aesthetic point of view but, contrary to von Neumann's contention, it does not embody the full limitation of size doctrine. Even though (∗) establishes non-equinumerosity with the set V of all elements as a necessary and sufficient condition for elementhood, (∗) in itself does not tell us when a given set is equinumerous to V. For instance, the axiom of power-set, which clearly falls within the limitation of size doctrine, does not follow from (∗) and the other axioms of G [3]).

We have explicitly described VNB and G and mentioned the original systems of von Neumann–Bernays and Gödel. In all these systems essentially the same theorems are provable [4]). As we saw, one can arrive at set theory with classes starting from two different motivations. We started first from the motivation of replacing the metamathematical notion of condition by the mathematical notion of class. This motivation was introduced by Quine and Bernays [5]). Thus classes, or at least proper classes, are regarded as a kind of objects different from sets, and in some sense less real than sets. On the other hand, von Neumann's motivation regards classes and sets as objects of the same kind with the same claim for existence. The only difference between proper classes and sets is that, because of the antinomies, the proper classes cannot be members of classes whereas sets can. In the next two subsections, we shall continue these respective lines of thought, to arrive at new axiomatic systems for set theory with classes.

[1]) von Neumann 25, Axiom IV2.

[2]) See von Neumann 29 and A. Levy 68.

[3]) To see this, consider the model Π_1 of Bernays 37–54 VI, § 17, in set theory with the axiom of constructibility.

[4]) Provided, of course, that the same attitude towards individuals is adopted. Not all the necessary proofs have been carried out in the literature because the matter is trivial but lengthy.

[5]) Quine 63, Bernays 58.

7.5. **Classes Taken Seriously – the System of Quine and Morse.** Adopting von Neumann's point of view, let us, for a short while, call again the classes sets and the sets elements, and let us even use lower case variables for the classes. The inconsistent axiom of comprehension of §3.1 requires that for every given condition $\mathfrak{P}(x)$ there exists a set which contains exactly the sets x which fulfil the condition $\mathfrak{P}(x)$. Zermelo's axiom of subsets (our Axiom V) amends this axiom of comprehension by requiring that the condition $\mathfrak{P}(x)$ in it be of the form "$x \in a$ and $\mathfrak{Q}(x)$", where a is a given set (i.e., a parameter). On the other hand, von Neumann's way of avoiding the antinomies is to require $\mathfrak{P}(x)$ to be of the form "x is a member of *some* set, and $\mathfrak{Q}(x)$" or, equivalently, "x is an element, and $\mathfrak{Q}(x)$". Translating this back to our ordinary terminology and notation we obtain:

AXIOM (XII) OF IMPREDICATIVE COMPREHENSION. **There exists a class A which contains exactly those elements x which satisfy the condition $\mathfrak{P}(x)$, where $\mathfrak{P}(x)$ is any condition.**

This is more than what we have in Axiom XI, in which the condition $\mathfrak{P}(x)$ is not supposed to mention any class quantifiers; indeed, not all the instances of Axiom XII are provable in VNB [1]). Axiom XII has been suggested by Quine in 1940 as one of the axioms of a system of his which will be discussed in Chapter III, §4 [2]). The annexation of Axiom XII to VNB is due to A.P. Morse and Wang [3]). Let us denote by QM the theory which consists of the axioms of VNB, with Axiom XI replaced by Axiom XII [4]).

Let us now study QM, comparing it with VNB. From the point of view that classes are extensions of pure conditions, QM is plain false since, as we saw above (in p. 125), some instances of Axiom XII fail for extensions of

[1]) Provided that VNB is consistent – Mostowski 51, Kruse 63a. Stronger results can be proved by the methods of Kreisel–Levy 68 (Th. 11) – see footnote 4 on p. 139.

[2]) First edition of Quine 51.

[3]) Wang 49, Kelly 55 (Classification axiom-scheme, in the Appendix), Morse 65(see pp. xxi–xxii).

[4]) We should have added Axiom XII to G rather than to VNB, since G is more in the line of the von Neumann approach; yet we chose to add it to VNB for the purpose of direct comparison of QM with VNB, since VNB and G can anyway be regarded as notational variants of each other.

A somewhat more detailed exposition of QM is given by Stegmüller 62, who borrows most of the technical features from Bernays 58. A very detailed development of a system strongly related to QM is carried out in Morse 65. This system has the unorthodox feature of identifying the notions of formula and term, thereby identifying the notions of class and statement (a statement is equal to the null-class O if it is false and to the *universal* class V if it is true; a class is true if it contains O). The strict equivalence of Morse's system with QM has been verified by Tarski and Peterson (Morse 65, p. xxiii). For a system which comprises QM, see Bernays 61a.

pure conditions. Therefore, let us compare QM with VNB as to its desirability as a system of axioms for set theory, from von Neumann's point of view mentioned above. The main argument in favor of QM is that once we agree, as von Neumann did, to avoid the antinomies not by forbidding the existence of large classes, but by denying their elementhood, there is no reason at all why we should stop at classes defined by conditions which do not involve class quantifiers and not admit classes defined by other conditions. Another argument in favor of QM is that when we define in VNB a class $\{x \mid \mathfrak{P}(x)\}$ we always have to check whether $\mathfrak{P}(x)$ is equivalent to a condition of the primitive language without class quantifiers. This requires some bookkeeping if one uses many defined notions of set theory [1]). Nothing of this sort is needed in QM where $\{x \mid \mathfrak{P}(x)\}$ is always a class.

A particularly embarrassing fact about VNB is that in VNB, unlike QM, one cannot prove all the instances of the induction schema, "If 0 fulfils the condition $\mathfrak{Q}(x)$ and, for every natural number n, if n fulfils $\mathfrak{Q}(x)$ then $n+1$ fulfils $\mathfrak{Q}(x)$ too, then every natural number fulfils $\mathfrak{Q}(x)$" [2]). In VNB the standard proof of the induction schema proves it only for conditions formulated in the primitive language without class quantifiers [3]) (and for other conditions which are equivalent to such conditions). In practice, induction is indeed almost always used for such conditions only, but to be on the safe side one has to keep track of uses of class quantifiers not only for the definition of classes, but also for the application of induction.

Another aspect in which VNB and QM differ is that VNB can be given by a finite number of axioms, as we saw in §7.2, whereas QM cannot [4]). The fact that VNB can be given by a finite number of axioms is not without

[1]) E.g., Gödel 40 keeps track of it by means of the metamathematical concept of a *normal* notion.

[2]) Assuming that VNB is consistent – Mostowski 51; (see also footnote 4 below).

[3]) The standard formulation of induction in VNB is by the single statement "For every class Q, if $0 \in Q$ and if, for every n in Q, $n+1$, too, is in Q, then Q contains all the natural numbers," – Bernays 37–54 II, §6. By Axiom XI this clearly implies the schema of induction for conditions $\mathfrak{Q}(x)$ without class quantifiers.

[4]) Let VNB* be the system obtained from VNB by adding the induction schema above as an axiom (where $\mathfrak{Q}(x)$ varies over *all* conditions). VNB* is a subsystem of QM since the induction schema is obviously provable in QM. VNB* is strongly semantically closed in the sense of Montague 61 since finite sequences of classes can be handled as suggested by R.M. Robinson 45 and, hence, no consistent extension of VNB* without new symbols, and in particular QM, which we assume to be consistent, can be determined by a finite number of axioms. Moreover, one can prove that no consistent extension of VNB* can be determined by a set of axioms in the language of VNB with a bounded number of class quantifiers. (This follows from Theorem 5 of Kreisel–Levy 68,

its aesthetic appeal, but it is not a serious advantage for the following two reasons. First, there is no simple and direct way of using Axioms XI1–XI8 for the run-of-the-mill theorems of set theory. If one adopts Axioms XI1–XI8, the only reasonable way to develop set theory seems to first present the cumbersome proof of Axiom XI from Axioms XI1–XI8 and then start using Axiom XI. Second, the notion of a *proof* in VNB is not simpler than the notion of a proof in QM, even though the notion of an axiom of VNB is (where we choose Axioms XI1–XI8 to take care of comprehension), since in the proofs one uses rules of inference which are like schemas in the sense that they apply to infinitely many possible statements. For instance, the rule of detachment – *modus ponens* – allows one to derive from the two premises $\mathfrak{P}$ and $\mathfrak{P}\rightarrow\mathfrak{Q}$ the conclusion $\mathfrak{Q}$, where $\mathfrak{P}$ and $\mathfrak{Q}$ are any statements out of the infinite set of all statements [1]).

We have not yet discussed the question whether in replacing Axiom XI by Axiom XII we introduce no contradictions. We shall now look into this question, as well as into the more general question of the comparison of the deductive power of VNB and QM. We know that, assuming that VNB is consistent, there is no convincing proof that if VNB is consistent so is QM. This is shown by means of Gödel's theorem on consistency proofs [2]). However, there is no reason why this should lead us to doubt the consistency of QM. We shall see below that if some reasonable set theories, which are formulated in terms of sets only, are consistent, so is QM.

We have already mentioned above that some instances of Axiom XII are not provable in VNB, assuming, as we shall do throughout the present paragraph, that VNB is consistent. In fact, one can prove in QM infinitely many statements which involve only sets (or, for that matter, only natural numbers), and which are not provable in ZF (or in VNB) [3]). Thus the transition from VNB to QM is of a different nature than the transition from ZF to VNB or

by means of a truth definition which uses the ideas of Mostowski 51 and A. Levy 65b). Finally, assuming the consistency of VNB*, not all the instances of Axiom XII are provable in VNB* – this follows from Kruse 63a, Th. 5.2. Whatever will be said about QM in the present paragraph and in the two following ones applies equally well to the weaker theory, VNB*.

[1]) In fact, axiom schemas can be considered to be rules of inference with 0 premises.

[2]) See p. 328; here one uses the fact that in QM one can prove the consistency of VNB – see Mostowski 51.

[3]) Kreisel–Levy 68, Theorems 10 and 11. A particularly interesting such statement is the negation of the Second Axiom of Restriction (p. 116), which is provable in QM (Kurata 64), but is not provable in VNB (if VNB is consistent – Shepherdson 51–53 III). On the other hand, if QM is consistent then the negation of the axiom of constructibility is unprovable, also in QM, (Tharp 66).

from VNBC to VNBC_σ, which did not add any new theorems about sets. Still, one has to bear in mind that, as far as it is now known, those new theorems of QM are of a metamathematical character and do not seem to be theorems one would consider in a development of set theory for the ordinary purposes of mathematics. At the same time, the facts expressed by Axiom XII are facts about classes which cannot be presented as facts about sets; one can construct a theorem of QM which is not implied in VNB by any statement which mentions only sets, unless the latter statement is refutable in VNB [1]).

Now that we know that QM is the source of infinitely many new theorems concerning sets which cannot be proved in ZF, we can ask, naturally, whether the notion of class is indeed essential for obtaining these theorems, or whether these will be the theorems of some natural set theory stronger than ZF but still formulated in terms of sets only. The latter happens to be the case. All the theorems of QM which mention only sets are also theorems of some set theory ZM, formulated in terms of sets only, and obtained by adding to ZF a certain axiom schema of strong infinity which implies the existence of many inaccessible cardinals [2]). As a consequence, if ZM is consistent so is QM. Actually, one can verify by a relatively simple argument the following stronger result. Let $\mathsf{ZF}^{\#}$ be obtained from ZF by adding to it as an additional axiom the statement that there exists at least one inaccessible number. If the theory $\mathsf{ZF}^{\#}$ is consistent then QM is consistent too. Obviously $\mathsf{ZF}^{\#}$ is much weaker than ZM [3]).

Adopting von Neumann's attitude of regarding classes as objects of essentially the same kind as sets, we went along with the consequences of this attitude which led us to the system QM [4]). Let us now go back and examine

[1]) Such a theorem is given by the proof of Th. 4 in Kreisel–Levy 68, where one has to use the truth definition for statements of ZF given by Mostowski 51, and the result on the equivalence of ZF and VNB.

[2]) This is a version of an axiom schema proposed, in essence, by Mahlo – see A. Levy 60.

[3]) Actually, all the theorems of QM which mention only sets are theorems of the theory ZF^* whose axioms (and theorems) are all the statements $\mathfrak{S}$ in the language of ZF which can be proved in ZF to hold in every model $R(\theta)$, where R is the function defined on p. 94 and θ is an inaccessible number. (This easily follows from the discussion in Shepherdson 51–53 II, § 3.7, provided we define an inaccessible number as in A. Levy 60, or else add the local axiom of choice to all the theories which we discuss here.) It is obvious that ZF^* is consistent if and only if $\mathsf{ZF}^{\#}$ is consistent. All the theorems of ZF^* are theorems of ZM (see A. Levy 60). On the other hand, the theorem of QM: "If there is an inaccessible number then $\mathsf{ZF}^{\#}$ is consistent" (which is proved along the lines of Mostowski 51) is not a theorem of $\mathsf{ZF}^{\#}$ (if $\mathsf{ZF}^{\#}$ is consistent), as follows from Gödel's theorem on consistency proofs.

[4]) One additional advantage of QM is that one can formulate in its language a strong axiom schema of infinity of set theory, which agrees neatly with QM – Bernays 61a.

again von Neumann's ideas. His addition of proper classes to the universe of set theory results from his discovery that it is not the existence of certain classes that leads to the antinomies, but rather the assumption of their elementhood, i.e., their being members of other classes; therefore he introduces these classes as proper classes which are not members of classes. This is not a completely satisfactory solution of the problem of the existence of collections as objects, since now, even though proper classes are real objects, collections of proper classes do not exist. The existence of real mathematical objects which cannot be members of even finite classes is a rather peculiar matter, even though in actual mathematical treatment such classes are rarely needed, and can, in case of need, be represented by some classes of sets [1]). One cannot just blame the antinomies for this peculiar situation; we shall see that if one proceeds carefully enough then the assumption that proper classes can be members of classes or of other objects can be seen not to cause any contradiction.

One example of a system in which proper classes are members of some objects is as follows. We introduce a new kind of objects – hyper-classes – which are like classes except that, unlike classes, they can also contain classes as members. We shall use Greek letters as variables for hyper-classes, with the understanding that every class and set is a hyper-class and every set is a class. (We shall not carry out the rather trivial task of describing the formal framework in which this is done.) In addition to all axioms of QM, the system contains the following axioms.

(a) *Every member of a hyper-class is a class* (or a set).

(b) (Schema) *There is a hyper-class* α *which consists of all classes (and sets)* X *which fulfil the condition* $\mathfrak{P}(X)$, where $\mathfrak{P}(X)$ is any condition.

This system can be seen to be consistent if the system $\mathsf{ZF}^{\#}$ mentioned above is [2]), and yet such comprehensive classes as the class of all sets or the class of all ordinal numbers are members of some objects, namely of hyper-classes [3]).

Another, more extreme example of a system where comprehensive classes can be members of classes, is the following system ST_2. The only variables of ST_2 are class variables. Its primitive relations are the binary relation of membership – $X \in Y$ – and the unary predicate of sethood – $S(X)$ (read:

[1]) See, e.g., R.M. Robinson 45.

[2]) The proof is essentially the same as the proof of the (relative) consistency of QM (see footnote 3 on p. 141).

[3]) In this system one can formulate an axiom which asserts the existence of a two-valued measure defined on all classes, which is α-additive for every ordinal α (see, e.g., Scott 61a or Keisler–Tarski 64). This is a strong axiom of infinity, which, in a very natural sense, is much stronger than that mentioned in footnote 3 on p. 141.

X is a set). Lower case variables are introduced as defined variables which vary over sets. The axioms are:

(a) A sethood axiom: *Every member of a set is a set.*

(b) All the axioms of QM (these axioms contain upper and lower case variables). The axiom of replacement is formulated as:

If F is a function and a a set then there is a set which contains exactly the values $F(x)$ which are sets, for all members x of a which are in the domain of F.

(c) The axioms of ZF with all variables replaced by upper case variables.

This is a two-tier set theory with sets in the bottom tier and classes in the upper tier. For instance, the class V of all sets is a class, as well as its power-class **PV**, which consists of all subclasses of V, its power-class – **PPV**, etc. A natural model of this theory can be given in the system $\mathsf{ZF}^{\#}$ as follows: We understand by "set" a member of $R(\theta)$, where R is the function defined on p. 94, and θ is a fixed inaccessible number, and by "class" we understand any set. Thus, assuming the consistency of $\mathsf{ZF}^{\#}$, the system ST_2 is also consistent. Moreover, every statement about sets which is provable in ST_2 is a theorem of ZM [1]).

Before we continue our discussion of the possible ways of handling classes in set theory, let us look into the use of classes in category theory, which is a new branch of algebra. Mathematicians working in that theory have found that even a set theory like QM is insufficient for their needs. Categories are classes, which may also be proper classes, and category theory also deals with functions defined on classes of categories and with other kinds of objects which are unavailable in QM [2]). Let us refer to the informal framework in which category theorists are working as *category theory*. Let us denote the informal power-class operation with **P**, i.e., for a class A, **P**A is the class of all subclasses of A (including the proper subclasses of A). Category theory involves only objects which are members of the classes V, **PV**, **PPV**, ..., $\mathbf{P}^n$V, where V is the class of all sets and n is some fixed finite number.

Some category theorists proposed to develop category theory within a system of set theory which is, essentially, ZF together with an axiom which asserts the existence of arbitrarily large inaccessible numbers [3]). In this system, the sets $R(\theta)$, where θ is any inaccessible number, are called by the category

[1]) The proof is along the same lines as the proof of the (relative) consistency of QM in footnote 3 on p. 141.

[2]) For the part of category theory which can be developed in VNB or in QM see MacLane 61 and Isbell 63.

[3]) Grothendieck, Sonner. See Kühnrich 66 and Kruse 66, where further references are given.

theorists universes. We shall refer to these sets as *subuniverses*, out of respect to the real universe. The idea is now not to deal with categories related to the universe of all sets, but to deal with categories related to a subuniverse $R(\theta)$. The latter categories are just subsets of $R(\theta)$, or of some $R(\theta+n)$, where n is a finite number; therefore one has sets of such categories, functions over such categories, etc.

A similar, but simpler way of dealing with categories in set theory is to assume only the existence of at least one inaccessible number, to choose for θ a fixed inaccessible number and to agree to talk only about categories related to the subuniverse $R(\theta)$. The reason why this way was not adopted by category theorists seems to be that they did not want to deprive the sets which are not members of the fixed subuniverse $R(\theta)$ from the blessings of category theory. Since they assumed the existence of arbitrarily large inaccessible numbers, given any set x, there is, by the axiom of foundation, a subuniverse $R(\theta)$ which contains x, and the category theory for this subuniverse applies to x. Looking deeper into the matter we see that even the assumption that every set belongs to some subuniverse is not sufficient for all conceivable needs of category theory; the results of category theory will concern in each case only the members of a subuniverse $R(\theta)$, but not *all* sets. To give an example, suppose $\mathfrak{S}(x,y)$ is some binary relation between groups, and suppose, for the sake of simplicity, that this relation holds or does not hold between x and y independently of the universe (or subuniverse) to which we refer in defining this relation. If we use category theory to prove the existence of a group g such that $\mathfrak{S}(g,h)$ holds for every group h, then with respect to the subuniverse $R(\theta)$ this means that there is a group $g\in R(\theta)$ such that $\mathfrak{S}(g,h)$ holds for every group $h\in R(\theta)$; but this does not establish in set theory the existence of a group g such that $\mathfrak{S}(g,h)$ holds for *every* group h, irrespective of which subuniverse it belongs to.

The best way of developing category theory within set theory seems to be to use a system of set theory like the system ST_2 described above. In ST_2 we can use categories related to *all* sets. This seems to be close to the way category theorists think about it when they are caught unawares, since in this approach categories are classes, not necessarily sets, yet, in almost all respects, classes can be treated like sets in ST_2. This can still be criticized as follows. Since the classes behave like sets this may mean that the classes of ST_2 are really sets, even though they are called classes. Thus, when we say here "all sets" we exclude the proper classes, which should have been included too. Thereby we seem to have arrived again at the situation discussed above where we dealt with a single fixed subuniverse (here we have a universe of classes and a subuniverse of sets). This criticism can be easily overcome by considering ST_2 only as a *façon de parler* for proving theorems about sets in the set theory ZM. We have mentioned above that every theorem of ST_2 which concerns sets only is also a theorem of ZM. Since ZM deals with sets only, when we say in ZM "all sets" we mean just that. Categories are, generally, objects of the system ST_2, but theorems about sets which are proved by means of categories are also theorems of ZM.

A different variant of ST_2 is obtained if we adhere to the limitation of size doctrine by requiring that a small class, e.g., a finite class, be a set even if its members are proper classes. In this variant we drop the sethood axiom (a) of ST_2, and amend the axiom of replacement VIII^c in (b) to read: *If F is a function and a is a set, then there is a set which contains exactly the values $F(X)$, which may be sets or classes, for all members X of a which are in the domain of F*. It should be noted that the axiom of union in (b) is: *For every set a, there is a set b whose members are exactly the members of the sets which are members of a* [1]). A natural model of this system is obtained in $\mathsf{ZF}^{\#}$ by interpreting 'class' as set, and 'set' as 'set of cardinality $<\theta$' [2]).

Stronger and more elaborate systems, which follow the basic ideas of the simple systems given here, are discussed in the literature; yet all such systems turn out not to yield any information concerning sets which is not contained, in one way or another, is some set theory of the ZF type [3]), which for the systems considered here is $\mathsf{ZF}^{\#}$ or ZM. This process of adding bigger classes and hyper-classes has to stop somewhere; and we have to decide where to do so. QM is a good place to stop at for reasons of convenience and neatness, yet, apart from these considerations, this choice is as arbitrary as any other. This arbitrariness is, to some extent, due to the antinomies and hence unavoidable; however, it is also due in part to the decision, originating with von Neumann, to admit classes other than sets as real mathematical objects. As mentioned in §7.4 above, we can use VNB without adopting von Neumann's point of view. More consistent and radical solutions to the problem of devel-

[1]) This is the correct way of reading the symbolic version of the axiom of union in §3.3; the verbal version of this axiom in §3.3 is refutable in the present system.

[2]) In the system G* of Oberschelp 64, too, proper classes can be members of sets. However, due to an axiom of comprehension weaker than the one given here, the system G* is essentially equivalent to QM. (One can prove that for a suitably formulated version H of QM which admits individuals, and for the natural translation of the statements of H to the language of G*, a statement is a theorem of H if and only if its translation is a theorem of G* – the proof is essentially given in Oberschelp 64; a more constructive version can be obtained by the method of Paul Cohen mentioned in footnote 2 on p. 132.

[3]) Takeuti 61 and 69; Solovay 66.

oping a system of set theory in which classes occur for the sake of convenience, while sets are still believed to be the only mathematical objects that exist, are handled in the next subsection.

7.6. **Classes not Taken Seriously – Systems of Bernays and Quine.** In §7.1 we introduced classes in VNB, in order to be able to neatly handle extensions of pure conditions. If we do not intend to regard classes as real mathematical objects, then a second look suggests that the system VNB might be too strong. We know that VNB is not too strong in the sense that we can prove in it false theorems about sets, or even about classes, yet VNB is too strong in the sense that, as we shall see, the machinery it contains for handling classes is much more than needed for the desired streamlining of set theory.

Bernays introduced [1]) an axiom system B which differs from VNB, in addition to purely technical matters, as follows. While B retains the full formalism that VNB has for handling sets and also retains the class variables, its language does not admit quantifiers over class variables. To partially compensate for this loss the language of B also admits, as primitive notions, the class abstracts $\{x \mid \mathfrak{P}(x)\}$, where $\mathfrak{P}(x)$ is any condition on x (of the language of B). Also, equality of classes is not primitive in B but is defined. The axioms of B are the axioms of VNB with the following changes:

(a) the universal class quantifiers in front of the axioms of subsets ($\mathrm{V^c}$) and replacement ($\mathrm{VII^c}$) are dropped,

(b) the axiom of extensionality (X) is dropped, and

(c) the axiom of comprehension for classes (XI) is replaced by the schema: *y is a member of the class $\{x \mid \mathfrak{P}(x)\}$ if and only if y fulfils the condition $\mathfrak{P}(x)$.*

Free class variables in a formula are interpreted as if they are universally quantified; e.g., the formula $\forall x(x \in A \leftrightarrow x \notin \{x \mid x \notin A\})$ is read "For every class A and for every element x, $x \in A$ if and only if $x \notin \{x \mid x \notin A\}$".

Even though the apparatus of B is more economical than that of VNB, it is enough for a streamlined approach to classes, since all the statements of VNB which are ordinarily used in mathematical arguments can also be expressed in B. B is closer to ZF than VNB. For instance, the proof that every statement of ZF which is provable in B is also provable in ZF is rather trivial (this should be compared to the corresponding proofs for VNB – both of which use deep theorems of logic) [2]). When one compares B and VNB one

[1]) Bernays 58.

[2]) The completeness theorem of the first-order functional calculus for the proof given on p. 131, and the cut elimination theorem (or the ϵ-theorem) for the proofs mentioned in footnote 2 on p. 132.

has to notice that the primitive symbols $\{x \mid \mathfrak{P}(x)\}$ of B are defined symbols of VNB. It is obvious that every theorem of B is also a theorem of VNB; but also the converse is true, i.e., every statement of B which is a theorem of VNB is also a theorem of B [1]). It is also worth noting that the embarrassing situation in VNB, viz. that there are conditions for which induction over the natural numbers cannot be proved, is avoided in B since all those conditions of VNB involve class quantifiers.

An even more radical attitude is advocated by Quine [2]). He proposes a system which is just ZF, except that statements of the language of B which involve classes are considered to be shorthand versions of statements of ZF. The translation from shorthand (in the language of B) to longhand (in the language of ZF) is as follows. (The shorthand may sometimes be longer than the longhand.) The language of B contains, in addition to what is in the language of ZF, free class variables and class abstracts – $\{x \mid \mathfrak{P}(x)\}$. The free class variables are interpreted as metamathematical variables for class abstracts $\{x \mid \mathfrak{P}(x)\}$ in which the condition $\mathfrak{P}(x)$ has no class parameters; e.g., the statement

(*) $\exists z \forall x (x \in z \leftrightarrow x \in A \wedge x \in \{u \mid u \notin B\})$

is understood to be a schema, where A and B stand for arbitrary class abstracts, i.e., the schema

$$\exists z \forall x (x \in z \leftrightarrow x \in \{v \mid \mathfrak{P}(v)\} \wedge x \in \{u \mid u \notin \{v \mid \mathfrak{Q}(v)\}\}) .$$

From this point of view the single statements contain no class variables, but they may contain class abstracts. These class abstracts can occur in such a statement only in a context like $y \in \{x \mid \mathfrak{P}(x)\}$ [3]). $y \in \{x \mid \mathfrak{P}(x)\}$ is taken to be shorthand for $\mathfrak{P}(y)$. Thus the statement (*) above is really shorthand for the schema $\exists z \forall x (x \in z \leftrightarrow \mathfrak{P}(x) \wedge \neg \mathfrak{Q}(x))$, where $\mathfrak{P}(x)$ and $\mathfrak{Q}(x)$ are any conditions of ZF. Simple checking shows that all the axioms of B are, in the present interpretation, theorems or theorem-schemas of ZF. It can easily be seen that also the logical axioms of B are interpreted as theorems or theorem-schemas of ZF, and that the logical rules of inference used in B lead us here from theorems and theorem-schemas of ZF to other theorems and theorem-schemas of ZF. Therefore all the theorems of B are also theorems of Quine's system, even though they are there interpreted somewhat differently. On the other

[1]) There are two proofs for this fact, both very similar to the proofs mentioned in the previous footnote.

[2]) Quine 63.

[3]) In fact, they may also occur as $\{x \mid \mathfrak{P}(x)\} = \{y \mid \mathfrak{Q}(y)\}$ or as $\{x \mid \mathfrak{Q}(x)\} \in y$, etc., but these are *defined* expressions of the language of B (see the definitions on pp. 122 and 128) and can therefore be eliminated.

hand, there are theorems of Quine's system which cannot be proved in B, since there is a statement $\mathfrak{S}(A)$ of B, with a free class variable A, such that when interpreted as a schema, all its instances are provable in ZF, and yet $\mathfrak{S}(A)$ is refutable in QM [1]).

7.7. **The System of Ackermann.** In the systems ZF and VNB the guiding principle for choosing the axioms was the limitation of size doctrine. Adherence to this doctrine prevented the occurrence of the known antinomies in those systems. In 1956, Ackermann [2]) proposed a system of axioms for set theory which is based on a completely different approach and which retains as axioms only the weakest consequences of the limitation of size doctrine, i.e., that a member of a set and a subclass of a set are sets. It is rather surprising that, as it turned out later, essentially the same theorems are provable in Ackermann's system as in ZF. It is also possible to formulate Ackermann's system for set theory with individuals [3]), but we shall follow our earlier practice of considering only set theory without individuals..

In Ackermann's system, which we denote with A, the universe consists of objects with a membership relation between these objects, denoted by the symbol $\in$. Two objects which have exactly the same members are equal. Therefore it stands to reason to refer to these objects as *classes*, which we do from now on. Notice that these are not classes in the sense of extensions of pure conditions as in §7.1, or, for that matter, extensions of any conditions. These are classes in the vague sense for which we also use the words 'collection', 'aggregate', etc., i.e., objects completely determined by their membership. Some of the classes are said to be *sets*. Unlike the system G of §7.4, not every class which is a member of some class is a set. The language of A is based on the first-order predicate calculus with equality. There is just one kind of variables – class variables; we shall use for them capital letters. The primitive predicate symbols are the binary membership symbol $\in$ and a unary predicate symbol M. We shall read $M(A)$ as "A is a set". We shall use lower-case letters as variables for sets. Lower-case variables are not primitive symbols of the language, we just say "for all x ..." instead of "for every class X if $M(X)$ then ..." and similarly for the existential quantifier.

DEFINITION. If A and B are classes such that every member of A is a member of B, we write $A \subseteq B$ and say that A is a *subclass* of B.

[1]) Such a schema was given by Mostowski – see A. Levy 60, §5. To see that it is refutable in QM use the method of Mostowski 51.

[2]) Ackermann 56.

[3]) Ackermann 56, §3.

In symbols, $A \subseteq B =_{\mathrm{Df}} \forall X(X \in A \to X \in B)$.

The axioms are:

α) THE AXIOM OF EXTENSIONALITY. **If the classes A and B have exactly the same members then they are equal.**
In symbols, $\forall X(X \in A \leftrightarrow X \in B) \to A = B$.

β) THE AXIOM OF COMPREHENSION FOR CLASSES. **There exists a class A which contains exactly those *sets* x which satisfy the condition $\mathfrak{P}(x)$,** where $\mathfrak{P}(x)$ is any condition of A.

The class A of this axiom is assumed to contain only the *sets* x which satisfy $\mathfrak{P}(x)$, rather than all classes X which satisfy $\mathfrak{P}(X)$, in order not to obtain here the contradictory axiom schema of comprehension of §3.1 (with sets replaced by classes). As in §7.1 we denote with $\{x \mid \mathfrak{P}(x)\}$ the class of *sets* x which satisfy $\mathfrak{P}(x)$; the existence of this class being guaranteed by Axiom β. Now, if we repeat the proof of Russell's antinomy we get that the class $\{x \mid x \notin x\}$ is not a set.

γ) THE AXIOM OF HEREDITY. **If Y is a member of the set x then Y is a set, too.**
In symbols, $Y \in x \to M(Y)$.

δ) THE AXIOM OF SUBSETS. **If Y is a subclass of the set x then Y is a set too.**
In symbols, $Y \subseteq x \to M(Y)$.

Let V be the class $\{x \mid M(x)\}$ of all sets. We saw above that the subclass $\{x \mid x \notin x\}$ of V is not a set, hence, by Axiom δ, V too is not a set.

ϵ) THE AXIOM OF COMPREHENSION FOR SETS. **If the only classes X which satisfy the condition $\mathfrak{P}(X)$ are sets then there exists a *set* w which consists exactly of those sets X which satisfy the condition $\mathfrak{P}(X)$,** where $\mathfrak{P}(X)$ is any condition *which does not involve the unary predicate M and which has no parameters other than set parameters.*
In symbols, $\forall x_1 \forall x_2 \ldots \forall x_n [\forall X(\mathfrak{P}(X) \to M(X)) \to \exists w \forall X(X \in w \leftrightarrow \mathfrak{P}(X))]$, where $\mathfrak{P}(X)$ does not involve M and has no parameters other than $x_1, \ldots, x_n$.

This is the main axiom of comprehension for sets in A (γ and δ, too, are axioms of comprehension for sets). Unlike the axioms of comprehension for sets in ZF and VNB, and unlike Axioms γ and δ, Axiom ϵ is not motivated by the limitation of size doctrine. Before we discuss the motivation of Axiom ϵ, let us first study the axiom from a technical point of view. If we lift the restriction that the condition $\mathfrak{P}(X)$ should not involve the predicate M, then by applying Axiom ϵ to the condition $M(X) \wedge \mathfrak{P}(X)$ we would obtain that the class $\{x \mid \mathfrak{P}(x)\}$ is a set, for every condition $\mathfrak{P}(X)$. In particular, the class $\{x \mid x \notin x\}$ is a set, which immediately yields Russell's antinomy. Thus the restriction that the condition $\mathfrak{P}(X)$ in Axiom ϵ does not involve the unary predicate M is necessary for the consistency of A. Also the second restriction, viz. that $\mathfrak{P}(X)$ may only have set parameters, is necessary in order to avoid Russell's antinomy. Were it not for the restriction on parameters in Axiom ϵ, we could have chosen, for $\mathfrak{P}(X)$ in Axiom ϵ, the condition $X \in Y$, and thereby we would have had "for every class Y, if all the members of Y are sets then the class $\{x \mid x \in Y\}$ is a set". If we substitute for Y the class $\{x \mid \mathfrak{P}(x)\}$, where $\mathfrak{P}(x)$ is any condition, we get that every class $\{x \mid \mathfrak{P}(x)\}$ is a set. This again, immediately yields Russell's antinomy.

As a consequence of Axiom ϵ, we get that if $\mathfrak{Q}(x)$ is any condition with no parameters other than set parameters, and which does not involve the predicate M, then $\mathfrak{Q}(X)$ cannot be equivalent to $M(X)$. This is shown as follows. Suppose $M(X)$ is equivalent to $\mathfrak{Q}(X)$, then if we take $\mathfrak{Q}(X) \wedge X \notin X$ for $\mathfrak{P}(X)$ in Axiom ϵ, we get that the class $\{x \mid x \notin x\}$ is a set, which, we know, is a contradiction. In other words, $M(X)$ cannot be defined in A by means of the membership relation ϵ (unless A is inconsistent).

Ackermann justifies Axiom ϵ as follows. Let us consider the sets to be the "real" objects of set theory. Not all the sets are given at once when one starts to handle set theory – the sets are to be thought of as obtained in some constructive process. Thus at no moment during this process can one consider the predicate $M(X)$ as a "well-defined" predicate, since the process of constructing the sets still goes on and it is not yet determined whether a given class X will eventually be constructed as a set or not. As a consequence, a condition $\mathfrak{P}(X)$ can be regarded as "well-defined" only if it avoids using the predicate M. Also, parameters are allowed in such a condition $\mathfrak{P}(X)$ only to the extent that they stand for "well-defined" objects, i.e., sets.

Ackermann's justification of Axiom ϵ is clearly insufficient. While one is not allowed to have in the condition $\mathfrak{P}(X)$ of this axiom a parameter Y which stands for a class which is not a set because membership in such a class is not "well-defined", one is allowed to use quantifiers over *all* classes in $\mathfrak{P}(X)$, i.e., $\mathfrak{P}(X)$ may contain expressions like "for all classes Y ...". If a single class is not "well-defined", why is the totality of all classes "well-defined"? It is possible to refine Ackermann's justification by some subtler arguments which may overcome the difficulty outlined here. However, taking into consideration all justifications known to the authors, Axiom ϵ is still far from having the intuitive obviousness of, say, the axiom of replacement of ZF. Thus, what makes the system A interesting and trustworthy is not the arguments brought forth in favor of its axioms but rather the beauty of the proofs in A, and the fact that A turned out to be equivalent in a strong sense to ZF, as we shall see below.

The last axiom of A is:

ξ) THE AXIOM OF FOUNDATION. **If y is a non-empty set then y has a member u such that $u \cap y = 0$**[1]).
In symbols, $y \neq 0 \rightarrow \exists u(u \in y \wedge u \cap y = 0)$.

When we come to compare A with ZF, we write the variables of ZF as lower-case variables, since in ZF all objects are sets. Every statement of (the language of) ZF is therefore also a statement of A. On the other hand, statements of A that mention classes which are not sets are not statements of ZF and have no natural translation into statements of ZF. Therefore, the question which we now ask is: Which statements of ZF are provable in A? It turns out that *the statements of* ZF *provable in* A *are exactly the theorems of* ZF. The proof that all the statements of ZF provable in A are theorems of ZF uses metamathematical arguments[2]). In order to prove the converse, namely that all the theorems of ZF are provable in A, it is enough to prove in A all the *axioms* of ZF. Here we shall prove in A all the axioms of ZF other than the axiom schema of replacement. The proof of that axiom schema makes use of Axiom ξ of foundation and involves metamathematical arguments[3]). The other axioms of ZF are proved as follows.

I. *The axiom of extensionality* of ZF follows immediately from Axiom α.

II. *The axiom of pairing*. Given sets b and c we consider the condition "$X = b$ *or* $X = c$". This condition does not mention the predicate M, it has only set parameters, and every X satisfying it is a set. Therefore, by Axiom ϵ, there is a set whose members are exactly b and c.

III, IV. *The axioms of union and power set* are proved similarly by applying Axiom ϵ to the conditions "*X is a member of a member of b*" and "*X is a subclass of b*" and using Axioms γ and δ, respectively.

V. *The axiom of subsets*. If b is a set and $\mathfrak{P}(X)$ is any condition, not even necessarily in the language of ZF, then by Axiom β there is a class $\{x \mid x \in b \wedge \mathfrak{P}(x)\}$, and by Axiom δ this class is a set. If we knew that there exists at least one set then the existence of the null-set would follow from the

[1]) This is Axiom IX* of §5.1. It is shown in Levy–Vaught 61 that the schema IX of §5.1 is now provable in A, and that if A without Axiom ξ is consistent then A with Axiom ξ is consistent. Axiom ξ was not proposed by Ackermann 56, but we need it here for the purpose of comparing A with ZF.

[2]) A. Levy 59.

[3]) Reinhardt 70.

axiom of subsets, as in p. 39 (or directly from Axioms β and δ). However, here we have to use Axiom ϵ to prove that there exists a set at all, so we may as well apply Axiom ϵ to the condition $X \neq X$ and prove right away the existence of the null-set.

VI. *The axiom of infinity*. It is rather surprising that this axiom can be proved in A since, unlike the systems of set theory discussed till now, none of the axioms of A directly mentions the existence of an infinite set. Let us prove the strongest version (VIc) of the axiom of infinity in §3.6 which is "There is a set a which contains a memberless set (i.e., the null-set) and such that, for all members y and z of a, $y \cup \{z\}$, too, is a member of a". Let us consider the following condition $\mathfrak{P}(X)$ on X: "X is a member of every class B such that B contains a memberless class and such that for all members Y and Z of B there is a member U of B which consists of Z and of the members of Y (informally, $U = Y \cup \{Z\}$)". The class V of all sets satisfies the requirements for B in $\mathfrak{P}(X)$, since there is a memberless set and since, by the axioms of pairing and union which we proved above, if Y and Z are sets then also the class $Y \cup \{Z\}$ is a set. Thus every class X which satisfies $\mathfrak{P}(X)$ is a member of V, and hence a set. Therefore the condition $\mathfrak{P}(X)$ satisfies the assumption of Axiom ϵ, and Axiom ϵ implies the existence of a set a which consists exactly of the sets X which satisfy $\mathfrak{P}(X)$. This set can be easily seen to be as required by the axiom of infinity above.

IX. *The axiom of foundation*, and in particular version IX*, follows immediately from Axiom ξ.

Having presented the result that the same statements of (the language of) ZF are provable in ZF and in A, let us draw some conclusions. First, we get that A is consistent if and only if ZF is consistent, since a contradictory statement, say, "There exists a set x which is both a member of itself and not a member of itself", is provable in A if and only if it is provable in ZF. Second, we get that if $\mathfrak{S}$ is a statement of ZF, then $\mathfrak{S}$ is consistent with A (i.e., not refutable in A) if and only if $\mathfrak{S}$ is consistent with ZF. Therefore, if A is consistent then A is consistent with the axiom of constructibility and, a fortiori, with the axiom of choice and the generalized continuum hypothesis. Also, almost all the independence results of §4 and §6 still hold when ZF is replaced in them by A.

Till now we have compared A with ZF; let us now say something about the relationship of A with the system QM of §7.5. As in the case of A and ZF a natural translation exists only from QM to A. In this translation the set variables of QM are translated as set variables of A while the class variables of QM are translated as class variables of A restricted to classes which consist only of sets. Using the proofs given above of the axioms of ZF it is easily seen that the translation of all the axioms of QM other than Axioms $\mathrm{VII}^{\mathrm{c}}$ of replacement are theorems of A. The translation of Axiom $\mathrm{VII}^{\mathrm{c}}$ is not provable in A; in fact, it is an axiom of strong infinity for A [1]. If A is consistent then (the translation of) Axiom $\mathrm{VIII}_{\sigma}^{\mathrm{c}}$ of global choice is consistent with A since, as we mentioned above, if A is consistent then A is consistent with the axiom of constructibility, which implies Axiom $\mathrm{VIII}_{\sigma}^{\mathrm{c}}$.

The central objects of A are the sets, and, as we saw, we know quite a bit about the sets in A. As to classes which are not sets, Axiom β asserts the existence of such classes whose members are sets. No axiom handles directly classes which have members which are not sets, yet by means of Axiom ϵ one can obtain many results about such classes. As a particularly simple example let us prove that there is a class A not all of whose members are sets. If there were no such class then a class X would not be a member of any class A unless X is a set. On the other hand, by the axiom of pairing, which we proved above, every set X is a member of some class. Therefore the property of being a set would be equivalent to the property of being a member of some class; but we proved above, by means of Axiom ϵ, that the property of being a set is not equivalent to any condition which does not mention the predicate M, and thus we have a contradiction. One can also prove much stronger results about classes whose members are not necessarily sets. For instance, by means of the axiom of foundation ξ one can prove the existence of the class $\{V\}$ whose only member is V $(=\{x \mid x=x\})$, and of the class $\mathbf{P}V$ which consists of all the subclasses of V, and of the class $\mathbf{PP}V$, etc. [2].

[1]) This can be shown by the methods of A. Levy 59 and Levy–Vaught 61 – see A. Levy 59, p. 157.

[2]) Levy–Vaught 61. On the other hand, one cannot prove in A any statement about classes which cannot be proved in ZF about sets – A. Levy 59.

CHAPTER III

TYPE-THEORETICAL APPROACHES

§1. THE IDEAL CALCULUS

We do not believe that there exists at this moment a single classification of the various approaches to the foundations of set theory which is decisively simpler and more "natural" than its competitors. We therefore make no such claim for the classification adopted here. The various approaches presented together in one chapter will, of course, have many common features, but the degree of communality will differ from chapter to chapter and the reader will notice that, occasionally, an approach dealt with in one chapter will have less in common with another approach treated in the same chapter than with one of the approaches mentioned in another chapter. As we see it, there exists a multi-dimensional continuum of possible attitudes out of which those that were historically chosen form a sometimes rather arbitrary-looking selection; of these systems those described in this book form an additional, perhaps not quite so arbitrary, selection.

In order to get a good preview of the features that distinguish the approaches to be treated in this chapter from those described in the preceding chapter, let us start with the presentation of a certain calculus that might be regarded as an adequate formalization of one "naive" approach: According to this approach, all the entities, if not within the whole universe, at least within the "universe of discourse", are essentially alike in their status, and it therefore makes always sense, to claim that one of these entities is a member of any other entity, even of itself, though such a claim may be wrong, sometimes perhaps even absurdly wrong. All the variables of the calculus in which this view is to be formalized would then be of the same kind, and one formation rule of this calculus would state, for instance, that any expression of the form '...∈– – –', or of the form '... = – – –', where the dots and dashes are replaced by occurrences of variables (different or identical), is a formula, and that for certain versions of such a calculus, such expressions are the only atomic formulae, from which the other formulae are formed with the help of the customary connectives, quantifiers, and parentheses.

In addition, according to the same approach, all those entities which fulfil a given condition always constitute another entity, a class, and only one class, since any two classes which have all their members in common are identical. This means that the corresponding calculus will contain, first, an *axiom (-schema) of comprehension* that might be symbolized as

$$(\ )\quad \exists y \forall x [x \in y \leftrightarrow \varphi(x)]\ ,$$

where '$\varphi(x)$' represents any formula in which 'x' is free and the initial pair of parentheses is meant to indicate the *universal closure* [1]) of the following formula, i.e. to stand for the string of the universal quantifiers binding all the remaining free variables of '$\varphi(x)$', if any; and, second, an *axiom of extensionality* that might be symbolized as

$$\forall x \forall y [\forall z (z \in x \leftrightarrow z \in y) \rightarrow x = y]\ .$$

The calculus, whose outline has been given here, will be called – following Hermes-Scholz [2]) – the *ideal calculus* and denoted by 'K'. This calculus may be extended by adding one or both of the axioms of infinity and choice (see Chapter II).

K and its extensions, though indeed constituting an ideal formalization of the naive approach in many aspects, have unfortunately one drawback: *they are inconsistent*. It is very easy to derive in them a formalized counterpart of Russell's antinomy. Let us do this in some detail.

Taking '$x \notin x$' for '$\varphi(x)$' in the axiom of comprehension, we get

$$(1)\qquad \exists y \forall x (x \in y \leftrightarrow x \notin x)\ .$$

According to a standard theorem of the first-order predicate calculus (which is again taken as the logic underlying the ideal calculus) we have

$$(2)\qquad \forall x (x \in y \leftrightarrow x \notin x) \rightarrow (y \in y \leftrightarrow y \notin y)\ .$$

According to a standard rule of the predicate calculus,

1) For the term, see Church 56, p. 228; for the symbol, see Carnap 37, p. 94.

2) Hermes-Scholz 52, pp. 57 ff.

(3) $\exists y \forall x (x \in y \leftrightarrow x \notin x) \rightarrow \exists y (y \in y \leftrightarrow y \notin y)$

is derivable from (2). (1) and (3) yield together, by *modus ponens,*

(4) $\exists y (y \in y \leftrightarrow y \notin y)$

from which follows, according to another standard theorem,

(5) $\exists y \neg (y \in y \leftrightarrow y \in y)$.

On the other hand,

(6) $y \in y \leftrightarrow y \in y$

is a theorem of the predicate calculus, from which, according to another standard rule of this calculus,

(7) $\neg \exists y \neg (y \in y \leftrightarrow y \in y)$

can be deduced. (5) and (7), however, are obviously a pair of contradictory statements. A calculus in which both are provable is inconsistent.

We may look upon the approaches discussed in the preceding chapter and to be discussed in the present chapter as so many different attempts to modify the ideal calculus in order to overcome its shortcomings. The "axiomatic" attitudes of the preceding chapter do not essentially change the language: the two kinds of variables in the systems VNB and QM of §7 in the previous chapter are not indispensable – we presented (on pp. 136–137) a system G which has exactly the same language as the ideal calculus K, but is essentially equivalent to VNB, and one can also easily extend it to a system essentially equivalent to QM. The decisive changes are made in the axiom of comprehension: in Zermelo's system it is replaced, on the one hand, by the much weaker *axiom (-schema) of subsets,* whose symbolization, adapted to our present purposes, is

( ) $\forall z \exists y \forall x\, [x \in y \longleftrightarrow (x \in z \vee \varphi(x))]$,

and, on the other hand, by a certain number of specific cases of the original unrestricted axiom of comprehension, i.e. by a certain number of axioms in which the original '$\varphi(x)$' is replaced by certain specific formulae; in one case, for instance, '$\varphi(x)$' is replaced by '$x \subseteq z$', i.e. by '$\forall w(w \in x \rightarrow w \in z)$', and we get the *axiom of power-set,*

$$\forall z \exists y \forall x\, [x \in y \leftrightarrow \forall w(w \in x \rightarrow w \in z)] \ .$$

A less far-reaching weakening of the axiom of comprehension consists in introducing a different conjunctive component into the right side of the equivalence, viz. the formula '$\exists z(x \in z)$. – instead of Zermelo's '$x \in z$' – getting thereby

$$(\ ) \quad \exists y \forall x\, [x \in y \leftrightarrow (\exists z\, (x \in z) \wedge \varphi(x))] \ .$$

This axiom-schema [1]), originating with von Neumann, at least in essence, assures the existence of a class comprising all and only those entities that fulfil a given condition and are members of *some* class or other, in short, of a class comprising all and only those *elements* that fulfil the given condition. (Zermelo's axiom of subsets guaranteed the existence of a class answering a given condition only if its prospective members were already members of a *given* class.) It is still weak enough not to allow the reproduction of the argument leading to contradiction in the ideal calculus. The reader would do well to check this by himself. He will find that instead of arriving at a contradiction he will be able to prove only the harmless theorem that the class of those entities that are not members of themselves is a non-element.

The common attitude characteristic for the approaches treated in the preceding chapter can then be summarized by saying that its deviation from the naive approach consists in repudiating the assumption that to any given condition there always corresponds a (membership-eligible) class, viz. the class of all and only those entities that fulfil this condition.

There exists a different attitude, going back in essence to Russell, that attempts to overcome the antinomies by tampering not with the axioms but rather with the language of the ideal calculus. By stipulating, e.g., that the string of symbols '$x \in x$' be not a formula, the antinomy is overcome in the rather trivial sense that its counterpart can no more be formulated in the calculus.

However, simply stipulating that no string of symbols of the form '...∈– – –', where the dots and dashes are replaced by (occurrences of) the same variable, be a formula would not do. It can easily be shown that a

1) This is Axiom XII of p. 138 formulated in the language of the system G of p. 136, with small letters instead of capital letters.

Russell-type antinomy can be generated from the string '$x \in y \wedge y \in x$' that would not be disqualified by the mentioned stipulation. In order to do away with all antinomies of this type, a more systematic and more radical deviation from the language of the ideal calculus seems to be required.

§2. THE THEORY OF TYPES

Instead of having only one kind of variables, as K does, the calculus we are now going to outline – we shall call it *type theory* and denote by 'T', in honor of its relationship to Russell's type theory (see §8) – contains a denumerable hierarchy of *levels* of variables. Each variable belongs to one level and one only. The *level-numbers*, 1, 2, 3, ..., will be indicated by right superscripts. The only atomic formulae of T are again formulae of the form '... ∈–––', and of the form '... = –––', but now in formulae of the first kind the level number of the left-hand variable must be lower by exactly one than the level number of the right-hand variable, i.e., $x^i \in y^j$ is a formula if and only if $j = i + 1$, and in formulae of the second kind the level numbers of both variables must be the same, i.e., $x^i = y^j$ is a formula if and only if $i = j$.

Since T contains infinitely many levels of variables, the single axiom of extensionality of K has to be replaced by an axiom-schema of extensionality

$$\forall x^i \forall y^i [\forall z^{i-1} (z^{i-1} \in x^i \leftrightarrow z^{i-1} \in y^i) \rightarrow x^i = y^i]$$

and the axiom-schema of comprehension

$$(\ ) \quad \exists y^{i+1} \forall x^i [x^i \in y^{i+1} \leftrightarrow \varphi(x^i)]$$

has become even more "schematic" in its new version.

We have already seen that Russell's antinomy is not reproducible in T. Let us check now that Cantor's antinomy cannot be reproduced in it either. In order to obtain Cantor's paradox we must prove the existence of the set v of all sets. In T we cannot even express that a set v contains all sets, since each variable of T belongs to exactly one level, and all we can say in T is that every entity of level i is a member of v. This is written as $\forall x^i (x^i \in v)$, and therefore v must belong to level $i + 1$. By the axiom schema of comprehension (where we take $x^i = x^i$ for $\varphi(x^i)$), there is indeed such a v^{i+1}. In K one obtains Cantor's paradox by showing that the power-set $\mathbf{P}v$ of v is a subset of v; in T the power-set of v^{i+1} is easily seen to be v^{i+2} (which is the set which contains all entities of level $i + 1$), and v^{i+2} is not a subset of v^{i+1}.

It can finally be shown that none of the customary arguments leading up to the other known logical antinomies are reproducible within T. As to the semantic antinomies, we shall waive their treatment for the time being, and take it up only later on.

How good is T? This question is meant in two senses: (a) how much of naive set theory, and how much of classical mathematics in general, can be developed on its basis; (b) how sure can we be of its consistency?

With regard to question (a), the following can be said. Let us supplement T by an axiom of infinity to the effect that the number of different objects of level 1 is $\aleph_0$ – where an object is regarded as belonging to a certain level if it belongs to the range of a variable of that level – and by axioms of choice for the various levels, and denote the resulting system by 'T*'. T* is essentially equivalent to (a simplified version of) the system of Whitehead and Russell, which we shall denote by PM, presented in their celebrated work *Principia Mathematica* [1]), in which counterparts of all basic theorems of classical set theory, arithmetic, and analysis have been rigorously proved and which is, therefore, almost generally considered to have provided a sufficient foundation for these disciplines [2]). However, the avoidance of the logical antinomies, which inspired the transition from K to T, has been achieved only at the price of certain drawbacks, some of which will be mentioned here.

Cardinal numbers can no longer be simply and uniquely defined as classes of equinumerous classes (see p. 96 and *T*, p. 59), in the tradition created by Frege, but, according to the level of the equinumerous classes, we get infinitely many cardinal numbers in each level (from two upwards). Each level has its own (quasi-)universal class, its own null-class, and the complement of a class becomes a quasi-complement, containing not all non-members of that class

1) Whitehead-Russell 10–13.

2) This essential equivalence is rather surprising in view of the fact that T is notationally so much inferior to PM, as it contains neither propositional nor relational variables. However, propositional variables can be replaced by schemata, a procedure introduced by von Neumann. Two-termed relations can be replaced by classes of ordered pairs, and ordered pairs in their turn by certain classes. If the ordered pair is homogeneous, i.e. of the form $\langle x^i, y^i \rangle$, it can be replaced by $\{\{x^i\},\{x^i,y^i\}\}$, according to the well-known method of Wiener-Kuratowski (cf. p. 33). Otherwise, i.e. if of the form $\langle x^i, y^j \rangle$, with $i \neq j$, this pair has first to be "homogenized", in a self-explanatory fashion, prior to the application of the Wiener-Kuratowski transformation. Many-termed relations, finally, can be replaced by classes of ordered *n*-tuples, which in their turn are reducible to ordered pairs in a way exemplified by the replacement of $\langle a, b, c \rangle$ by $\langle a, \langle b, c \rangle \rangle$, after homogenizing, if required.

but only those non-members that belong to the same level as the members of that class.

Many mathematicians will find this reduplication repugnant, not only for intuitive reasons but also, perhaps mainly, because it severely restricts their accustomed freedom of expression and notation and often requires complicated technical operations in order to overcome the unwanted notational restrictions.

Some of these technical drawbacks could be overcome by utilizing a principle of *typical ambiguity*, in adaptation of a procedure of *Principia Mathematica*. Instead of proving each theorem with the variables (and constants introduced by definition) carrying with them general superscripts and provided with occasionally long-winded clauses stating the relationships between these superscripts [1]), one could decide not to use those superscripts from the beginning but assume that each theorem is preceded by the clause 'provided that all terms belong to the appropriate levels'. The resulting formalism [2]) would look once more like that of K and one could think that it would be notationally as easily manageable.

In practice, however, it would sometimes be quite difficult to keep track of all the tacit provisions [3]). In addition, special care would have to be taken in the formulation of the rules of formation so that expressions leading to antinomies should not inadvertently be reestablished as formulae. Though, for instance, in a calculus embodying a principle of typical ambiguity, each of the two expressions '$x \in y$' and '$y \in x$' will separately be regarded as a formula, their conjunction '$x \in y \;\&\; y \in x$' must most definitely be denied this character, on pain of reestablishing a Russell-type antinomy.

1) Another procedure to the same effect would be to prove the theorem with the smallest possible superscripts attached to all occurring terms and proving once for all a meta-theorem stating that each theorem remains valid when all superscripts have been raised by the same amount.

2) This formalism is called PM4 in Quine 51a. Quine regards it as a certain version of PM. Beneš 52 points out that PM4 differs from PM – or rather from T – in containing a universal class and in completely abolishing the segregation of entities into types, the last deviation making PM4 "a set theory rather than a type theory". Without trying to diminish the extent of the deviations, the issue seems to be a verbal one. Russell's original type theory differed from Zermelo's original set theory in countenancing an infinite hierarchy of universes of discourse as against a single universe with Zermelo, as well as in declaring certain formulae as meaningless regarded as meaningful by Zermelo (in addition, of course, to many other differences). In PM4, the first difference is indeed abolished, but the second subsists. Cf. also below, pp. 191 ff.

3) For a case where the masters themselves, Russell and Whitehead, failed, see Gödel 44, pp. 145–146.

The required rules of formation are best stated in terms of *stratification* following Quine's usage. An assignment of levels to variables is said to *stratify* a formula of the form ...∈––– if it assigns two consecutive levels to its left-hand and right-hand variables, and such an assignment is said to stratify a formula of the form ... = ––– if it assigns the same level to both its variables. A general formula ψ of K is called *stratified* if there is an assignment which assigns levels to all the variables which occur in the expanded form of the formula ψ, i.e., in the form it assumes when expanded to primitive notation exclusively, in such a way that it simultaneously stratifies all the atomic formulae in ψ (all occurrences of the same variable being, of course, assigned the same level). A formula of K will then, finally, be admitted as a formula of our calculus if and only if it is stratified.

Though the careful specification of the language of this calculus would keep the logical antinomies out, working with it would require additional care in view, for instance, of the fact mentioned above that no longer can a conjunction of two formulae be automatically taken to be a formula itself.

In addition, adoption of typical ambiguity does not help to overcome other intuitively repugnant features of T*, such as the reduplication of the universal and null classes, of the cardinal numbers, etc.

With regard to question (b), i.e. the consistency of T*, let us be satisfied for the moment with stating that T* is consistent if the system of the axioms I–VI is; more will be said about this topic in Chapter V. Let us only remark at this point that the consistency of T* does not depend on the fact that the *language* of T* is restricted by the decree that the expression $x^i \in y^j$ is not a formula unless $j = i+1$; T* would stay equally consistent if $x^i \in y^j$ would be taken to be a formula for all i and j ($x^i \in y^j$ being false if $j \neq i+1$) [1]). What makes T* consistent is that the axiom of comprehension of T* asserts only the existence of sets of level $i+1$ which contain members of level i only.

§3. QUINE'S NEW FOUNDATIONS

The mentioned shortcomings of T* induced quite a few logicians to look for ways and means by which to avoid them, without – as much as possible – endangering the relative security of T*. One of the most interesting attempts in this direction was made by Quine, who tried to combine Zermelo's approach, which retained the language of K and excluded the antinomies, by

1) See pp. 191–192 below for a formulation of type theory with variables which range over all objects of all levels.

giving up the unrestricted use of the axiom schema of comprehension, with Russell's approach, which saw the culprit rather in the use of "meaningless" phrases. Would it not suffice to leave the language of the ideal calculus alone, i.e., not to discard the unstratified formulae, but rather to sterilize their harmful effects by restricting the '$\varphi(x)$' of the axiom of comprehension to stratified formulae?

The resulting system was described in a paper *New Foundations for Mathematical Logic* (1937) and became subsequently known simply as *New Foundations*. We shall denote with NF the system whose axioms are

THE AXIOM OF EXTENSIONALITY,

$$\forall x \forall y [\forall z (z \in x \leftrightarrow z \in y) \rightarrow x = y],$$

and

THE AXIOM OF COMPREHENSION,

$$(\ ) \quad \exists y \forall x (x \in y \leftrightarrow \varphi(x)),$$

where $\varphi(x)$ is a stratified formula in which y does not occur free. NF differs from *New Foundations* only in the handling of equality, and this difference has no effect on what we whall say about NF. NF is, in some respects, convenient to handle, and many of the shortcomings of T* are indeed overcome. According to NF, there are unique universal and null sets, each set has a complement, the cardinal numbers are unique, etc. It is also interesting to notice that the axiom schema of comprehension of NF can be replaced by a finite number of axioms which are quite similar to Axioms XI1–XI8 (pp. 129–130) [1]).

What about the antinomies? Since the decisive formulae in the derivation of the customary logical antinomies are unstratified, the existence of the corresponding antinomic sets cannot be proved directly by means of the axiom of comprehension of NF. This in itself is, however, no guarantee against establishing the existence of the antinomic classes by an indirect proof [2]).

1) Hailperin 44.

2) A simple example of a set whose existence is provable in NF, even though it is determined by a non-stratified condition, is the set $z \cup \{z\}$. Using the stratified condition $x \in z \vee x = u$ for $\varphi(x)$ in the axiom of comprehension one can prove that, for all sets z and u, $z \cup \{u\}$ exists, and hence, in particular, that the set $z \cup \{z\}$ exists. However, this example is rather misleading. In fact, the stratification requirement on the formula $\varphi(x)$

As a set theory NF has serious drawbacks. First, one can prove in it theorems which contradict some simple and obvious facts of classical set theory. One can prove in NF that there is a set y which is not equinumerous to the set $\{\{x\}|x \in y\}$ of its singleton subsets – the universal set, e.g., is such a set. One can also prove that there is a set which is equinumerous to its power-set – here, again, the universal set is such. Does this not prove that Cantor's antinomy can be reestablished in NF? No, it does not; Cantor's theorem (T, p. 70) does not hold in NF. In its customary proof there occurs an unstratified formula, as the reader will verify for himself. Serious difficulties arise in connection with the fact that mathematical induction holds only for stratified formulae – in NF one cannot even prove that for every finite cardinal number n there are exactly n cardinal numbers less than n (provided NF is consistent) [1]). Also, the axiom of choice is not compatible with NF[2]). Finally, there is a certain property $\mathfrak{P}(\alpha)$, expressible in the language of NF, such that one can prove in NF that there is an ordinal α which has this property, but there is no least ordinal which has this property, i.e., the natural ordering of the ordinals is not really a well-ordering [3]).

The anomalies just mentioned can be overcome, in one way or another; Rosser indeed wrote an extensive textbook in which he shows how to develop a large part of mathematics in NF or in simple extensions of it [4]). Mathematical induction for *all* formulae can be added as an additional axiom schema to NF. As to the other examples mentioned above of strange behavior of sets in NF, we notice that the sets which behaved strangely were only very large sets, such as the universal set or the set all ordinal numbers. One can formulate notions of "relatively small" sets, *prove* that some of the above mentioned anomalies, e.g., the failure of Cantor's theorem, do not hold for those sets, and *enforce*, by means of additional axioms, that the other anomalies, too,

in the axiom of comprehension of NF can be restricted to the variable x and the *bound* variables of $\varphi(x)$; for any such condition $\varphi(x)$ the corresponding instance of the axiom schema of comprehension is anyway a theorem of NF, as we saw in the case of $z \cup \{z\}$.

1) Cf. Rosser 53, Ch. XIII and Orey 64.

2) This, at the time rather surprising, result was found by Specker 53.

3) Rosser-Wang 50, p. 117. $\mathfrak{P}(\alpha)$ is the property "the set of all ordinals less than α has an order type less than α" which is expressible by a non-stratifiable formula; NF admits transfinite induction for stratified formulae only (cf. Rosser 53, Ch. XII). Rosser and Wang put it as follows: In every model of NF the set of all ordinals is not well-ordered by the natural ordering and hence no model of NF is "standard".

4) Rosser 53. Yet Curry 54a strongly questions the wisdom of choosing an NF-type system as a basis for a textbook of logic for mathematicians.

will not hold for those sets [1]). For example, one may add to NF a new axiom which asserts that all "relatively small" sets can be well-ordered [2]). Let us also mention that the existence of an infinite set is provable in NF [3]). Therefore one can develop in NF, without any additional axioms, all the mathematical theories which can be developed in T*, such as the theories of the natural and of the real numbers, the theory of real functions, etc.

From the point of view of the philosophy and the foundations of mathematics the main drawback of NF is that its axiom of comprehension is justified mostly on the technical ground that it excludes the antinomic instances of the general axiom of comprehension, but there is no mental image of set theory which leads to this axiom and lends it credibility. This could be forgiven only if NF would be a set theory which allows for a trouble free development of mathematics, at least to the extent that ZF does; in this case the elegance of basing set theory on a single syntactically simple form of the axiom schema of comprehension might compensate for the loss of a convincing intuitive justification for this particular form. However, we saw that if one aims at a reasonable development of mathematics in NF, one must add to it additional axioms. Those axioms, unlike the axiom of comprehension of NF, are motivated by direct mathematical needs, and thus the axioms of comprehension of the resulting system will not be based on one fundamental idea, and the simple elegance of NF will be lost in that system. Also the special attention given to the "relatively small" sets, and the possible special axioms concerning them, reintroduce the limitation of size doctrine (see p. 32) through the back door. Since only the "relatively small" sets can be expected to be well-behaved, the reason for dealing with other sets becomes now esthetical rather than mathematical, e.g., the universal set and the complements of "relatively small" sets exist, but those are sets which cannot participate in ordinary mathematics.

When we come to discuss the practicality of NF as a basis for mathematics, an additional point comes up. When we dealt with ZF, we specified exactly what the underlying formal language was, but this was not really necessary for the development of mathematics from it. Zermelo, in his original system, did not specify the underlying language without thereby incurring any undesirable results. Specifying the language is absolutely necessary for metamathematical investigations of set theory and apparently avoids the seman-

1) Such "relatively small" sets are the *Cantorian* and the *strictly Cantorian* sets – see Rosser 53, pp. 347, 486.

2) Quine 63, p. 296.

3) Specker 53, or cf. Quine 63, p. 299.

tical antinomies, but it is of no advantage to a mathematician who is not interested in the metamathematics of set theory and who is not worried about the semantical antinomies. The same remarks apply also to the theory QM (p. 138) which admits sets and classes. The reason why such a casual consideration of the language is tenable is because the axiom schemata of ZF and QM admit *all* formulae of the language (up to a natural restriction necessary in order to avoid the use of one variable for different things) and that even the use of reasonable stronger languages does not seem to endanger the consistency of those systems. On the other hand, in NF it is essential that the language be explicitly specified, since we have to be able to tell which formulae are stratified in order to know which applications of the axiom schema of comprehension are admissible in NF. For this purpose one has also to keep track of the status, with respect to stratification, of all the new operation and relation symbols which are defined in set theory. This puts a considerable burden on a mathematician to whom this bookkeeping does not seem related at all to the subject matter studied by him. We already voiced such criticism (on p. 139) with respect to the system VNB but it applies still more in the present case since there are few branches of mathematics in which formulae with bounded class variables would be used, while unstratified formulae are used much more often.

The practical difficulty of getting a mathematical rather than a syntactical criterion as to which instances of the axiom of comprehension are admissible in NF has been somewhat eased by the "model" of Specker [1]), which we shall now describe. Let TS be the system obtained from the type theory T by adding to it a unary operation symbol f and the following axiom schema which asserts that f is an automorphism of the universe which for every i maps the objects of level i on the objects of level $i+1$.

$$(*) \qquad \forall y^{i+1} \exists x^i (f(x^i) = y^{i+1}) \wedge \forall x^i \forall y^i (x^i = y^i \leftrightarrow f(x^i) = f(y^i))$$
$$\wedge\ \forall x^i \forall y^{i+1} (x^i \in y^{i+1} \leftrightarrow f(x^i) \in f(y^{i+1}))$$

The rules for computing the level of a term are such that if the level of a term t is j then the level of $f(t)$ is $j{+}1$, e.g., the levels of $f(x^i)$, $f(y^{i+1})$, and $f(f(x^i))$ are $i{+}1$, $i{+}2$ and, $i{+}2$, respectively. Thus the axiom (*) does indeed satisfy the formation rules of type theory. In TS, formulae $\varphi(x^i)$ in which the symbol f occurs are not allowed in the axiom schema of comprehension of type theory (p.158). The system TS is a "model" of NF in the sense that if φ is any

1) Specker 58 and 62.

sentence of the language of NF, and ψ is a sentence of the language of TS obtained from φ by putting level superscripts on the variables and by applying the symbol f any number of times to the different occurrences of the variables (ψ has, of course, to satisfy the formation rules of type theory), then φ is a theorem of NF if and only if ψ is a theorem of TS. TS is somewhat easier to work with than NF, at least to mathematicians used to ZF, since the restrictions on the use of its axiom of comprehension are simpler and more natural than those of NF; in TS, formulae used in the axiom of comprehension, as well as all other formulae, have to satisfy the formation rules of type theory but, unlike stratification, those formation rules are very natural in the context of type theory.

When we come to ask whether NF is at all consistent, we notice first that, by Gödel's theorem on consistency proofs (Chapter V, p. 313) if NF is consistent then we cannot expect to find a proof of the consistency of NF which uses no means beyond those of NF and, a fortiori, we cannot find such a proof which uses only the means of Peano's number theory. Therefore we have to look for a proof of the consistency of NF in some set theory such as ZF, or to look for a possibly finitary proof (Chapter V, p. 305) that if some set theory, say ZF, is consistent then so is NF. Nothing of the kind has been found yet, and the search for such a proof is still a major task of the metamathematics of set theory. As a consequence of this state of affairs and because of the lack of a clear mental image of NF as a set theory, one cannot take the consistency of NF for granted. A little light is shed on the question of the consistency of NF by noticing that the following is an immediate corollary of what was said above concerning the provability of φ and ψ in NF and TS. If we take for φ a sentence $\varphi' \vee \neg\varphi'$, then we can take ψ to be of the form $\psi' \wedge \neg\psi'$, and we get that NF is consistent if and only if TS is. This does not establish the consistency of NF since we do not know whether TS is consistent, but it somewhat increases our confidence in the consistency of NF since, if NF and TS were both inconsistent, one could expect that a contradiction would be exhibited faster in TS, TS being a more natural theory from a mathematical point of view [1]).

Let us close with a few remarks concerning individuals in NF. Quine chose to introduce individuals as sets x which are equal to their singleton, $\{x\}$. If NF is consistent then it remains consistent when one adds to it an axiom asserting that there are no individuals as well as when one adds to it an axiom asserting

1) An investigation of the consistency of NF, from a different angle, is carried out in Grišin 69.

that there is at least one such individual [1]). Quine chose to introduce individuals in this way, rather than in the straightforward way used in the previous chapter [2]), for reasons of syntactic elegance [3]); in particular, he wanted to retain the full axiom of extensionality. As it turned out, the effects of this choice go far beyond questions of elegance; if Quine were to formulate NF so as to allow individuals which do not coincide with their singletons, and if he, consequently, changed the axiom of extensionality, he would get a much weaker theory NF′ in which, unlike NF, the existence of infinite sets could not be proved and which, again unlike NF, can be shown to be consistent by a proof which uses no means beyond those of Peano's number theory [4]).

§4. QUINE'S MATHEMATICAL LOGIC

NF served as a basis for the formulation of another system of set theory by Quine in his book *Mathematical Logic* [5]). This system, too, became known by the name of the book; we shall denote our version of it by ML. In ML, Quine follows von Neumann [6]) in distinguishing between classes which can be members of classes, and are therefore called *elements*, and classes which cannot be members of classes. Since we want to compare ML with the other set theories which we have discussed, and in particular with NF, we shall refer to the elements as *sets* and to the classes which are not elements as *proper classes*; as in Chapter II, § 7, we shall use capital letters for class variables and small letters for set variables. The difference between the system G (or VNB) of Chapter II and ML is that in the former system the sets behave as they behave in NF. The axioms of ML are

THE AXIOM OF EXTENSIONALITY,

$$\forall A\, \forall B\, [\forall x (x \in A \leftrightarrow x \in B) \rightarrow A = B],$$

THE AXIOM OF COMPREHENSION BY A SET,

$$(\ ) \quad \exists y\, \forall x\, (x \in y \leftrightarrow \varphi(x)),$$

1) Scott 62.
2) See footnote 1 on p. 30.
3) Quine 63, pp. 31–33.
4) Jensen 69.
5) Quine 51.
6) See system G of Chapter II, § 7.4.

where $\varphi(x)$ is any *stratified* formula with set variables only, in which y does not occur free [1]), and

THE AXIOM OF IMPREDICATIVE COMPREHENSION BY A CLASS,

$$(\ ) \quad \exists Y \forall x (x \in Y \leftrightarrow \psi(x)) ,$$

where $\psi(x)$ is *any* formula in which y does not occur free.

The axiom of impredicative comprehension by a class is exactly Axiom XII of impredicative comprehension for classes of QM (p. 138). When we compare both axioms of comprehension of ML, we see that stratified formulae which mention only sets determine sets, whereas arbitrary formulae determine classes. In VNB it is exactly the "small" classes which are sets, but this is not the case here. The largest class of them all, viz. the universal class, is a set since it is determined by the stratified formula $x=x$. The class of all sets x which are not members of themselves, which is a subclass of the universal set, is a proper class, as can be easily shown by repeating the argument of Russell's antinomy. The property of being a proper class is not even necessarily confined to large classes – even the class of all natural numbers (according to its usual definition in ML) cannot be shown to be a set [2]).

The axioms of NF, written with set variables, are obviously theorems of ML – the axiom of extensionality of NF follows directly from the corresponding axiom of ML, the axiom of comprehension of NF coincides with the axiom of comprehension by a set of ML. These are the only facts about sets mentioned in the axioms of ML; in addition to these the axioms of ML consist of that "part" of the axiom of extensionality which concerns proper classes, and this part can be essentially regarded as a definition of equality (p. 122), and of the axiom of comprehension by a class, which says nothing about the existence of sets. As a consequence one can prove that if $\mathfrak{S}$ is any statement containing set variables only, then $\mathfrak{S}$ is a theorem of ML if and only if it is a theorem of NF. In particular, by choosing for $\mathfrak{S}$ a contradictory statement we get that ML is consistent if and only if NF is. The proof of this is similar to the proof that the corresponding relationship holds between the

1) In 1940, in the first edition of Quine 51, Quine permitted also bound class variables in $\varphi(x)$. That version of ML was shown to be inconsistent by Rosser 42 and Lyndon , since it implies the Burali-Forti paradox; in that system one can, essentially, carry out transfinite induction for the property $\mathfrak{P}(\alpha)$ of p. 163 and of footnote 3 on that page. The present version of this axiom is due to Wang 50a.

2) Rosser 52.

set theories VNB and ZF, given in the previous chapter (pp. 131–132) [1]). Notice that this relationship holds only between ZF and VNB, VNB having only an axiom of *predicative* comprehension by a class, but fails to hold between ZF and QM, QM having an axiom of *impredicative* comprehension by a class. This relationship holds between NF and ML, even though ML contains an axiom of impredicative comprehension by a class, for the following reason. In QM we can prove statements $\mathfrak{S}$ of ZF which cannot be proved in ZF, by first using the axiom of impredicative comprehension by a class to obtain certain classes and then using those classes in the axiom of subsets (V^c) to obtain certain sets. Nothing of the kind is possible in ML which has no axiom like the axiom of replacement (VII^c) which makes the existence of sets depend on the existence of classes. This feature of ML, which guarantees the consistency of ML if NF is consistent, is not an unmixed blessing. It is just the absence from ML of such axioms that casts serious doubts on the role of the classes in ML. If the proper classes do not affect the sets at all, perhaps one could do without proper classes at all?

Starting from his system NF, Quine went on to propose ML and develop it. He was motivated, to a large extent, by the lack of full mathematical induction in NF [2]). In ML, full mathematical induction is available, but at a very high price. When we recall that every statement $\mathfrak{S}$ which contains only set variables and which is a theorem of ML is also a theorem of NF, we see immediately that for the notion of natural number of NF, or for any other notion of natural number definable in NF, we cannot have full mathematical induction in ML unless we already had it in NF. Thus if we expect ML to help us with induction we must use in ML a notion of natural number unavailable in NF. Naturally, we take the natural numbers of ML to be sets rather than proper classes. The formula $N(x)$ of ML which asserts that x is a natural number will contain quantifiers over class variables (since otherwise $N(x)$ would have been available already in NF). Let us denote by $N'(x)$ the formula of NF which asserts that x is a natural number in the sense of NF. The

1) Wang 50a, or Rosser 55, pp. 45–47. The only difference is that now we define *all* subsets of u to be model-classes. If ML is consistent, we cannot expect to obtain a method which directly produces a proof of $\mathfrak{S}$ in NF from a proof of $\mathfrak{S}$ in ML, as we had for VNB and ZF on p. 132 and in footnote 2 on the same page; if what we mentioned here about the provability of $\mathfrak{S}$ in ML and NF were provable even in the theory of natural and real numbers (i.e., in second order number theory) then it would be provable also in ML, and we would obtain that ML is inconsistent, as in footnote 4 on p. 170.

2) Quine 55, p. 165 and Quine 63, p. 300. As a consequence ML is not finitely axiomatizable (the proof of this is like the proof, in footnote 4 of p. 139, that QM is not finitely axiomatizable).

relationship between N(x) and N′(x) is that N(x) implies N′(x) in ML but N′(x) does not imply N(x) in ML [1]) (unless ML is inconsistent). Thus, in ML we use a notion of natural number which is possibly more restricted than that of NF.

Since N(x) contains quantifiers over class variables we cannot use the axiom of comprehension by a set to prove that the class $\{x|N(x)\}$ of all natural numbers is a set; indeed, if ML is consistent, this class cannot be shown in ML to be a set [2]). This makes N(x) a very poor formal counterpart of the intuitive notion of natural numbers. From a conceptual point of view, having two kinds of collections – sets and proper classes – can be tolerated as long as all the collections which play an active role in mathematics are sets, while only very large, or otherwise exceptional, collections are proper classes. The collection of all natural numbers is so central in the development of mathematics that it must be a *set* in any set theory worth its name. This state of affairs in ML is a drawback not only from the conceptual point of view. Even though ML includes the set theory NF, which is a fairly strong set theory in certain respects, nothing of the machinery of NF can be applied to the class $\{x|N(x)\}$ of all the natural numbers of ML. Thus, while one can define real numbers in ML as classes and develop the arithmetic of real numbers [3]), this is done in exactly the same way as one develops the theory of real numbers in second-order number theory, and is completely divorced of the "set theoretic" part of ML. As a consequence, not even the theory of real functions can be developed in ML [4]).

These drawbacks can be overcome in two different ways. First, one can strengthen ML by some additional axioms [5]), in particular, by an axiom which asserts that the class $\{x|N(x)\}$ is a set. Thus strengthened, ML does indeed become a reasonably strong set theory in which we can develop at

1) Cf. the next footnote.

2) Rosser 52, p. 241. This proves, of course, that N(x) is not equivalent in ML to any stratified formula which contains only set variables. In Quine's original 1940 version of ML, $\{x|N(x)\}$ is directly shown to be a set, but that version is, alas, inconsistent.

3) Quine 55, §§51–52.

4) Let us denote by Con(NF) and Con(ML) some arithmetical statements which assert, in a natural way, the consistency of NF and that of ML, respectively. We mentioned on the previous page and in the first footnote there Wang's proof of Con(NF)→ Con(ML). That proof can be carried out in the theory of real functions (i.e., in third-order number theory); however, as we shall see, if ML is consistent then Con(NF) → Con(ML) is not a theorem of ML. By a theorem of Wang (Rosser 52, Lemma 2), Con(NF) is a theorem of ML; if one could also prove Con(NF) → Con(ML) in ML then one could prove Con(ML) in ML which by Gödel's theorem on consistency proofs (p. 313) would mean that ML is inconsistent.

5) Orey 55 and 56.

least as much of classical mathematics as can be developed in the type theory T*. However, if one recalls that ML was proposed as an improvement of NF because ML has full-fledged mathematical induction whereas NF has induction for stratified formulae only, one may prefer to stick to NF since, once one starts adjoining to ML axioms as mentioned, one might as well add an axiom of full mathematical induction to NF. On p. 164 we voiced some criticism against adding new axioms to NF – the same criticism also applies to ML. The second cure to the maladies of ML, to be adopted by whoever prefers not to add axioms to ML, is to retreat, i.e., to abandon the definition $N(x)$ of the notion of natural numbers in favor of the definition $N'(x)$ used in NF. With the latter definition we can develop in ML all the classical mathematics one can develop in NF, but we have mathematical induction restricted to stratified formulae which contain set variables only.

Finally, let us remark that in the set theories with classes discussed in the previous chapter we had two kinds of collections – sets, which are the collections which play an active role in mathematics, and proper classes. We had a somewhat similar distinction in NF between the "relatively small" sets and the "relatively large" sets. In ML, even if strengthened by additional axioms, we have a threefold classification – relatively small sets, relatively large sets, and proper classes. While the main advantage of NF over ZF lies in the elegance of its axioms, ML, like NF, does not allow a simple and troublefree development of mathematics, without sharing this elegance.

§5. THE HIERARCHY OF LANGUAGES AND THE RAMIFIED CLASS CALCULUS

Before we turn to the presentation and discussion of the other trend noticeable in the type-theoretical approach, and as a partial preparation for this discussion, let us deal with a question which must have by long arisen in the mind of the careful reader: type hierarchies and stratification seem to work well enough for the avoidance of the logical antinomies, but what about the semantic antinomies? Indeed, it seems on first sight as if a Richard-type antinomy can be derived within a calculus embodying a type rule. For the purpose of descriptional simplicity, let us choose for illustration not the calculus T treated before, but a certain variant P which has natural numbers as individuals (rather than as certain classes of classes) and a few more axioms corresponding to the well-known Peano axioms for the arithmetic of natural numbers (p. 48, footnote 1); the reader will easily verify that the transition to this variant is not essential for the validity of the following argument but only serves to simplify it considerably. There are in P denumerably many expressions determining classes of natural numbers. Let these classes be enumerated as, say, r_1^1, r_2^1, r_3^1, ...; then the axiom of comprehension seems to ensure the existence of a class, say r_0^1, that contains all and only those natural numbers n that do not belong to the class r_n^1. r_0^1, being a class of natural numbers determined by a well-formed expression, must be identical with some r_n^1, say r_m^1. What about the relation between m and r_m^1? If m belongs to r_m^1, then from the identity of r_0^1 and r_m^1 and the definition of r_0^1 it follows that m does not belong to r_m^1; if m does not belong to r_m^1, it belongs by definition to r_0^1, hence to r_m^1. The responsibility for this contradiction cannot be charged to a violation of the type rule.

It can however be argued, and has been argued, that the condition '$n \in r_n^1$' playing a decisive role in the above argument cannot really be formulated in the language of P. 'r_n^1' is short for 'the n-th class of natural numbers determined by expressions of P (according to some enumeration)', hence is not (an abbreviation for) an expression of P itself but rather of the metalanguage of P. Strict adherence to the distinction between object-language and metalanguage is enough to remove the ground from under the argumentation leading up to Richard's antinomy (and the other semantic antinomies). By implementing the hierarchy of levels within one language by a *hierarchy of languages* [1]), the last threats presented by the classical antinomies seem to have been successfully repulsed.

Once more, there is a price to be paid for this success. The strict distinction between the language layers looks counter-intuitive to many thinkers, to about the same degree as the strict distinction between the levels. In addition, the question must be raised whether the layering of languages excludes the detrimental possibility that a given object-language might contain expressions which – though not identical with metalinguistic expression – would still stand in some kind of isomorphic correspondence to them, thereby allowing for reestablishment of the semantic antinomies, if only in a roundabout way. This is a very serious possibility indeed, and in Chapter V the subject will be taken up again.

Yet, if one wants to make sure that the semantic antinomies do not turn up in a roundabout way, one can use a different implementation of the type hierarchy considered so far. We recall that the "Richardian" class of natural

1) This insight, prepared by Ramsey's distinction between the logical and the "epistemological" antinomies and Hilbert's distinction between mathematics and metamathematics, is due mainly to Tarski; see Tarski 56, VIII. Cf. also Carnap 37, pp. 211 ff.

numbers was determined through reference to a totality, i.e. the totality of all such classes determined by expressions of P, of which it turned out to be a member itself. There have been, and still are, quite a few thinkers who regard such a type of determination, if "essential", i.e. not replaceable by a determination not exhibiting this feature [1]), as illicit, since viciously circular. No object, they would say, should be regarded as belonging to a certain totality if the very existence of this object can be shown only by the use of an expression referring to that totality. By outlawing expressions purporting to refer to such over-inclusive totalities as the totality of all expressions, an illicit application of the axiom of comprehension can be avoided and the existence of the "Richardian" class and other antinomic classes no more proved. This aim can be achieved, e.g., by changing the language of T, to which calculus we now return after P has served its purpose, such that each variable will have to carry along with it not one but two indices, one, say a right superscript, indicating its level as before, the other, say a left superscript, indicating its "order". The formulation of the axiom of comprehension is now changed to:

> If the highest order-index of any bound variable of level $i+1$ occurring in '$\varphi({}^k x^i)$' is j then
>
> $$(\;)\quad \exists {}^{j+1}y^{i+1} \forall {}^k x^i \, ({}^k x^i \in {}^{j+1}y^{i+1} \leftrightarrow \varphi({}^k x^i))$$
>
> (the order-index of 'y^{i+1}' being 1, if no such bound variable occurs in '$\varphi(x)$' altogether).

In the resulting *ramified class-calculus*, RT, the semantical antinomies are indeed eliminated. Trying to reproduce, e.g., the argumentation leading up to Richard's antinomy we are immediately impressed by the fact that no longer does there exist in RT a counterpart for 'for all classes of natural numbers of level i' but only for 'for all classes of natural numbers of level i and order k'. Forming the Richardian (class) corresponding to an enumeration of the classes of natural numbers of order k determined by the formulae of RT, we notice that its order is $k+1$, so that we no longer have any right to claim that it is identical with one of the classes of order k.

The arguments which we have just given show that the semantical antinomies cannot be directly reproduced in RT. Moreover, the semantical antinomies, or any other antinomy, cannot even turn up indirectly in RT since

1) This is the type of determination that is known as "impredicative". Cf. Chapter II, p. 38 and below, § 10.

the consistency of this calculus is demonstrable (from certain relatively weak assumptions) [1]. However, as might be expected, this very desirable trait has its price. On the basis of RT only a fraction of classical mathematics can be reconstructed. This will be easily understood, without our having to delve into details, as soon as we notice, e.g., that the least upper bound of a class of real numbers, constructed in RT in any natural way, is of an order higher than these real numbers [2].

In order to counterbalance the deplorable loss of strength incurred by the ramification, one could stipulate an *axiom of reducibility* whose effect would be to ensure, corresponding to each class of a certain level and any order, the existence of a class of the same level and order 1 containing the same members as that class. These two classes would nevertheless not be regarded as identical since otherwise the effect of the introduction of the axiom of reducibility would be nothing but the exact cancellation of the ramification; but any two classes containing the same members *are* identical, according to the axiom of extensionality. This axiom would then have to be sacrificed. This, however, would now have repercussions of its own, which we shall not pursue here any further [3].

1) Such a demonstration has been given many times, both model-theoretically, as in Fitch 38, and proof-theoretically, as in Lorenzen 51 and Schütte 52. For these two methods of consistency proofs, see Chapter V, § 4.

2) This difficulty was discussed already by Weyl 18, p. 23.

3) Cf. below, p. 204. The situation concerning the axiom of reducibility is still quite confusing. The objection raised by Ramsey, Waismann and others, as if this axiom were of an empirical character rather than of a logical one, hence if true only factually so and, so to speak, by lucky coincidence – recalling a similar criticism against the axiom of infinity (see below, p. 185) – seems to have little justification. As a matter of fact, the axiom of reducibility is incomparably weaker than the unresticted axiom of comprehension in the ideal calculus; this last axiom, however, is rejected by intuitionists and other constructivist authors not because it is regarded as empirically false or doubtful but because its acceptance does not square with a constructivist attitude. For a Platonist, the axiom of reducibility is no less logical than the axiom of comprehension, and if he rejects either or both, it is done because the logical system incorporating them is either contradictory or in some other sense inefficient. In an appropriate metalanguage, the logical character ("analyticity") of the axiom of reducibility is easily demonstrable. See especially Quine 36, Copi 50, Church 51, Copi 54 (Appendix B) and Church 56, pp. 354–356.

Altogether, the real choice, as many have observed, seems to be between the simple theory of types, appealing to the Platonist, and the ramified theory without the axiom of reducibility, appealing to the constructivist. The latter has then the choice between the "heroic", or rather, "quixotic" course taken in the twenties by Chwistek, for instance, of countenancing finite types only and being therefore obliged to sacrifice large parts of classical analysis, or working with types of transfinite order, in the line of Wang and Lorenzen to be discussed in the following sections.

The ramified calculus can be credited with being a forerunner of Gödel's notion of constructibility, which plays such an important role in the metamathematics of set theory (see pp. 60 and 108ff) [1]. In introducing the constructible sets, only the notion of order is retained, while the notion of level is not used.

§6. WANG'S SYSTEM Σ

Reinforcing ramified type theory through an axiom of reducibility never enjoyed much popularity, and is now generally regarded to have led into a blind alley. Ramified type theory as such, however, seems to be undergoing a process of rejuvenation at the hands of such able logicians as Wang and Lorenzen. It is not so much its capability of eliminating the semantic antinomies which makes it so attractive – this aim can be attained more simply through the method of language hierarchies outlined above – as its connection with a "constructivist" philosophy of mathematics. Postponing the philosophical discussion let us turn to the presentation of a system that aims at preserving the desirable features of a ramified type theory but avoids its undesirable traits without recourse to such dubious means as an axiom of reducibility. The major point of this system is that it restores to a sufficiently high degree the freedom of use of such phrases as 'for all real numbers' instead of the awkward 'for all real numbers of order i'. The system we are going to present, Hao Wang's system Σ, exists so far only in outline [2]) but this outline looks so interesting that at least a short discussion is indicated.

The hierarchy of objects in Σ differs in many respects from that of RT. First, the messy two-dimensional array of RT is replaced by a one-dimensional array that manages to combine the level and order distinctions through allowing for "mixing of types". The lowest (or 0-th) *layer* – we shall use this term instead of Wang's own 'order' so as to keep it apart from the "orders" of RT – consists of some denumerable totality of objects (which may be taken to be, for instance, the positive integers or all the finite sets built up from the empty set). The first layer contains all the objects of the 0-th layer and, in addition, all those sets of these objects which correspond to conditions that contain no bound variables ranging over objects of the first or higher layers; in general, the $n+1$-st layer contains all the objects of the n-th layer together with all such sets of these objects as are determined by conditions whose

1) For this relationship, see Gödel 38 and 44.
2) Wang 54.

bound variables range over objects of the n-th layer at most. We have here a combination of a conception of *cumulative* layers [1]) according to which all objects of a certain layer belong also to all higher layers – a rather serious deviation from the more customary conception of *exclusive layers* – with a strict embodiment of the *vicious-circle principle* [2]), according to which no entity determined by a condition that refers to a certain totality should belong to this totality. For any two objects, there exists therefore a layer (hence, infinitely many ones) to which they both belong; the lowest layers, on the other hand, to which they belong separately may of course be different.

Secondly, the hierarchy of layers is continued beyond the finite ordinals [3]). Layer ω, for instance, is the union of all finite layers, and layer $\omega + 1$ contains, in addition, also such sets as are determined by conditions whose bound variables range over the objects of layer ω at most. Attempts to extend the type hierarchy beyond the finite ordinals were made before [4]), but in Wang's system this device is exploited to its utmost. Though the naive freedom of expression is not yet fully restored by it and expressions corresponding to 'for all real numbers' are still not reestablished, we are already entitled to use 'for all real numbers of finite layers', though only from the partial system $\Sigma_{\omega+1}$ onwards, that deals with the entities of layer ω, and not in any system with lower index. That this is indeed a far-going improvement can be realized from the fact, for instance, that the least upper bound of a set of real numbers that belong to layer n is, in general, an entity that belongs to layer $n + 1$, hence different from all these numbers, according to the vicious-circle principle and the exclusive-layer conception; in Wang's system, however, the original real numbers and the upper bound of their set do belong to a common layer, indeed to all layers from $n + 1$ upwards. This clears the ground for a proof of the least upper bound theorem pretty much along the traditional lines. And the situation is similar with respect to other basic theorems in analysis, such as the Bolzano-Weierstrass and the Heine-Borel theorems that proved to be stubborn obstacles in the way of any constructivist reconstruction of analysis.

Wang also forms the union, Σ, of all the partial systems Σ_α, where α is any

1) The conception of cumulative layers (or orders) is not original with Wang. Cf., e.g., Quine 53, p. 123.

2) For an especially penetrating discussion of this principle, see Gödel 44, pp. 133 ff.

3) Whitehead and Russell explicitly reject this procedure though their reasons are none too convincing. See Whitehead-Russell 10–13 I, p. 53.

4) See, e.g., L'Abbé 53, especially footnotes 2, 3, 4 and 15 in which the relevant literature is referred to; cf. Andrews 65. The set theory ZF itself is viewed by many logicians as an extension of type theory beyond the layer ω – see, e.g. Kreisel 65.

"constructive" ordinal of the second number class, but this procedure raises many moot questions with regard to the exact characterization of the term 'constructive' in this context as well as with regard to the legitimacy of this procedure in general; these questions will not be discussed here. It is clear, however, and admitted by Wang, that Σ itself is no more a formal theory in the sense in which its partial systems Σ_α are.

With this proviso in mind, we shall now describe the structure of Σ in outline. Σ is based on standard predicate calculus, meant to hold for entities of any layer. It seems that Wang intends to regard all entities as sets – as in the axiomatization of Chapter II – since he defines identity in terms of equal extension. (But he might also have in mind an adaptation of Quine's treatment described above.) Leibniz' principle is then taken as an axiom, in an appropriate form. Wang claims that all the sets of layer α are enumerable by a function E_α (i.e. a certain relation, hence a certain set – cf. p. 43) of layer $\alpha + 2$, that a truth definition for Σ_α can be formulated in $\Sigma_{\alpha+2}$, and that the consistency of Σ_α can be proved in $\Sigma_{\alpha+2}$. (The transition to $\alpha + 2$ comes about through the fact that in these definitions and proofs sets have to be used that correspond to formulae which contain bound variables ranging over entities of layer $\alpha + 1$.) All sets of any subsystem Σ_α of Σ are therefore enumerable in some other subsystem of Σ, and the consistency of each subsystem provable within some other subsystem, hence – in a sense which does not yet seem to be quite clear – all the sets of Σ are enumerable in Σ and the consistency of Σ provable in Σ itself; the Gödel method of constructing undecidable statements (see Chapter V) is not directly applicable to Σ. All this is achieved with the help of powerful *axioms of limitation* whose effect is to ensure that there are no entities in Σ except those enumerated by the functions E_α. Since the functions E_α, by enumerating all the entities of layer α, automatically well-order them, certain theorems corresponding to the axiom of choice are provable in Σ. The continuum hypothesis, at any rate, is not independent of the other axioms but becomes either provable or refutable, according to whether the equinumerosity between sets of layer α is defined by the existence of a one-one mapping (cf. p. 44) between these sets within $\Sigma_{\alpha+2}$ (or higher layers) or within Σ_α itself.

It is probably advisable to postpone judgment on the system Σ until a more detailed and rigorous exposition is available. But the partial system $\Sigma_{\omega+1}$ is interesting on its own account and looks indeed promising enough. True, many of the attractive (but also perhaps treachorous) features of Σ, such as the inclusion of its own consistency proof, the immunity towards Gödel's incompleteness arguments, and – for some thinkers at least – the enumerability of all its sets, do not hold for $\Sigma_{\omega+1}$, but $\Sigma_{\omega+1}$ seems, on the other hand, to be quite sufficient as a foundation for large parts of analysis, with additional parts becoming reconstructible in, say, Σ_{ω^2+1} or $\Sigma_{\omega^\omega+1}$.

Whereas Σ_ω is roughly equivalent to a ramified class-calculus (without an axiom of reducibility) [1]) – neither containing full-fledged counterparts of such decisive phrases as 'for all real numbers of finite order' – $\Sigma_{\omega+1}$, by containing bound variables ranging over all entities of layer ω, i.e. over all entities of any finite layer, allows for the formulation of just such counterparts and therefore takes care of all those procedures for the smooth operation of which the axiom of reducibility was introduced. Whereas, however, the axiom of reducibility all but destroys the constructive character of the ramified class-calculus, this character is completely preserved in $\Sigma_{\omega+1}$, thereby enabling Wang to present the outlines of a model-theoretic consistency proof similar to that given by Fitch [2]) for the ramified class-calculus. Incidentally, Wang also claims to be able to give a proof-theoretic consistency proof for $\Sigma_{\omega+1}$ (as well as for any partial Σ_α) analogous to those given by Lorenzen and Schütte [2]) for the ramified class-calculus.

And here we come to the one immediately apparent drawback of $\Sigma_{\omega+1}$ (and of Σ), if indeed a drawback it is: Cantor's theorem (*T*, p. 70), according to which – among other things – the power-set of a denumerable set contains *absolutely* more members than this set, is not valid, and we saw that, on the contrary, any infinite sets can be enumerated in an appropriate partial system of Σ. That part of Cantor's theory which countenances indenumerable sets is thrown overboard, as a consequence of the repudiation of certain formulae which play a decisive role in the proof of this theorem, though his theory of transfinite ordinals (at least within the second number class and perhaps only insofar as they are "constructive") is kept intact. However, there is a number of mathematicians and logicians who would not regard this consequence as a serious loss and would even welcome it as materializing their constructivist intuitions. If the price for providing mathematics with sound foundations is no more than renouncing a certain part of the paradise into which Cantor had led the mathematicians at the end of the 19th century, then some thinkers would not regard this price as too high [3]).

1) As is the essentially equivalent system L_ω of Quine 53, pp. 124 ff.

2) See footnote 1 on p. 174.

3) A good semi-formal description of Wang's system is given in Stegmüller 56–57, pp. 66 ff.

§7. LORENZEN'S OPERATIONIST SYSTEM

Wang's approach is admittedly strongly influenced by that of Lorenzen who in the early fifties developed his own version of constructivism in the foundations of mathematics. Lorenzen's work culminated in a book *Einführung in die operative Logik und Mathematik* [1]), published in 1955. As evident from the title, Lorenzen now uses the adjective 'operationist' – probably the most adequate English rendering of the German term '*operativ*' – in addition to the adjective 'constructivist'. This is more than a mere verbal matter. According to Lorenzen, 'contructivist' in mathematics should be reserved for those methodological attitudes that insist on restricting their investigations to the effectively calculable or describable, whereas 'operationist' indicates rather that the entities investigated are schematic operations. (There exists therefore a connection with the "operational" methodology of Bridgman [2]) and an ever closer contact with Dingler's [3]) views.) For Lorenzen, the main (though not the only) subject of mathematics is the treatment of calculi – this should by no means be misunderstood as a claim that mathematics *is* a calculus, which Lorenzen would very definitely reject – where a calculus is understood to be a system of rules for schematic operations with figures, which may but need not be marks on paper; they might as well be pebbles (*calculi*) or any other physical objects. In addition to this precisification of the subject-matter of mathematics – and only slightly connected with it – Lorenzen stipulates that the methodical frame be as wide as compatible with the *conditio sine qua non* that all mathematical statements be *definite*. Due to the fact that Lorenzen – who in this respect shows a certain affinity to intuitionistic thinking (Chapter IV) – cares little for exact formalization, insisting that precisification is by no means identical with axiomatization or formalization, the exact meaning of this central term is not altogether clear. It certainly has nothing to do with the term 'definite' as used by Zermelo (p. 36) and is clearly wider than the term 'definite' as used by Carnap [4]). In Carnap's usage, a statement containing a regular (unlimited, or unrestricted) quantifier, such as '$\forall x\varphi(x)$', is indefinite – and so is a language whose rules of formation allow for such sign-sequences as the formulae of the ideal calculus – whereas it is definite in Lorenzen's usage, more precisely refutation-

1) Lorenzen 55; 62 and 65 contain simplifications in the foundations of logic and analysis, respectively.

2) Bridgman 27.

3) Dingler 13, 31a.

4) For details, see Carnap 37, p. 45 and pp. 160 ff.

definite (*widerlegungsdefinit*), on condition that '$\varphi(x)$' is definite, and this because '$\forall x\varphi(x)$' is refutable through a derivation of an instance of '$\neg\varphi(x)$'; being a derivation is a definite characteristic, since it is decidable through the performance of schematic operations on the figures that make up the lines of the derivation.

Lorenzen's system is then much more liberal than constructivist systems of the most rigorous brand that either contain no quantifiers at all or, at most, only *limited* quantifiers [1]) in which quantified statements with a constant limit are equivalent to conjunctions or disjunctions, respectively, of finite length. Nevertheless, it still excludes impredicative concept formations. The limitations thereby imposed are overcome, just as with Wang, by a utilization of the transfinite layers [2]). Lorenzen, however, is able to go into much greater details in the justification of his hope that almost all of classical analysis will remain essentially intact, though certain reformulations will be required. Absolutely indenumerable sets don't exist any more, of course; in those mathematical fields, such as topology or the theory of integration (Lebesgue measure theory is an immediate example), where the contrast: denumerable – indenumerable seems to play a decisive role, the contrast: primary – secondary – where a set is called *primary* if it belongs to a layer whose index is less than some arbitrary limit-number θ_1 and *secondary* if the index of the smallest layer to which it belongs is greater than θ_1 but less than some other fixed (non-problematic, i.e. "constructive") limit-number θ_2 – can in general serve as a satisfactory substitute. However, should it turn out that certain classical theorems will not be reproducible on the operationist basis, there remains the possibility of paying serious heed to the words of Skolem: "We ought not to regard all that's written in the traditional textbooks as something sacred".

The logic of dealing with schematic operations turns out to be the intuitionistic logic (Chapter IV) in which the *tertium non datur* is not generally valid. By adding this law to the effective predicate calculus, Lorenzen gets the "fictional" predicate calculus, corresponding to classical functional logic. For the justification of this transition, Lorenzen uses an argumentation which is strongly related to the ideas developed by van Dantzig in connection with the so-called *stable* statements [3]).

The fact that Wang and Lorenzen, starting from quite different back-

1) I.e., quantifiers of the form 'for all [some] x up to y'. For a theory of the generalized notion of restricted quantification, see Hailperin 57.

2) These, however, are no longer needed in Lorenzen 65.

3) van Dantzig 47.

grounds, the one from a Russell-Zermelo tradition, the other from a constructivist Hilbertism (Chapter V), converge to systems which have so much in common cannot but bring their streamlined version of the ramified type theory back into the race. The old animadversions against this theory do not hold any more.

§8. THE LOGICISTIC THESIS

Though we might have thought that metaphysical considerations should be of no relevance to the problem of providing a secure foundation for set theory (or mathematics in general), so long as one judges by what logicians and mathematicians *say* about this problem, this is definitely not so. Let us pause for a while to see this at some detail.

Metaphysical convictions certainly make no difference – with the possible exception of some very tough-minded intuitionists – for the application of mathematics to science and technology. No director of research in some industrial outfit would inquire into the metaphysical beliefs of the mathematician he is about to hire. There seems to exist no correlation between these beliefs and the performances in which the director of research is interested. In the attempt to solve some set of differential equations, all mathematicians will peacefully cooperate though they might thereafter, during a lunch hour conversation, disagree violently about "the nature of mathematics". We make these commonplace observations, not in order to disparage discussions about the nature of mathematics – this whole book is dedicated to the discussion of one aspect of this problem – but to put them in their right place.

Most present-day thinkers – though by no means all [1]) – would agree that there are at least two different kinds of true scientific statements: the empirical truths on the one hand and the logical and mathematical truths on the other. We shall take here this approach for granted and investigate only into the relation between the respective sets of the logical and the mathematical truths. Five views are theoretically possible concerning the relationship between these two sets: they are identical; the set of mathematical truths is a proper subset of the set of logical truths; the set of logical truths is a proper subset of the set of mathematical truths; the two sets are mutually exclusive; the two sets are overlapping. In addition, one could of course also be less committal and subscribe to any disjunction of at most four of these theses.

1) For a forceful presentation of the minority view, see Quine 53, especially essay II. Cf. also the remarks on significs in Chapter IV, pp. 218–220.

Looking at the literature, one could indeed find expositions of each of the five fundamental views as well as of many of their disjunctions, if one were to take these expositions at their face value. This would, however, probably be a mistake since very often the terms 'logical' and 'mathematical' are used in different, sometimes in very different, senses by different authors (and occasionally by the same author in different publications), so that it is often very hard to distinguish between purely verbal and real disagreements, or to decide that a verbal agreement is an indication of a substantial agreement.

We mention all this because the type-theoretical approach to the foundations of set theory, to the various variants of which this chapter is devoted, has often been identified with the *logicistic* approach to the foundations of mathematics, i.e. with the thesis that mathematical truths form a proper subset of the logical truths. This identification, however, is definitely misleading and due to nothing more but the fact that there is a union between the inventor of type theory and the foremost logicist of our time in the person of Bertrand Russell. Logicism, however, was created by Frege long before type theory was invented, and there are now many formulations of type-theoretical systems that do not embody the logicistic thesis. Let us turn now to an exposition of this thesis.

The last quarter of the 19th century saw the triumph of the so-called *arithmetization of mathematics*. By explicit, though occasionally quite complicated, definitions, the various kinds of numbers, up to complex numbers and beyond, and the operations upon them were introduced on the basis of the natural numbers and the operations upon them. Almost all mathematicians were satisfied with this achievement and regarded with suspicion any further inquiry into the reducibility of natural numbers to other entities or into the deducibility of the arithmetical theorems from other theorems [1]). Frege, however, had no great difficulties in showing [2]) that all the interpretations given in his time to the calculus of natural numbers were gravely deficient and that the last refuge of the mathematicians in this respect, i.e. leaving the calculus without interpretation, could not be seriously upheld since in certain contexts numerical signs must be regarded as interpreted, e.g. in "every quadratic equation has exactly two roots". By his own interpretation, the natural numbers were the cardinal numbers [*Anzahlen*] of certain con-

1) Let us recall Kronecker's much-quoted dictum: *"Die ganzen Zahlen hat der liebe Gott gemacht, alles andere ist Menschenwerk"*. (God made the integers, everything else is human creation.)

2) In his masterfully written Frege 1884. However, at least up to the turn of the century he met with the antagonism of most mathematicians.

cepts, with 'the cardinal number of the concept *F*' defined as short for 'the extension of the concept equinumerous-with-the-concept-*F*'; finally, the statement 'the concept *G* is equinumerous with the concept *F*' was regarded as short for 'there exists a one-one-correlation between the objects falling under the concept *F* and the objects falling under the concept *G*'. This last expression, Frege was able to show, could be reduced to purely logical expressions. With infinite care and meticulousness, using a strange geometrical symbolism, Frege continued to define one basic arithmetical term after the other, proving, as he went along, the fundamental arithmetical theorems governing these terms. For this purpose, he almost recreated formal logic, gave the first complete axiom system for the propositional calculus, and greatly expanded the predicate calculus.

Even if Frege had been entirely successful in reducing arithmetic to logic – which he was not in view of the emergence of antinomies in his system [1]) –, it is not always sufficiently realized today and has not been sufficiently realized by Frege himself that this would still not have meant that all of analysis were reducible to logic, in spite of the already accomplished arithmetization of analysis. This is so because the arithmetization means reduction to "integers and finite or infinite systems of integers", in Poincaré's terms quoted above (p. 14), hence not only to integers but also to what we would now call *sets of integers*. The reduction could therefore be regarded as completed only if, in addition to arithmetic, also the general theory of sets – or at least the theory of sets of integers and, probably, of sets of sets of integers etc. – were "reduced" to logic. This, however, was never done nor even attempted by Frege. This really final step was attempted only by Russell (and Whitehead). It was when dealing with Cantor's set theory in order to reduce it to logic that Russell ran afoul of his antinomy. This accident forced him to a revision of what logic amounted to and made him give it the specific shape of the Ramified Type Theory. Historically speaking, type theory is then an accidental by-product of an attempt to implement the logicistic thesis, but systematically their connection is very slight. In many current variants of type theory, given by Gödel, Tarski, Carnap, and others, the logicistic thesis is abandoned and logic and arithmetic are simultaneously developed.

We already mentioned above that the usual formulations of the logicistic thesis are far from being precise and universally accepted. The following formulation, e.g., seems to be precise enough: All specific mathematical terms are definable on the basis of the logical vocabulary, and for the proof of all mathematical theorems no axioms beyond the axioms of logic and no rules of inference beyond those accepted in logic are needed. The precision of this thesis, however, is not as great as it appears to be on first sight. There exists no universal agreement on the meaning of most of the terms occurring

1) See above, pp. 2–3.

in it, such as 'logic', 'axioms of logic', 'rules of inference' or 'logical vocabulary'. Depending on the specific interpretation of these terms as well as on a precisification of the term 'mathematics' itself, e.g. whether geometry is or is not to be regarded as mathematics (a decision that depends of course in its turn on the interpretation given to 'geometry'), the logicistic thesis runs the whole gamut between an uninteresting truism through a pious hope, whose basis is not quite clear, to an almost obvious falsity.

We have no intention of dealing here with the problem of the relationship between logic and mathematics in all its generality. (Cf. also Chapter IV.) At the moment, we are concerned only with set theory. Unfortunately, this does not reduce by much the vastness of the problem, nor is it thereby appreciably simplified. The situation is by no means such that there exists a clear conception of what logic on the one hand and set theory on the other hand are, allowing a dispassionate, though perhaps still difficult, investigation into their relationship; on the contrary, different apriori conceptions of this relationship color the conceptions of these disciplines themselves, forming one complex of problems whose disentanglement is a formidable task indeed.

The axiomatic approach to set theory, treated in Chapter II, is inherently neutral with regard to our problem. To construct set theory, or any theory, for that matter, as an uninterpreted axiom system based upon some smaller or greater fragment of logic – in Chapter II, this fragment consisted of the first-order predicate calculus (with equality) – means, by definition, that the question of interpretation of this axiom system is so far not taken up. It is by no means excluded that one possible interpretation should be a logical one, in the sense that the primitive terms of this axiom system – for most systems discussed in Chapter II just $\in$ – is defined on the basis of the vocabulary of either the underlying fragment of logic or of some more extensive fragment in such a way that the axioms become logical theorems. Of course, he who strongly believes in the existence of such a logical interpretation and moreover does not believe in the existence of other useful interpretations may regard the erection of an independent axiom system as a waste of time and prefer to present his theory from the beginning as a part of logic. This is what Frege and Russell, and many others after them, have done. By interpreting '$\in$' as denoting class-membership – which they take to be a logical notion though perhaps one that belongs to a "higher" part of logic – they identified from the very start the (mathematical) theory of sets with the (logical) theory of classes.

This identification, however, meets with certain difficulties. They arise only when the realm of finite sets is left and therefore were of greater concern to Russell who intended to reconstruct all of set theory to its complete

Cantorian extent as part of logic than to Frege who, a contemporary of Cantor, was interested in the theory of sets mainly insofar as it was needed for the arithmetic of the natural numbers. Whereas most axioms of ZF are easily provable in logic after this identification, the counterparts of the axioms of infinity and of choice could not be proved in PM. For the counterpart of the first versions of the axiom of infinity of ZF this is immediately obvious since their formulation violates the rule of types adopted in PM, according to which the membership of any class must be homogeneous, whereas the classes whose existence is asserted by these axioms have a heterogeneous membership. But neither can any statement to the effect that the number of individuals, i.e. of the entities belonging to the lowest level, is denumerably infinite be derived from the axioms of PM. For each statement provable on the basis of logic plus this axiom, a theorem in the form of an implication whose antecedent is the axiom and whose consequent is the statement is derivable from logic alone. Whitehead and Russell were therefore able to prove, corresponding to each mathematical theorem T for whose derivation the axiom of infinity AxInf was needed, not T itself but rather AxInf$\rightarrow T$. But this meant, of course, that any such mathematical theorem T by itself could not be shown to be a theorem of logic unless AxInf was taken to be an axiom of logic. The authors of PM, however, were very reluctant about taking this step since the content of this axiom, i.e. the existence of an infinity of individuals, had a definite factual look, indeed so much so that not only its logicality but even its truth was in doubt: whether the universe was composed of a finite or infinite number of ultimate particles – taking these particles to be the "individuals" of the system of PM, which was indeed one of the explicitly admitted interpretations – seemed to be a question which could be answered, if at all, only by physics, and this in spite of the fact that InfAx was formulated in logical terms exclusively. This situation was disturbing even for the authors of PM themselves and caused others to reject this "reduction" of mathematics to logic.

The situation is different, but no less disturbing, with regard to the Axiom of Choice, or the Multiplicative Axiom – MultAx, for short – as Russell called one of its variants (see p. 57). Again there were large parts of higher classical mathematics whose derivation required the use of MultAx. But the authors of PM could not persuade themselves to treat it as a logical truth on a par with the other logical axioms. They had an intuitive notion of logical truth, from which they were unable to derive, to their own satisfaction, either the truth or the falsehood of MultAx. Afraid that MultAx might eventually be shown to be false, they had to content themselves once more, with regard to those classical mathematical theorems T in whose proof MultAx was commonly used, to prove MultAx→T. Though they likened the situation to that with which the geometrician is confronted by the Axiom of Parallels, it is doubtful whether this analogy is to be taken seriously: had they known – something that was proved only decades later – that MultAx is independent of their other logical axioms, i.e. that neither it nor its negation are derivable from them, it rather seems that they would have been ready to stretch their intuitive notion of logical truth – this notion was admittely never quite definite [1]) – so as to include MultAx rather than its negation. At any rate, no attempt was made by them to develop mathematics on the assumption of the truth of the negation of MultAx – in analogy with the development of non-Euclidean geometries.

But with regard to MultAx, the situation should probably be looked upon not so much as throwing doubt on the reducibility of mathematics to logic as on the adequacy, reliability, and determinateness of our logical intuitions. There is certainly nothing specifically mathematical in its formulation. Whether it should, might, or should not, be accepted as an axiom, now that it is demonstrably independent of the other axioms, is a moot question within the philosophy of logic, into which to delve at any length and depth it is not feasible without a detailed discussion of this discipline, something that is beyond our reach here [2]).

It seems, then, that the only really serious drawback in the Frege-Russell thesis is the doubtful status of InfAx, according to the interpretation intended by them. But is this interpretation under which the individuals are certain factual entities like particles, events, etc., an integral part of the thesis? This is another moot question, in the discussion of which it is very difficult not to fall into verbal traps. But, whether regarded as an essential deviation from this thesis or an inessential reinterpretation of it, there exist logical systems which embody a unified simultaneous development of logic and arithmetic – the expression 'reduction of mathematics to logic' could be applied to this development only in a rather stretched sense – thereby managing to sidestep the disturbing problematic status of InfAx within PM.

These systems, of which Gödel's system P mentioned above (p. 171) is one of the better known, are couched within what might be called, following Carnap, *coordinate-languages* [3]). In such languages, in contradistinction to the more customary *name-languages*, the objects of the fundamental domain are not designated directly by proper names but indirectly by systematic

1) See, e.g., Russell 19, p. 205.

2) See, e.g., Mostowski 67, Bernays 67, and Körner 67.

3) See Carnap 37, p. 12.

positional coordinates, i.e. by symbols that show the place of the objects in the system and thereby their positions in relation to each other. Instead of saying, then, that entity *a*, i.e. the entity whose proper name is '*a*', is blue, as customary in name-languages, one says in a coordinate-language that the entity occupying such and such a position is blue. In some of the languages specifically constructed and treated by Carnap, the basic domain of positions is taken to be a one-dimensional series with a definite direction whose initial position is designated by '0', the succeeding position by '0′', etc.; in other words, the natural numbers are treated as the coordinates of these languages. Without going into any more detailed description of these languages, let us only notice that through adopting some Peano-type axioms the infiniteness of the basic domain becomes provable. But – and this is the decisive point – the fact that the infiniteness of the domain is part of the logic of the system is by no means as disturbing as within PM. No longer does the statement of infinity assert the existence of infinitely many different particles or other physical entities but rather the fact that the one-dimensional series of positions has no last member, leaving the answer to the question how many of these positions are occupied by physical entities entirely to extra-logical science.

Those authors who have voiced strong feelings against existence assumptions in logic, whether of an infinity of "entities" or even of at least one "entity", seem to have had name-languages in mind; these objections look rather pointless when directed against coordinate-languages, where these "entities" are nothing but positions that might well be unonccupied. It appears that the strengthening of the logicistic attitude towards the foundations of mathematics entailed by the shift from name to coordinate-languages has not yet been sufficiently taken into account by most authors in this field.

Let us notice in this connection that Carnap [1]) has succeeded in proving that the axiom of choice is analytic – where 'analytic' was understood to embody a certain refinement of the pre-systematic concept of logical validity – in his Language II, one of the mentioned coordinate-languages; for his proof, he assumed that an axiom of choice held in the metalanguage of Language II. This assumption does not make the proof circular in any formal sense but, on the other hand, it does not contribute to the solution of the problem whether such an axiom is to be regarded as a logical principle. Carnap himself, at least in the thirties, regarded this problem not as one referring to the material truth of the principle of choice but rather to its expediency. Since, within Language II, an axiom of choice is expedient and allows for the reconstruction of a mathematical calculus adequate for science

1) Carnap 37, pp. 122 ff. Cf. already Ramsey 26.

and since it is unlikely that this language will turn out to be inconsistent, there is very little to be said against the admission of the axiom of choice as a logical principle, if one accepts Carnap's ontology-free standpoint.

§9. TYPES, CATEGORIES, AND SORTS

The idea that not all objects are of one kind, in the sense that there are properties which can be meaningfully predicated of certain objects but not of others, is an old one and had originally no connection with the problem of antinomies. This conception can be traced back to Aristotle [1]), Schröder drew level distinctions in 1890 [2]), and indications to this effect are to be found in Frege's writings [3]). The philosopher E. Husserl dealt at length with categories of meaning [*Bedeutungskategorien*] [4]) and exerted much influence on the Polish school of logicians.

Russell himself was never quite satisfied with the ontological validity of his own type distinctions though he remained convinced that some sort of hierarchy is necessary [5]). In one of his later publications, he was even ready to admit – perhaps somewhat rashly – that his definition of types was wrong, insofar as he had originally distinguished different types of *entities* whereas he should have made these distinctions rather with regard to *symbols* [6]).

Among Russell's followers, we can distinguish between those for whom type distinctions were a matter of intuitive coercion, a philosophical necessity, and those who regarded them as a necessary evil, something to be grudgingly admitted so long as no better way of avoiding the antinomies was in view. The foremost proponent of the first conception is probably the Polish logician Stanisław Leśniewski [7]) for whom, as a matter of linguistic intuition, (almost) all expressions of any language, natural or artificial, belonged to exactly one *semantic category* out of a potentially infinite and highly ramified hierarchy of semantic categories; this hierarchy consisted of two *funda-*

1) *Categoriae*, ch. 3.

2) See Church 39.

3) But no more than indications. Frege's hierarchy of *Stufen* was a consequence of certain formal considerations. He almost explicitly rejected a theory of types. See Church 39.

4) Husserl 21–28 II 1, pp. 317 ff. Cf. also Bar-Hillel 57, 67.

5) Russell 44, p. 692.

6) *Ibid.*, p. 691.

7) Leśniewski 29; for the most thorough treatment of Leśniewski's conceptions, see Luschei 62.

mental categories, that of *names* and *statements* (always including open names such as 'the father of x' and open statements, respectively), and an infinity of *functor categories*, distinguished according to the category and number of the argument-expressions of the functors belonging to them and to the category of the expression resulting from the application of these functors to their arguments.

Among the simpler members of this hierarchy, we have the category of functors that out of a name form a statement, that out of two names form a statement, ..., that out of a statement form a statement, that out of two statements form a statement, ..., that out of a name form a name, that out of two names form a name, ..., that out of a statement form a name, ... Using a convenient quasi-arithmetical notation originating with Ajdukiewicz [1]), these categories could be denoted by

$$s/n,\ s/nn,\ ...,\ s/s,\ s/ss,\ ...,\ n/n,\ n/nn,\ ...,\ n/s,\ ...,$$

respectively. This symbolism also indicates how the more complex members of this hierarchy could be characterized and denoted: the category of a functor, e.g., that out of a functor (as argument) that out of a name forms a statement forms a functor (as value) that out of a name forms a statement would be denoted by '$s/n//s/n$'. A few illustrations taken from English (but to be evaluated *cum grano salis*) and the set-theoretical symbolism adopted in this book will help:

Category	English	Set theory
n	John, poet	x, y
s	John is a poet	$F(x), x \in y$
s/n	... is-a-poet	$F(\ ), ... \in y$
s/nn	... is-a ---	$... \in ---,\ ... \subseteq ---$
s/s	It-is-not-the-case-that ...	$\neg(\ ), \forall x(\ )$
s/ss	... if-and-only-if ---	$... \leftrightarrow ---$
n/n	The-father-of ...	$C(\ )$
n/nn	The-product-of ... and ---	$... \cap ---$
n/s	whether ...	$\{x \mid ...\}$
s/ns	... doubts-whether ---	$\forall ... (...)$
$s/n//s/n$	gracefully (in 'John gracefully retired')	
$s/nn//s/nn$	gracefully (in 'John gracefully greeted Mary')	/ (in '$\notin$')

1) Ajdukiewicz 35, further elaborated in Bocheński 49, Bar-Hillel 50, 53, Geach 70. This notation should be compared with those developed in Carnap 37, § 27 (who uses 'functor' in a sense which is roughly that of the expression 'name-forming functor' used here), and Church 40 for types of expressions.

For Leśniewski, transgressing the limits set by the *Theory of Semantic Categories* is a sin against logic, occasionally though not necessarily punished by the arisal of antinomies. The validity he claimed for his theory was quite apart from the fact that by adherence to it some of the logical antinomies could not be formulated [1]) and was based rather, as stated above, upon an intuitive insight into the conditions of meaningfulness of natural language expressions, in elaboration of Husserl's views. Leśniewski, in his turn, was followed by a host of Polish philosophers and logicians, the most important among whom were Kotarbiński, Ajdukiewicz, and Tarski. How strong the influence of Leśniewski among his pupils was, can be judged, e.g., from a passage written by Tarski in the early thirties where he claims that

> the theory of semantical categories penetrates so deeply into the fundamental intuitions regarding the meaningfulness of expressions that it is scarcely possible to imagine a scientific language in which the sentences have a clear intuitive meaning but the structure of which cannot be brought into harmony with the above theory [2]).

Later on, however, Tarski lost faith in the intuitive necessity of Leśniewski's theory of semantic categories, probably when realizing the grave strictures imposed by it upon the structure of languages, and started investigating language systems in which this theory was not obeyed [3]). It seems that Wang's recent studies (cf. above, § 6) form a direct continuation of this line of thinking.

Whatever the importance of the Theory of Semantical Categories as a logico-philosophical doctrine, there can be little doubt as to the importance for linguistics of its syntactical counterpart, the *Theory of Syntactical Categories* [4]), especially in the form of the notational calculus it was given by Ajdukiewicz [5]).

1) Not all logical antinomies are due, according to Leśniewski, to a confusion of categories. Some result from a violation of certain rules of definition. In addition, Leśniewski claims that the current conception of the term 'class' confuses two quite different notions, namely that of "distributive class" and that of "collective class" (roughly the same distinction already made by Russell 03, pp. 68 ff, in terms of "class-as-many" and "class-as-one") and traces in particular the origin of Russell's antinomy to this confusion. Cf. Sobociński 49–50.

2) Tarski 56, p. 215.

3) *Ibid.*, VIII, Appendix

4) Bar-Hillel 64, 67.

5) See footnote 1 on p. 189; grammars built on these principles are now called "categorial grammars"; cf. Bar-Hillel 64.

Typical for the second, opportunistic conception of the theory of types is Reichenbach [1]) who regarded the fact that this theory made language consistent as its best possible justification.

We shall not deal here with the various philosophical objections that have been launched against the theory of types, its intuitive plausibility, its applicability to natural languages, or its consistent formulability. Whatever the force of these objections with regard to natural languages, there can be no doubt that calculi embodying type distinctions can be constructed according to the highest standards of rigor and that it can even be proved of some of these calculi that they are consistent. Many mathematicians have nevertheless found the theory of types repugnant as a foundation for mathematics. There are probably many reasons behind this reaction. It might often be due to no more than an unanalyzed idiosyncratic aversion, leaving us nothing more to say about it besides acknowledging the fact, but the type division has also certain technical drawbacks some of which we have discussed above.

Let us deal here with what is probably the most serious disadvantage of the Theory of Simple Types. Developing set theory on the basis of a logic incorporating a type stratification of its variables means that this logic is no more the predicate calculus of the first order but rather the predicate calculus of order ω, i.e. one that contains quantifiable variables of any finite level whatsoever. Set theory as developed in *Principia Mathematica*, in contradistinction to ZF, is therefore no more an *elementary* theory and no longer enjoys the desirable features of such a theory, the most important of which is that the *proof procedures of its underlying logic are complete* (see Chapter V, §4). In an elementary theory, all variables have the same range and there exists therefore just one single universe of discourse. The set theory of *Principia Mathematica* is not the only non-elementary one; so is the system VNB of von Neumann-Bernays in which there are two disjoint universes, that of the classes and that of the sets.

Now, there exists a technique of converting any *many-sorted* theory, i.e. any theory containing more than one universe of discourse, into an equivalent *one-sorted* theory [2]). Applying this technique to the set theory of *Principia Mathematica* we arrive at a system that was called by Quine [3]) the *Standardized Theory of Types* and which represents an interesting transition between the simple theory of types, or rather its still more simplified version T* of §2, and Zermelo's original system. It is still a type theory in the sense

1) Reichenbach 44, p. 38.
2) See Schmidt 51, Wang 52a, Hintikka 55.
3) Quine 56, and 63, §37.

that its ontology comprises individuals, classes of individuals, classes of classes of individuals etc., and that these classes remain homogeneous. But it differs from T* in that its variables are not merely typically ambiguous, i.e. ranging in a given context over a definite, though unspecified type, but fully general, i.e. always ranging over all types.

Without going into the details of the highly elegant treatment given by Quine to this transformation, let us briefly describe one resulting system of axiom-schemata for the standardized theory of types. Let 'T_0' be the predicate that holds for all and only the individuals, 'T_1' the predicate true of all and only the classes of individuals, etc. (Quine shows that all these predicates can be defined in terms of '$\in$' alone and need therefore not be added to the primitive notation.) The axiom-schemata of comprehension and extensionality receive now the following form:

$$(1)\quad (\;)\quad \exists y[T_{n+1}(y) \wedge \forall x(T_n(x) \rightarrow (x \in y \leftrightarrow \varphi(x)))],$$

$$(2)\quad (\;)\quad T_{n+1}(x) \wedge T_{n+1}(y) \wedge \forall w(T_n(w) \rightarrow (w \in x \leftrightarrow w \in y)) \rightarrow x = y\,.$$

To these axioms a new set of axioms has now to be added through the axiom schema:

$$(3)\quad (\;)\quad x \in y \rightarrow (T_n(x) \leftrightarrow T_{n+1}(y)).$$

(The function of the T-clauses in (2) is to prevent the identification of the individuals with each other and with the null-sets of the different types as well as the identification of the various null-sets among themselves.)

Notice that according to (3) any formula of the form '$x \in y$' where x and y do not belong to consecutive ascending types (and one of them belongs to some type), hence especially the formula '$x \in x$', is *false* and not meaningless as in the original system. This is a straightforward effect of the standardization. That this deviation from the original approach has no technically pernicious consequences should have some impact on the evaluation of those views that saw in the introduction of the category "meaningless" in addition to "true" and "false" the great philosophical achievement of the theory of types.

For those adherents of the theory of types who are ready not only to trade meaninglessness for falsity (or truth, as the chips may fall) in order to achieve standardization but even to tamper with the original ontology for the sake of technical simplicity, Quine shows that a considerable increase in simplicity arises if the null-sets of all types are identified. A different but equally considerable simplification results from the identification of the individuals with each other and with the null-set of the lowest type, as in ZF.

Half a century after Zermelo and Russell published their theories, independently of each other and starting from seemingly totally different and even contrary approaches, an almost complete reunion of these theories is now in full view. True, the deviations from the original attitudes through which this rapprochement is brought about are quite considerable and some of them might well be regarded as militating against the spirit of both Zermelo and Russell themselves. Many a Russellian will find it hard to swallow the notion of cumulative types. Those, however, for whom a satisfactory foundation of set theory means not so much a construction that corresponds closely to their intuitions but rather a system from which classicial analysis can be effectively reconstructed and which is either demonstrably consistent or at least within which the customary arguments leading up to the standard antinomies cannot be reproduced, will welcome these recent developments [1]).

§10. IMPREDICATIVE CONCEPT FORMATION

It is now time to deal at some detail with another most interesting, though also highly obscure and controversial, ingredient of Russell's original attempt at overcoming the antinomies: the recognition that *impredicative* concept formation is the root of all evil [2]) to be eradicated by strict adherence to the *vicious-circle principle*. (The connection of this conception with the constructivist approach to the foundations of mathematics will be discussed on pp. 196–197.)

Part of the obscurity is caused by the fact that Russell uses, in the various formulations he gives the vicious-circle principle, at least three different terms which he seems to regard as synonymous: first, '*definable*' in "If, provided a certain collection had a total, it would have members only definable in terms of that total, then the said collection has no total" [3]); second, '*presupposes*' in "... it is possible to incur a vicious-circle fallacy at the very outset, by admitting as possible arguments to a propositional function terms which presuppose the function" [4]); third, '*involves*' in "A function is what ambigu-

1) Wang-McNaughton 53, Klaua 64, Kreisel 65 and Shoenfield 67. More on the subject is found on page 89.

2) Skolem 52 continues to regard this recognition as a natural one and accordingly recommends further study of the ramified type theory.

3) Whitehead-Russell 10–13 I, p. 37.

4) *Ibid.*, pp. 37–38.

ously denotes some one of a certain totality, namely the values of this function; hence this totality cannot contain any members which involve the function since, if it did, it would contain members involving the totality, which, by the vicious-circle principle, no totality can do" [1]). Gödel shows [2]) that we have before us three different principles rather than three formulations of the same principle, that the second and third of these principles are much more plausible than the first, and that it is only the first which obviates the derivation of mathematics from logic and militates against procedures in constant use in classical mathematics.

The presently existing confusions around the concept of impredicativity are enhanced by the fact that one now used the adjective 'impredicative' as a modifier for nouns denoting linguistic entities such as 'definition', 'statement', or 'axiom', non-linguistic entities such as 'concept', 'property', 'set', and even 'procedure' or 'concept-formation', without taking the necessary care in establishing the interconnections between these various uses.

Of late, it has been shown that there exists an important difference in the conceptions of what forms an impredicativity between the two first and foremost proponents of the prohibition of impredicative concept formation in (logic and) mathematics: Poincaré and Russell [3]). This difference is of more than historical interest since it might well throw light on the currently much discussed question where the borderline between the harmful and innocent impredicativities lies.

We shall gain in clarity, though we shall lose somewhat in generality, if we restrict the discussion of impredicativity to systems of the kind treated so far. This enables us to focus our attention to the axiom (-schema) of comprehension (in its various formulations and variants, including the axiom of subsets). We shall say that a certain version of this schema, whose core is – as we recall –

$$(\ )\quad \exists y \forall x (x \in y \leftrightarrow \varphi(x))$$

(where, according to the variant, '$\varphi(x)$' is required to fulfil various conditions, and where the superscripts occurring in some variants are omitted as irrelevant in our context), is *impredicative* if '$\varphi(x)$' contains a bound variable of the same kind (type, layer, level etc.) as 'y'; in other words, and in a somewhat less formal and rigid formulation, if the class whose existence is guaran-

1) *Ibid.*, p. 39.

2) Gödel 44, p. 235.

3) For the early history of the objections against impredicative procedures, see Fraenkel 28, pp. 247 ff. Cf. also Church 56, p. 347, footnotes 573 and 574.

teed by this axiom-schema belongs to the range of a bound variable occurring in the determining condition. Derivatively, we shall call *impredicative* also those conditions $\varphi(x)$ themselves which contain bound variables of higher level than that of x [1]) and also the corresponding classes and the process of forming them, i.e. of proving their existence.

Usually, one formulates the problem of impredicativity in terms of "impredicative definitions" rather than in terms of "impredicative concept formation". This is partly due to the fact that PM – and other systems – have no axiom of comprehension but instead an (explicit or implicit) rule admitting '$\{x \mid \varphi(x)\}$' – i.e., "the class of all x, for which $\varphi(x)$ holds" – as an abstract-expression substitutable for a variable and an axiom (-schema) of conversion: $y \in \{x \mid \varphi(x)\} \longleftrightarrow \varphi(y)$ [2]). In such systems, the principle of comprehension becomes a derivable theorem. (Indeed, from the tautology $\varphi(x) \longleftrightarrow \varphi(x)$ we get $x \in \{x \mid \varphi(x)\} \longleftrightarrow x \in \{x \mid \varphi(x)\}$ by conversion, $\forall x[x \in \{x \mid \varphi(x)\} \longleftrightarrow x \in \{x \mid \varphi(x)\}]$ by universal generalization, $\exists y \forall x[x \in y \longleftrightarrow x \in \{x \mid \varphi(x)\}]$ by the above-mentioned rule and existential generalization, finally $\exists y \forall x[x \in y \longleftrightarrow \varphi(x)]$ again by conversion [3]).) If now a single symbol is introduced by definition as an abbreviation of such an abstract-expression and '$\varphi(x)$' is an impredicative condition in the sense clarified above, this symbol is introduced by an "impredicative definition". Since this definitional abbreviation is, however, only a minor technical point, we shall continue to see the gist of the problem of impredicativity in the impredicative concept formation, whether through an application of the axiom-schema of comprehension or through the introduction of abstract-expressions.

The vicious circle principle requires that measures be taken to disqualify impredicative concept formation. As for its justification, we must clearly distinguish between four quite different kinds of argumentation. The first is an argument from pragmatic effectiveness: the application of the principle overcomes the semantic antinomies. Its critics will then point out that other means are known – such as the theory of language hierarchies [4]) – which are equally effective in this respect but less destructive in their effect on the reconstruction of mathematics. The second argument refers to the self-referential character of impredicatively defined terms. After over half a century of detailed discussion it should, however, be clear that self-reference as such without some additional qualification, though certainly circular in a sense, is by no means always viciously circular [5]).

1) Here we disregard the requirements which concern the parameters of $\varphi(x)$, i.e., the free variables of $\varphi(x)$ other than x – see Chapter V, p. 330.

2) Such is also the system of Bernays 58 – see system B on p. 146.

3) Cf. Hilbert-Ackermann 28/49, 2nd ed. p. 125, 3rd ed. p. 137.

4) Cf. Pap 54.

5) See, e.g., the forceful and witty defense of the meaningfulness of self-referential expressions in Popper 54.

The third argument stresses that the applicability of an impredicatively formed concept in a given concrete case is not decidable and that any attempt in this direction must result in an infinite regress. To use an example of Carnap's [1]), modified to apply to system P (mentioned in §5): Consider the condition on x^1,

$$\forall z^2[7 \in z^2 \wedge \forall y^1(y^1 \in z^2 \rightarrow Sy^1 \in z^2)] \rightarrow x^1 \in z^2$$

(where 'Sy^1' denotes the successor of y^1), which may be read as 'x belongs to all those sets that contain 7 and together with any natural number also its successor'. The axiom of comprehension guarantees the existence of a set containing all those natural numbers that satisfy this condition. Trying to find out whether 5, for instance, belong to this set, one has to check whether 5 belongs to *all* sets of natural numbers that contain 7 etc., but in order to do this, one has, among other things, also to check whether 5 belongs to that set of natural numbers whose existence is guaranteed by the axiom of comprehension – and we are very clearly running around in a vicious circle. Notice that the argument does not rely on the fact that the checking procedure requires the testing of infinitely many sets. There are indeed authors who object to such procedures in general, but then they reject not only impredicative conditions but all *indefinite* statements, i.e. statements containing essentially unlimited quantifiers of any type whatsoever. Here we are concerned with those who, though accepting indefinite conditions, reject impredicative ones. But if indefinite statements are not rejected, probably because a general statement is established not by a case-after-case test – which could indeed not be performed as it would consist of infinitely many operations – but rather by the construction of a proof, which is a finite operation, then the third argument is not very convincing either, since the applicability of an impredicative concept can often be established by a proof procedure. In the above-mentioned example, it is indeed very easy to prove that 5 does not belong to the set determined by the displayed condition.

The fourth argument, and certainly the strongest one, refers to the non-constructive character of impredicatively introduced objects. We can hardly be said to have a clear idea of a totality if the membership of a certain object in this totality is determinable only by reference to the totality itself. Not only is this blurred view unsatisfactory from a philosophical standpoint [2]), it

1) Carnap 37, pp. 163–164.

2) Intuitionism (see Chapter IV) rejects, of course, impredicative procedures. Cavaillès 47, p. 17, sees in this rejection the distinctive characteristic of the intuitionistic

might occasionally give rise to antinomies and will in any case cause great difficulties in the construction of models that would prove the consistency of the system. The last point seems to be the decisive one and the one that has in the last years focussed again the attention of the logicians on impredicative concept formations. Let us recall that the consistency of the ramified type theory (with an axiom of infinity but without the axiom of reducibility) has been proved long ago from relatively weak assumptions [1]), whereas proving the consistency of the type theory T*, which contains an axiom of infinity, requires much stronger assumptions [2]).

The systems of Wang and Lorenzen embody a vicious circle principle in the sense that an object determined by a condition referring to a totality of objects of layer α is not itself of layer α but rather of layer $\alpha + 1$. Since, however, mixing of layers is no longer forbidden, this does not exclude the possibility of there being a layer to which all objects of layer α as well as the object of layer $\alpha + 1$ determined by a condition on these objects belong. In these systems, this holds indeed for layer $\alpha + 1$ itself and for the infinitely many layers of a higher ordinal. The mixing of layers, however repulsive to thinkers of certain strong ontological convictions, does not reduce the constructiveness of the system, nor does the use of transfinite ordinals if one restricts oneself here to the "constructive" ordinals of the second number-class.

An entirely different, very interesting but so far rather abortive attempt to take the avoidance of impredicative concept formation seriously, was made by Hintikka [3]). Utilizing a suggestion of Wittgenstein [4]), later elaborated by Kolmogorov and Zich [5]), he exhibited a much weaker way of avoiding universal variables than the theory of orders or layers. He proposed to give bound variables an *exclusive* interpretation, in contradistinction to the customary *inclusive* interpretation. According to the customary interpretation, the statement, 'There are at least two different individuals having the property *P*', is symbolized by

program rather than in the rejection of the principle of excluded middle. This view is explicitly reaffirmed by Gilmore 56.

1) In fact, the statement that ramified type theory with an axiom of infinity, RT*, is consistent is a theorem of the type theory T*, and hence the consistency of RT* does not imply that of T* – see Chapter V, p. 328.

2) Fitch 38.

3) Hintikka 56.

4) Wittgenstein 22, 5.53–5.5352.

5) Zich 48 which contains also, in a footnote, a reference to Kolmogorov's relevant work.

$$\exists x\, \exists y\, [x \neq y \wedge P(x) \wedge P(y)]\,,$$

whereas in the exclusive interpretation of the bound variables this statement would be symbolized by

$$\exists x\, \exists y\, [P(x) \wedge P(y)]\,,$$

with the word ‘different’ being taken care of by the distinctiveness of the variables. Notice that in the second symbolization, under the exclusive interpretation, the variables are not universal since their ranges are not allowed to coincide. The restriction of the universality is, however, not incorporated once for all in a rigid system of type and order indexes but is induced rather by their interplay within any given specific context.

If we change the logical basis of the ideal calculus (of § 1), i.e. the first-order predicate calculus, such as to accord with the exclusive interpretation of the bound variables, then the standard derivation of the antinomies will not come through. The derivation of Russell’s antinomy, for instance, will fail because step (2) [1]), which is in accordance with the standard (inclusive) first-order predicate calculus, would no longer be legitimate in the revised (exclusive) calculus.

However, it is not necessary to revise the customary first-order logic in order to take account of the exclusive interpretation. The same end can be achieved without tampering with the usual formation rules of this logic. The desired exclusiveness can also be taken care of by explicit diversity conditions on the variables. The axiom schema of comprehension of the ideal calculus would now be revised in an entirely different way and look something like

C_{ex}: ( ) $\exists y\, \forall x\, [x \in y \leftrightarrow (x \neq y \rightarrow \varphi(x))]$

where each quantifier ‘$\exists z$’ in the original formulation of ‘$\varphi(x)$’ is replaced by ‘$\exists z(z \neq y \wedge ...)$’ and each quantifier ‘$\forall z$’ by ‘$\forall z(z \neq y \rightarrow ...)$’. Once more, the standard derivation of Russell’s antinomy would not work, though this time for a different reason. All we would arrive at, as the reader might check, would be the true formula

$$\exists y\, [y \in y \leftrightarrow (y \neq y \rightarrow \neg y \in y)]\,.$$

Of course, we may expect to have to pay a price for the interpretation of the axiom of comprehension along the exclusive line, an interpretation that

1) On p. 155.

seems to be so efficient in overcoming the antinomies. Indeed, we can avoid here Cantor's antinomy, for instance, only by becoming unable to derive Cantor's theorem (which should not surprise us any longer).

It may be claimed that C_{ex} is the simplest modification of the axiom of comprehension that accords with Russell's vicious circle principle, since the entity whose existence is asserted by C_{ex} is the only one which is excluded from the ranges of the bound variables that occur in it. This might well be the cheapest materialization of Gödel's hunch [1]) according to which it should be possible to carry out the idea of limited ranges of significance without having to divide up the objects into mutually exclusive ranges. The distinctness clauses would then correspond to the "singular points' which are mentioned by Gödel in his metaphor. Hintikka's system keeps our logical intuitions intact up to minor corrections; these minor corrections themselves are not even real deviations from our logical intuitions but embodiments of the exclusive interpretation which is the appropriate one in the case of the axiom of comprehension, though our intuitions are certainly not sharp enough to make this clear without the "help" of the antinomies.

The problem of distinguishing, within mathematical reasonings in their pre-formalization stage, between the occurrence of essential, vicious impredicativities and inessential, innocent ones has always been a rather pressing one, from the first discussions between Poincaré and Zermelo [2]) on this point. Nobody ever seriously questioned the legitimacy of such a concept as the maximum of a function in a given interval, though its standard definition as a greatest of the function values in this interval is clearly impredicative. Poincaré's own way of showing that this impredicativity is not essential is by redefining the maximum of a function as a greatest function value for all *rational* arguments rather than for *all* arguments; even if this function value itself turns out to be rational (and within the interval of the argument values), this is not entailed by the definition as such. Hintikka's way is clearly much simpler and relies on the obvious fact that we can redefine the maximum of a function (within a given interval) as any one of these function values which are not smaller than any of the *other* function values (in this interval), a condition to be formalized with the help of a distinctness clause.

Hintikka's attempt, though, has so far turned out to be abortive for a simple reason: C_{ex} still engenders contradiction, as Hintikka himself realized shortly after having made his proposal [3]). Though a stronger revision of the

1) Gödel 44, p. 150.

2) See, e.g., Fraenkel 28, pp. 250–251.

3) Hintikka 57.

axiom of comprehension than the one embodied in C_{ex} would avoid these new contradictions, the resulting system would, even if consistent, deviate from standard set theories in much more than no longer containing Cantor's theorem. As a matter of fact, no further attempt was made since 1957 to continue along these lines.

§11. SET THEORIES BASED UPON NON-STANDARD LOGICS

The reader will by now have come to appreciate the grave obstacles that stand in the way of safeguarding a set theory that will be both strong enough to allow for the derivation of classical analysis as well as weak enough not to allow for the derivation of antinomies. All the systems of set theory discussed so far have in common that they are based either on classical first-order predicate logic or coincide with certain natural extensions of this logic. Their adherents believe that it is either too liberal an interpretation of the term 'formula' or too liberal a use of the axiom of comprehension which is to blame for the occurrence of the (logical) antinomies.

It is, however, understandable that some authors should have come to blame the shortcomings of existing set theories rather on the underlying logic and should, in consequence, have tried to arrive at a more satifactory set theory by changing this basic logic. The whole next chapter will be dedicated to the most important of these heresies – the intuitionistic conception; let it, however, be added immediately that this conception, together with a denial of classical logic, rejects the whole view as if set theory, or mathematics in general, were based upon some underlying logic. In this section we shall discuss very briefly some of the less spectacular attempts.

11.1. **Leśniewski's Ontology**. There is probably no country which has contributed, relative to the size of its population, so much to mathematical logic and set theory as Poland. Leaving the explanation of this curious fact to sociology of science, we shall dedicate the next two subsections to a brief outline of the contributions of two of the most original Polish thinkers to the foundations of set theory: Stanisław Leśniewski and Leon Chwistek. Both of them perished during World War II. The fact that many of the most important writings of these authors were published in Polish, that Leśniewski wrote his German papers in a style that made no concessions to the reader – some of them are printed in an esoteric symbolism with hardly a word of explanation in ordinary language –, and that Chwistek often combined an equally excentric symbolism with obscure asides on metaphysics, philosophy of mathematics, and science in general, did not encourage Western logicians to make special efforts to master their writings, and their untimely death – in the case of Chwistek, together with his most gifted pupils – still increased the difficulties. There can be no doubt as to the high originality of the contributions of these thinkers to the foundations of mathematics, and now, after the recent appearance of various authoritative articles in Western languages expounding Leśniewski's teaching [1]), and, in particular, of Luschei's monograph,[2]), and after the appearance of an English translation of Chwistek's principal work [3]) and of an extremely illuminating review of it, also in English [4]), we seem to stand at the verge of a real revival of interest in the work of these two logicians that has already fertilized the thought of many a worker in the foundations of mathematics.

In this subsection, we shall give the barest outline of that part of Leśniewski's system which is most relevant to the theory of sets; the next subsection will do the same for Chwistek's system.

Leśniewski called that part of his system which dealt with the logic of the particle 'is', or rather of the Polish particle '*jest*', *ontology*. Though in a sense this theory dealt indeed with "the general properties of being", the term 'ontology' carries no metaphysical connotations. Other authors, in order to avoid explicitly the arisal of such connotations, prefer the term *calculus of names* [5]). Within Leśniewski's total system, ontology occupies an intermediate position: it is based on *protothetics*, a generalized propositional calculus, and serves, in its turn, as a basis for *mereology* which deals with the part-whole relationship. (Leśniewski has no need for predicate logic, as the predicational ingredient of this part of standard symbolic logic is taken care of in ontology while its quantificational ingredient belongs already to protothetics.)

Ontology has just one primitive term in addition to those of protothetics, just one additional axiom, and some additional rules of inference. This primitive term is – 'ϵ', which is however not meant to symbolize the class-membership relation but rather to serve as the symbolic counterpart of the natural language '*jest*' (in Polish) or '*est*' (in Latin), and to a lesser degree of 'is' (in English). Modern logicians have often been at great pains to point out that

1) Sobociński 49–50, Słupecki 53, 55, Grzegorczyk 55.

2) Luschei 62.

3) Chwistek 48. The introduction by Miss Brodie should be perused with some reservation.

4) Myhill 49.

5) This is done, e.g., in Kotarbiński 29.

the *copula* of traditional logic is highly ambiguous and that the English term 'is', for instance, stands for class-membership, class-inclusion, and identity, respectively (and fulfils perhaps still other functions on occasion) in such sentences as 'Socrates is wise', 'The dog is an animal', and 'Socrates is the husband of Xantippe'. Leśniewski took the contrary view that the Polish '*jest*' and the Latin '*est*' should be regarded as non-ambiguous particles, whatever the situation in English. (The difference is caused by the fact that English – in contradistinction to Polish and Latin – has articles, definite and indefinite, which are obligatory in the contexts 'Socrates is ... man', 'The dog is ... animal', 'Socrates is ... husband of Xantippe' – with sentences like 'Socrates is white' and 'Tully is Cicero' still more complicating the situation – whereas the corresponding Latin sentences would be '*Socrates est homo*', '*Canis est animal*', '*Socrates est coniunx Xantippae*' – and '*Socrates est albus*', '*Tullius est Cicero*'. We shall spare the reader the corresponding Polish sentences.)

Leśniewski is completely successful in providing, through his axiom, his own term 'ϵ with a meaning which is a kind of conglomerate of the three mentioned meanings, without falling into contradiction. On the contrary, the consistency of ontology relative to protothetics can easily be proved. Leśniewski's original single axiom of ontology, after some notational adaptations (which would however not have been approved by Leśniewski himself) is:

$$x \epsilon w \leftrightarrow \exists y(y \epsilon x) \wedge \forall y \forall z[(y \epsilon x \wedge z \epsilon x) \rightarrow y \epsilon z] \wedge \forall y(y \epsilon x \rightarrow y \epsilon w).$$

Since all symbols in this axiom, with the exception of 'ϵ', are supposed to have their standard interpretation, we see that a Leśniewskian ϵ-sentence is true only if to the left of the 'ϵ' stands a non-empty, singular name; this follows from the first and second conjunctive components of the right side of the equivalence, respectively. Accordingly, such sentences as '*Hamlet est albus*' or '*Canis est animal*' are treated as false, the first since '*Hamlet*' is empty (i.e. not denoting), the second since '*canis*' is a general name.

That Leśniewski's theory of semantical categories should by no means be regarded simply as a kind of generalized linguistic counterpart of a simple theory of types [1]) is obvious from the fact that the symbols flanking the Leśniewskian 'ϵ' – and similarly with their natural-language counterparts – belong always to the same semantical category, even when the right-hand symbol is a class-symbol: '*Socrates*' and '*homo*', in the

1) This is the impression one could get from a not too careful reading of Tarski 56, pp. 213 ff, or Sobociński 49–50 who often uses the expression 'catégories sémantiques (types logiques)' – and Tarski and Sobociński are two of the most outstanding pupils of Leśniewski. Cf., however, Tarski 56, p. 213 n, from where it should be clear that Tarski is fully aware of the real situation.

sentence '*Socrates est homo*' belong to the same semantical category, i.e. to the category of names, whereas for Russell the denotata of these two words, i.e. Socrates and the class of men, belong to different types.

Since Leśniewski's 'ϵ' is not meant to be a symbol for class-membership, it is preferable to regard his ontology not as a variant of set theory but rather as a rival of set theory for the foundations of mathematics. This view does not exclude that counterparts of many set-theoretical axioms turn out – under a certain notational transformation – to be ontological theorems or that the ontological axiom should be transformed into a type-theoretical axiom. The latter possibility is easily materialized by interpreting '$x \epsilon w$' as 'x is a unit-class of individuals, w is a class of individuals, and $x \subset w$'.

How important a rival of set theory ontology is, or could be made to be, is a question which it is still very difficult to decide. But Leśniewski has convincingly shown that the standard arguments leading to the logical antinomies cannot be reproduced, in any of the various plausible rephrasings, in his system; some of the counterparts of these arguments fail to comply with the theory of semantical categories, others require certain steps which are not viable in ontology [1]).

Let us finish this sketch by noting that what might be regarded as the ontological counterparts of the axiom of comprehension are treated at length by Leśniewski under the name of *pseudo-definitions* [2]).

11.2. **The Systems of Chwistek and Myhill.** After continuous struggles and many changes of mind, Chwistek – the highly original Polish logician whom we recall as the first influential critic of Russell's ramified type theory [3]) – finally constructed a system which, had it been presented in a less obscure language and a somewhat less forbidding terminology and symbolism, might well have exerted a decisive influence on the reconstruction of mathematics on a consistent "constructive" basis. This final system of Chwistek's was published posthumously in *The Limits of Science* in 1948. This book is so different from the Polish original, *Granice Nauki*, published in 1935, that it should by no means be regarded as a mere translation of it. Fortunately, Chwistek has now found an equally competent interpreter in the American logician Myhill who was able to bring Chwistek's ideas closer to the main present currents of thought. Even so it is still impossible in the available

1) For an exhaustive discussion of this issue, see Sobociński 49–50, Luschei 62.

2) For the significance of Leśniewski's use of pseudo-definitions for the formulation of the second-order predicate calculus, see Henkin 53.

3) See footnote 3 on p. 174.

space to describe Chwistek's approach in any serious detail. Let us therefore content ourselves with stating that Chwistek's set theory is a kind of "*inverted*" *ramified type theory*, in the sense that classes always have members of "higher" types than themselves. The types are, in addition, cumulative – as, e.g., in Wang's system Σ (§ 6) or in some of the systems discussed by Quine – but again with the direction inverted, each entity belonging to all types lower than any type to which it belongs. By adding to this ramified type theory a certain variant of the axiom of reducibility, Chwistek obtains what he calls a *pure theory of types* which can easily be proved to be consistent; by adding another variant he gets a *simplified theory of types* which, however, combined with a certain powerful metasystem is inconsistent. Chwistek's pure theory of types, in its last stages, looks sufficiently similar to PM to make it plausible that classical analysis can be constructed on its basis. A completion of this program would therefore mean a proof of the consistency of classical analysis from "nominalistic" assumptions whose soundness can hardly be doubted in any serious sense.

Myhill himself, in an unfinished series of papers [1]), tried to complete the program. His own system is an inverted cumulative type theory with an all-inclusive type 0 but with no highest type. Vicious-circle paradoxes are avoided by arranging that no expression containing a quantified variable be a value of that variable. (In Myhill's – as in Chwistek's – system, expressions, not what the expressions denote, are values of variables; this curious feature is consistently carried through without involving him – or Chwistek, contrary to the general opinion – in fallacious thinking due to the "confusion between use and mention of signs".) He uses a *non-finitary* consequence relation according to which certain formulae are regarded as consequences of appropriate classes of formulae. This means, of course, that the system, though nominalistic in a sense, is anything but constructivistic. Its consistency can be shown, along the pattern of a proof by Fitch (p. 174, footnote 1), by transfinite induction (up to ϵ_0, at least). A certain analogue of the axiom of reducibility can be shown to hold in the system. He is finally able to derive in the system (type-restricted) analogues of Bourbaki's axiom system for set theory [2]) (plus an axiom of replacement) including the axioms of choice and infinity, but excluding the axiom of extensionality. Interestingly enough, Myhill believes it to be very unlikely that analysis needs a principle of extensionality, probably being influenced in this by Chwistek who regarded it as a product of metaphysical idealism. It seems, however, that so far Myhill has

1) See Myhill 49, 51, 51a.

2) Viz., the system presented in Bourbaki 49.

not been able to accomplish the proof that a non-extensional Bourbaki set theory contains a model of extensional Bourbaki set theory [1]).

11.3. **Fitch's System.** After years of constantly refining earlier attempts. F.B. Fitch published a textbook on *Symbolic Logic* [2]) in which he developed a new, and in many respects heterodox, approach to this topic, promising to publish a sequel dealing with the detailed derivation of the more important theorems of mathematical analysis from his system [3]). There is no need to deal here at length with its many notational and pedagogical innovations. Suffices it to state that Fitch's system contains no variables, either free or bound – e.g., the class of entities greater than 2, customarily symbolized by '$\hat{x}(x>2)$' or $\lambda x(x>2)$', is rendered by Fitch as (3/3>2) or even (Caesar/Caesar>2) – and no type restrictions but, on the other hand, no general rule of excluded middle or law of extensionality, the last feature backed up by an extremely narrow treatment of identity, according to which an identity sentence is valid if and only if its two sides, in their disabbreviated form, are notationally the same, in other words, by avoiding the occurrence of notationally different but logically synonymous sequences of primitive terms. (This requires, among other things, to treat one of the just mentioned expressions '(3/3>2)' and '(Caesar/Caesar>2)' as an "abbreviation" of the other, or both of them as "abbreviations" of a certain arbitrarily chosen expression of the form '(.../...>2)'.)

Antinomies are avoided either through the fact that the relevant derivations fail to satisfy the mentioned restrictions on the proof procedures – this is the way in which Curry's paradox [4]), a variant of Russell's paradox of particular interest insofar as it does not use negation, is overcome – or through the simple device of regarding each such antinomy as a proof of the fact that the proposition involved fails to satisfy the excluded middle. Russell's paradox, e.g., in its original version but in Fitch's notation, deals with the class (or attribute) $(x/\neg[x \in x])$ (where 'x' is *not* a variable) – to be denoted by 'Z' – and winds up by proving that $[Z \in Z] \leftrightarrow \neg[Z \in Z]$. But

1) This was one of the aims he set himself in Myhill 51a, p. 135.

2) Fitch 52.

3) Among Fitch's later papers that contain further developments of his system of Basic Logic – without yet amounting to a fulfilment of the promise –, we mention only Fitch 56, footnote 1 of which contains further references.

4) This paradox was created in Curry 42. The role of Russell's class of all classes that do not contain themselves as members is taken over there by the class of all classes such that if they contain themselves as members, an arbitrary statement is true. For a full formal derivation of the paradox, see Fitch 52, pp. 107–108.

this equivalence is not paradoxical by itself, it becomes so only under the assumption that the proposition $Z \in Z$ satisfies the excluded middle, i.e. that $[Z \in Z] \vee \neg[Z \in Z]$ holds; in that case the really paradoxical proposition $[Z \in Z] \wedge \neg[Z \in Z]$ can be easily derived from the equivalence. Since Fitch claims to have shown the consistency of his system, he is able to conclude that $Z \in Z$ is one of those propositions for which the principle of the excluded middle does not hold and which he terms *indefinite* – which should be clearly distinguished from Russell's term *meaningless* 1). The Liar antinomy is overcome analogously. What this antinomy shows is that the proposition expressed by the sentence: "this proposition itself is false", is indefinite. No structural criterium by which sentences expressing definite propositions would be distinguished from sentences expressing indefinite propositions is given. This is, of course, a drawback, though not so much with regard to natural language, for which no more effective criterium should be expected 2), as with regard to language systems; still it is not fatal, especially if such a system is demonstrably consistent, since in this case at least a sufficient condition for the indefiniteness of a proposition is provided, i.e. that from the assumption of its definiteness a contradiction can be derived. Incidentally, Fitch has a *rule of excluded middle for identity*, according to which every proposition of the form $[a = b]$ is definite.

Fitch's main reasons for rejecting Type Theory are that, on the one hand, certain sorts of self-reference are required for philosophical logic as well as for the development of the theory of real numbers and that, on the other hand, this theory cannot be stated without violation of its own requirements 3).

Fitch claims that his system is demonstrably consistent. This is achieved through certain restrictions on the proof procedures. These restrictions, however, whatever their intrinsic plausibility, have the effect that one cannot always conclude from the validity of an implication and of its antecedent to the validity of its consequent or from the validity of two statements to the validity of their conjunction but must, in general, take into account the way in which the premises of these inferences were obtained. Part of the meaning of a formula would then reside in the specific way of its derivation – which is no longer as esoteric an idea as it sounded in the past 4).

1) As well as, of course, from Carnap's term 'indefinite' mentioned above, p. 196.

2) Cf. Bar-Hillel 57a, 66.

3) This objection has often been made, e.g. by Weiss 28, and as often refuted. Though it certainly is not sound from a purely formal standpoint, it retains a certain heuristic value and is probably one of the strongest reasons why so many thinkers do not like type-theoretical systems. Cf. Gödel 44, p. 149.

4) It has become a central conception of generative-transformational grammar; cf. Chomsky 57.

In addition, the consistency proof provided by Fitch for his system seems not to be fully constructive, hence in a sense circular. In one decisive place, he uses, metalogically, an argument amounting to the transition from '$\forall x(p \vee \varphi(x))$' to '$p \vee \forall x\, \varphi(x)$', which is not constructively valid (and is rejected, for instance, in intuitionistic logic). This metalogical argument is, incidentally, used by Fitch in order to show that an analogous type of argument, *within* the system, is eliminable [1]).

Fitch's system is one of the many erected during the last decades to avoid the antinomies through a departure from the classical first-order predicate calculus. In Fitch's case, the departure takes the form of constructing the propositional calculus in such a way that the principle of the excluded middle is not generally valid in it. This, of course, recalls at once intuitionistic logic, to be treated at length in the next chapter. Fitch's logic, however, is by no means identical with intuitionistic logic as formalized by Heyting [2]).

This departure from classical logic does not make Fitch's logic three-valued, indefiniteness – in Fitch's sense – not being a third truth-value that could possibly be treated on a par with truth and falsity. Three-valued and, more generally, many-valued logics will be treated in the next subsection.

11.4. **Many-valued Logics.** It seems rather natural to look for the culprit in the arisal of antinomies in a still different direction: in most, if not all, antinomies the crucial contradiction has, or can be given, the form of an equivalence between a certain statement and its negation. The final line of a derivation of Russell's antinomy, for instance, is usually some variant or abbreviation of

$$[\hat{x}(x \notin x) \in \hat{x}(x \notin x)] \leftrightarrow \neg [\hat{x}(x \notin x) \in \hat{x}(x \notin x)],$$

the final line of a derivation of Grelling's antinomy is something like 'heterological' is heterological $\longleftrightarrow \neg$ ('heterological' is heterological), etc. Now, such statements embody contradiction if the logic of the language in which they are formulated is the classical two-valued logic, in which any statement of the form

$$p \leftrightarrow \neg p$$

1) All these criticisms were launched by Ackermann 52.

2) There are many resemblances between Fitch's system and that developed by Ackermann 50; cf. also Ackermann 52–53 and Grize 55.

is indeed self-contradictory. But in some many-valued logics – i.e. logics in which the meta-principle that every (closed) statement has exactly one of two truth-values does not hold – and under some natural interpretations of 'negation' and 'equivalence', statements of the mentioned form are no longer self-contradictory. In certain three-valued logics, such as the famous system L_3 of Łukasiewicz [1]), the negation of a statement with the "intermediate" truth-value has itself the intermediate truth-value so that in this case a statement is equivalent (in truth-value) to its negation.

Are then many-valued logics antinomy-free? This would be a rash conclusion. Though the derivation of such paradoxes as Russell's and Curry's would indeed not come through [2]), it can be shown that in a set theory allowing for unrestricted comprehension and based upon certain of these logics, e.g. on the mentioned L_3, a Curry-type paradox is derivable [3]).

The shift from the term 'antinomy' to 'paradox' that already occurred above, p. 205, in connection with the first mentioning of Curry's paradox, is not accidental. The term 'antinomy', as defined on p. 1, does not apply as such to negation-free or many-valued logics. Curry's paradox consists not in proving that two contradictory statements are equivalent – notions which obviously involve (two-valued) negation – but in proving any statement whatsoever by seemingly innocuous procedures, thereby proving the system to be (formally) inconsistent (see Chapter V, §4).

However, the avoidance of antinomies was by no means the only incentive that made logicians occupy themselves with many-valued logics. Many logicians treated such logics as pure calculi without insisting from the beginning on interpretation and application [4]). Others believed that such logics might be of use in the analysis of certain epistemologically puzzling situations such as the assignment of truth-values to future contingent events [5]). Still others thought that some of these logics were better suited to deal with certain phenomena in quantum physics than classical two-valued logic [6]).

No serious attempt has been made so far to construct a set theory or a theory of numbers on a many-valued logic [7]). There seems to exist no conclusive

1) See Tarski 56, Chapter IV, 3.

2) See, e.g., Bočvar 39, 43.

3) See Shaw-Kwei 54, Prior 55.

4) Cf., e.g., Rosser-Turquette 52, Introduction.

5) See Łukasiewicz 30 and, among recent publications, e.g. Prior 53.

6) See Reichenbach 44 and, out of the many writings by Destouches and Paulette Destouches-Février dealing with this topic, e.g., Destouches-Février 51 (with the sharply critical review of McKinsey-Suppes 54).

7) The construction of models of set theory in which the truth values belong to some Boolean algebra, as in Scott 67, Rosser 69, or Jech 71, should not be considered as such an attempt. Those Boolean-valued structures are used, as mentioned in Chapter II, to obtain the compatibility of certain set-theoretical statements with ZF and not in order to construct a system of set theory on a logical basis different from that of ZF.

apriori reason why certain many-valued logics should not prove to be fruitful and provide a secure foundation for set theory and mathematics. So long, however, as the proponents of these logics have not come forward with a full-fledged set theory or arithmetic, the *onus probandi* doubtless rests upon them, in view of the fact that these theories will certainly become much more complex than they are today. Until then, it is best to refrain from any final evaluation [1]).

11.5. **Combinatory Logic.** We shall finally just mention that there exists a method of developing logic which is radically different from the "standard" ones. The so-called Combinatory Logic was founded by Schönfinkel and Curry [2]) and is a highly interesting variant of modern symbolic logic. Since, however, its impact on the foundations of set theory does not seem to be very great, we shall here only call attention to a publication by Cogan [3]) in which a set theory is formalized from the point of view of combinatory logic, after an illuminating presentation of the essentials of this logic itself. The system of Cogan turned out to be inconsistent [4]), but Curry has expressed his conviction that the inconsistency can be avoided by a reformulation of the system.

1) A survey and bibliography of some of the recent work on this topic is given in Chang 65a.

2) Schönfinkel 24 and Curry 30.

3) Cogan 55, Curry-Feys 58.

4) Titgemeier 61. Other sources of inconsistency were found by students of Curry.

CHAPTER IV

INTUITIONISTIC CONCEPTIONS OF MATHEMATICS

§ 1. HISTORICAL INTRODUCTION. THE ABYSS BETWEEN DISCRETENESS AND CONTINUITY

The axiomatization of set theory (Chapter II) undertakes to avoid the antinomies that arise from the classical theory by *restricting the concept of set* or its mathematical use to such an extent as is suitable for that aim, but without fundamentally modifying the structure of the theory. Logicism (Chapter III) conceives the antinomies as a danger signal which concerns not only set theory but shows that something in the mathematical methods in general is out of order; therefore logicists attribute the defect to logic and its use in mathematics rather than to mathematics itself and propose a penetrating *reform of logic* which, as one of its incidental consequences, entails the elimination of the antinomies.

The attitude that is described in the present chapter is much more radical in its conception as well as in the consequences following from it. The mathematical schools which we will consider here maintain that traditional mathematics has misinterpreted and mismanaged the *concept of infinity*. They do not take the issue lightly, as most of them stress the fact that infinity is the very lifeblood of mathematics, to the extent that the part of mathematics dealing with finite concepts only is from the viewpoint of foundations almost trivial. With the criticism of the traditional treatment of infinity goes a thorough revision of concepts such as "existence", "proof" and "mathematical object". Yet analysis and geometry as developed since the 17th century and especially since the beginning of the 19th century have – so they argue – utterly disregarded the peculiar traits of infinity and their consequences for mathematics. The supposedly strict methods introduced into real number theory and calculus during the 19th century [1]) from Cauchy

1) For the interrelationship between infinity and the demand for strictness in mathematical proof during various periods, cf. Pierpont 28.

to Weierstrass and Cantor, far from reaching the desired goal, have rather raised to an elaborate system the erroneous tendency of treating infinity with methods created for finite domains.

According to this view the antinomies appearing at the turn of the century are but a secondary symptom, evolving at a rather accidental spot; actually they are caused by the brittleness of the foundations of mathematics itself, rather than just logic or set theory. The emergence of contradictions in set theory is due to the fact that in no other branch of mathematics has such an abundant and unlimited use of infinity been made. All the same, the contradictions are due to traditional dealing with infinity *in general* and not to its application in the degree involved by set theory. Therefore the significance of the antinomies as warning signals can be met only by a reform of mathematics as a whole; this will automatically exclude not only the antinomies actually found hitherto but any conceivable antinomy.

In addition, the concept of infinity in set theory assumes specific significance in connection with one of the oldest and most intricate notions of science in general, viz. *continuity*. True, the continuum is the very domain to which analysis and most of geometry refer. However, in these domains the continuum is presumed from the first as a *basis* while set theory ventures to *construct* the continuum in a way which appears as a special case of a more general method (power-set, diagonal method, Cantor's theorem); a method which belongs to the strongest and the most daring procedures of set theory. The starting-point of this procedure is a *discrete* (discontinuous) aggregate, e.g. the denumerable set of all integers or of isolated points of a line.

To be sure, the starting-point mentioned is itself an infinite set. But this type of infinity – in one or the other form of conceiving it, for instance, through iterated construction – is the very foundation of mathematics. Since it is the *conditio sine qua non* of mathematical reasoning the problem is not *whether*, but *how*, we should accept it, and various opinions and theories on this question are displayed in the present book: axiomatic, Platonistic, logicistic, intuitionistic, and metamathematical opinions or methods.

Bridging the gap between the domains of discreteness and of continuity, or between arithmetic and geometry, is a central, presumably even *the* central problem of the foundation of mathematics [1]). Cantor claimed to have bridged the gap, as claimed before by "classical" analysis; the sharpest criticism of these claims has been expressed by the intuitionistic schools.

To understand the nature of the problem one should stress the funda-

1) See Chevalier 29 where the problem is conceived in its generality, not only in its mathematical and physical aspects. Cf. also Jørgensen 32, Fraenkel 37.

mental difference [1]) between the *discrete, qualitative, individual* nature of *number* in the "combinatorial" domain of *counting* (arithmetic) and the *continuous, quantitative, homogeneous* nature of the points of *space* (or of time) in the "analytical" domain of *measuring* (geometry). Every integer differs from every other in characteristic individual properties comparable to the differences between human beings, while the continuum appears as an amorphous pulp of points which display little individuality.

Bridging the gap between these two heterogeneous domains is not only the central but also the oldest problem in the foundations of mathematics and in the related philosophical fields. The existence of incommensurable magnitudes discovered in the fifth century B.C. and for a while kept secret by the experts lest it might offend the public, had initiated the first crisis in the foundations of mathematics, especially in the Pythagorean school [2]). This crisis found an apparently satisfactory settlement through the Greeks themselves, while the second crisis, originating from the treatment of continuity in calculus and modern analysis in general, seemed to have been overcome in the sixties and seventies of the 19th century. Kronecker's fanatic challenge to recognize integers only and to purge mathematics of all other numbers (see below) remained than a voice in the wilderness.

As mentioned before, it is customary to assign the beginning of the third crisis in the foundations of mathematics to the turn of the century, with the appearance of antinomies in set theory. However, the real trouble began only a few years later, with the reactions to Zermelo's first proof of the well-ordering theorem (1904; cf. p. 75), and from 1907 on with the intuitionistic frontal attack against classical mathematics which stands in the center of the present chapter. By then, not only set-theoretical antinomies but contradictions on the whole were shrinking to mere *symptoms* of the inadequacy of classical mathematics, which naturally was first visible at the outskirts of the mathematical field; yet without the emergence of contradictions the situation would have been no less disastrous.

In the course of the discussions carried on since then it turned out more and more how closely these problems were related to those that seemed twice to have been solved, viz. the riddles of the Pythagorean and Eleatic schools and the difficulties which arose in the French and German centers of the theory of functions. Though the arguments have changed, the gap between

1) Cf. Freudenthal 32, Weyl 31; in the latter paper also a musical illustration is given.

2) Aristotle also points out the contrast between the discrete notions of reason (thought, counting) and the notions of the external world which seem to be continuous.

discrete and continuous is again the weak spot – an eternal point of least resistance and at the same time of overwhelming scientific importance in mathematics, philosophy, and even physics. The very same arithmetical theories of real number which a generation before seemed to be the solid basis of analysis now provoked the intuitionistic objections.

Incidentally, it is not obvious from the first which of these two regions, so heterogeneous in their structures and in the appropriate methods of exploring, should be taken as the starting-point. Certainly the discrete admits an easier access to logical analysis, and the tendency of arithmetization, already underlying Zenon's paradoxes, has been impressing its mark upon modern mathematics and may be perceived in axiomatics of set theory (Chapter II) as well as in metamathematics (Chapter V). However, the converse direction is also conceivable, for intuition seems to comprehend the continuum at once; mainly for this reason Greek mathematics and philosophy were inclined to consider continuity to be the simpler concept and to contemplate combinatorial concepts and facts from an analytic view. Whereas some traces of the latter attitude are still visible in some intuitionistic opinions and are not missing even in Brouwer's intuitionism (cf. below § 5), intuitionism in general considers arithmetic, say in the shape of mathematical induction, not only as the primary concept but as the origin of mathematics on the whole; it therefore tends to restrict analysis by imposing upon it "combinatorial" methods.

Such intuitionistic restriction of the concept of continuum and of its handling in analysis and geometry, though carried out in quite a variety of different ways by various intuitionistic schools, always goes so far as to exclude vital parts of those two domains. (This is not altered by Brouwer's peculiar way of admitting the continuum as a "medium of free growth".) On the other hand, intuitionists maintain that the restricted system covers all legitimate mathematical processes.

To establish the *necessity* of their restrictions, intuitionists critically analyze the methods of classical mathematics. In particular they claim that the dangers inherent in infinity, as emerging in 18th century's mathematics and further aggravated by Dirichlet's concept of an arbitrary function, have but seemingly been checked by the "classical" theories of real number, limit, continuity, integral, etc.; these theories are charged with logical circles which necessarily lead to contradictions. According to them, hardly any progress towards guaranteeing the solidity of classical mathematics has been achieved in the 19th century.

On the other hand, it is maintained that the restricted system is *sufficient* for the needs of what may justly be called 'mathematics' – a notion taken, to be sure, in a sense quite different from the traditional one. This claim shall be justified through basing arithmetic and the "meaningful" parts of analysis (including geometry) and set theory upon the mathematical constructions intuitionistically admitted.

Thus, in striking contrast with the "conservative" steps taken in Chapters II and III to avoid the sources of danger revealed by the set-theoretical antinomies, we deal here with a revolutionary trend meant to thoroughly alter the essence of mathematics and its methods, at least in comparison with the tradition of the last three hundred years. In passing it should be noted that in the hypothetical case that the intuitionist attitude should oust the classical view it might take generations to save, and to firmly base with intuitionistic methods, those parts of mathematics which do not become meaningless or false according to the new conception.

The group of mathematicians who call themselves *intuitionists* or *neo-intuitionists* [1]) is composed of various and widely diverging trends which, save for few exceptions dating back to the 19th century, arose with the beginning of the present century. The name does not derive from the intuitively creative nature of mathematical discovery or invention [2]) which is accepted universally but from the "primordial intuition" explained in § 5. One may trace back hints of this attitude to earlier periods of mathematical research, even to Greek antiquity [3]).

1) Various names have been used in the past, among them the now obsolete term 'neo-intuitionism'. We shall stick to the present terminology, which reserves 'intuitionism' for Brouwer's school. Older schools, especially the French school, are referred to as 'semi-intuitionism'.

In the older French literature the name 'pragmatist' (so Poincaré) or 'realist' is often found while the opponents are called 'idealists'; this terminology may lead to confusion in view of the Platonistic use of the term 'realist'.

The name *formalists* for the opponents of intuitionism, particulary employed for Hilbert's school of axiomatics and metamathematics, chiefly originates from polemic intentions and does not appropriately characterize that school. In fact "cantorians" or logicists are far more contrary to intuitionism than those "formalists", who in some regards are more finitistic than intuitionists.

2) A mathematician as remote from intuitionism as Hadamard beautifully described the intuitive character of mathematical creation (Hadamard 45).

3) Cf. Boutroux 20 (especially Chapter IV), and Hadamard's preface to Gonseth 26. According to the analysis of Aristotle's theory of science as given in Scholz 30 (cf. O. Becker 36), the attitude of this classic of ancient logic was rather neo-intuitionistic: he would have regarded intuitionistic mathematics as an επιστημη, mathematical fields beyond this sphere only as a δοξα. Also other Greek scientists, including Euclid, display an "intuitionistic" touch and strictly distinguish between constructibility and abstract existence; see, for instance, O. Becker 33.

Among the forerunners of modern intuitionism, I. Barrow, the teacher of Newton, may be mentioned in view of his criticism of Euclid's theory of proportions.

The number of scholars who adhere to intuitionistic principles, let alone those who actually adjust their methods of mathematical research to intuitionistic restrictions, has been but small so far and shows little tendency to increase. But it is highly remarkable that they include some of the outstanding mathematicians of the last generations from various countries, as is shown by the names of Kronecker, Poincaré, Borel, Lebesgue, Brouwer, Weyl, and Skolem [1]). To a large extent these and other mathematicians arrived independently at their daring ideas, forced as it were by an inability to adapt themselves to the traditional way of mathematical thinking; it is still more surprising that in spite of their relative isolation they showed themselves quite convinced of the final victory of their principles [2]). It is as if those ideas were at a certain time in the air where, however, only people with a proper scent could track them. The arguments are sometimes just intuitive and dogmatic, at other times based on philosophical motives, or again using strictly mathematical reasons. Accordingly, the various intuitionistic trends are connected by a loose affinity of their basic ideas only while considerably diverging in the detailed pursuance of these ideas.

For the first time in modern mathematics intuitionistic ideas appeared in Berlin in the seventies and eighties of the 19th century with Kronecker and a few of his pupils [3]). Their fight against the modern methods in the theories of real numbers and functions, as developed especially by Weierstrass and his school, proved on the whole unsuccessful, in spite of Kronecker's great authority; its principal and tragic effect was a considerable suppression of Cantor's ideas during two decades. The theories of real numbers, of functions, and of sets emerged victorious and, owing to their triumphal progress all over analysis, not even the alarm brought on by the antinomies of set theory at the beginning of the present century was felt initially, except at the very outskirts of mathematics.

However, with the first proof of the well-ordering theorem (1904) a general attack began mainly by French scholars, including a number of analysts who had themselves taken an active part in the application of set

1) The highly gifted British mathematician F.P. Ramsey (who dealt chiefly with foundational problems and died in 1930 at the age of 27) also took an inclination to intuitionism (according to Russell 31) after having opposed it in his earlier publications. Cf. furthermore Herbrand's attitude.

2) This already applies to Kronecker. Cantor however retorts: This is, so to speak, a question of strength . . . it has to turn out *whose* ideas are more powerful, more comprehensive, more productive, Kronecker's or mine; the success only will decide our conflict in due time. (Schoenflies 28, p. 12.)

3) See, in particular, Kronecker 1887. – Hölder's criticism of the arithmetical theories of the continuum dates from 1892; cf. Hölder 24, p. 194.

theory to the theory of functions, such as Baire, Borel, Lebesgue [1]); distinguished foreign mathematicians, e.g. Lusin, later joined this *Paris School* of intuitionists, while others, for instance Pasch, independently took a similar line. Poincaré even ventured, at one time, to reject set theory in general and to charge the methods of many classical branches of mathematics with the illicit use of impredicative procedures (pp. 181 f) [2]); despite the weakness of many arguments in his attacks on Zermelo and Russell, the authority of the foremost mathematician of his generation contributed to creating an atmosphere of alarm and to preparing the ground for the assaults of the Dutch school. The objections against the axiom of choice raised by almost all intuitionists – except for Poincaré whose view in this respect and in the question of mathematical existence in general is *sui generis* [3]) – attracted the attention of wider circles, even when many other scruples began losing ground in view of the successful axiomatization of set theory. The attitude of the members of the Paris School exhibits on many occasions examples of a penetrating analysis of foundational and set theoretical problems, though it shows a certain lack of consistency. Although their action is usually connected with the discussions concerning the axiom of choice, they also expressed their opinions on matters of a much wider philosophical bearing. A few examples may serve to illustrate some of their views.

(i) According to Lebesgue 05, p. 205, an object exists when it has been defined in finitely many words. Borel in this context speaks of "effectively defined" objects.

(ii) Lebesgue rejects the concept of an arbitrary number sequence, he only accepts those determined by a law.

(iii) Baire protests against the use of the power set of a given set as being well-defined.

Many of these issues were to be taken up later by Brouwer and his followers.

1) Yet most scholars of this trend, in contrast with intuitionists, were not eager to apply their principles in the practice of their own analytical researches.

2) However, the attitude taken by Poincaré (who died in 1913) was far less radical during the last years of his life. One should remember that later (see Poincaré 10) he gave a new proof for one of the strongest theorems of set theory, the non-denumerability of the continuum. (It is misleading that the posthumous collection of essays (Poincaré 13), some of which had appeared long before, bears the title *Dernières Pensées.*) – Here we may disregard the conventionalistic ingredient of Poincaré's mathematical philosophy, which is essentially restricted to geometry (and physics); cf., for instance, Rougier 20, Mooij 66.

3) Poincaré 08, Chapter V, part III.

At some points their criticism even surpasses Brouwer's, e.g. Borel in 47 p. 765 states that the very large finite involves the same difficulties as the infinite, thus anticipating ideas of Esenin-Volpin 61.

In 1907 the Doctoral Thesis of Brouwer marked the first step in a distinct intuitionistic direction for which he at first introduced the term *neo-intuitionism.* Besides several of his pupils in Amsterdam – hence the name *Dutch School* – a few others joined his circle, in particular (during the 1920's) Weyl, who before had taken an attitude of his own [1]) which in some respects resembled the Paris school. Especially in the decade beginning 1918 Brouwer unfolded his banner to an impetuous attack on traditional mathematics and to a thoroughly new foundation of analysis and set theory, which for some time deeply alarmed the mathematical world and provoked no less vehement reactions, occasionally producing something like Homeric talks; nevertheless quite a few of his opponents proved themselves considerably influenced by the new ideas. (While this is usually pointed out for the later works of Hilbert and his school from the 1920's on, one is in the habit of forgetting that Hilbert's earliest research in logic, which preceded Brouwer's Thesis and at that time remained almost unnoticed, might be regarded as an intuitionistic program more radical than intuitionism.) Brouwer's principles were far more revolutionary than those of earlier intuitionists, although in the theory of the continuum he was less rigid. His radical attitude is the reason why he found a good deal of both support and opposition in philosophical circles.

Describing separately the various trends of intuitionism and their implications with regard to the extent of "legitimate" mathematics would require a lengthy exposition. Therefore we shall mainly exhibit the *principles of Brouwer's intuitionism* and their consequences while including other trends in the literature and describing some of their chief ideas. The features in which Brouwer's intuitionism in principle differs from other intuitionistic trends are chiefly the following:

A) The conception of the *nature of mathematics* and mathematical existence, involving Brouwer's attitude towards the relation of mathematics to *language and logic* (§ § 2–4).

B) Brouwer's *theory of the continuum,* in particular the *choice sequences* (§ 5), based on his peculiar definition of set and species.

1) See Weyl 18 and 19 and the first part of 21. Later, however, Weyl swerved considerably from intuitionism, in particular from its intolerance of other opinions; cf. Weyl 26 (§ § 10–11) and his remarks at the end of Hilbert 28.

Literature on various aspects and modifications of intuitionism is abundant. Instead of listing all the relevant papers we refer the reader to the bibliographies of the following texts; Beth 59, Bockstaele 49, Fraenkel–Bar-Hillel 58, Heyting 55, Kleene 52, Kleene-Vesley 65, Mooij 66, Mostowski 66. However, literature pertaining to particular fields and problems, such as the *principle of the excluded middle* and *intuitionistic logic,* even when embedded in more general considerations, is given in the following sections. (For literature on the axiom of choice and its existential character, see Chapter II, § 4.)

Brouwer himself refused to acknowledge most interpretations of his attitude, including some of the Dutch school itself.

The prominent critics of intuitionism start either from logicistic or from formalistic principles; cf. Chapters III and V. Some other critical or polemic papers, partly of a philosophical nature, are given here [1]).

This historical exposition is a suitable opportunity for presenting a somewhat related attitude which, nevertheless, cannot be subsumed under intuitionism and which in many respects is even more radical than intuitionism. It dates back to about 1910 and has, not quite accidentally, the same geographical origin (Amsterdam) as intuitionism; its Dutch name is *Significa* (significs in English) and its character is largely relativistic and pragmatic. In its mathematical direction this attitude was developed by Mannoury [2]) but it was also applied to other sciences, including sociology and law. Some Dutch mathematical intuitionists, notably Brouwer and Van Dantzig, showed a close connection to signific ideas. However, these ideas have obtained practically no recognition outside the Netherlands, partly because of the rather obscure form of their exposition.

As shown in §§ 2 and 3, the intuitionistic criticism of traditional mathematics is directed against its treatment of *infinite* and *indefinite* aggregates, which form the chief part and the characteristic feature of mathematics on the whole; with respect to *closed finite* aggregates, even if their extent surpasses human imagination, the classical attitude is not altered.

1) Bernstein 19, Hölder 26, Ramsey 26 and 27, Cassirer 29 (in particular Part 3, Chapter IV refers especially to O. Becker), P. Levy 30, Ambrose 35 and 36. Curry 51 and Dieudonné 51 derive their arguments against intuitionism from *finitistic* attitudes.

2) Mannoury 25, 31, 34, Morris 38. The comprehensive expositions Van Dantzig 46, 49a and Vuysje 53 – the latter containing an extensive bibliography – give very clear surveys of significs in its original (Dutch) sense. The paper de Iongh 49 contains a criticism of intuitionism from the signific viewpoint and stresses the distinction between mental construction and its linguistic expression.

Therefore the conception of some mathematical statements as objective truths does not become meaningless in intuitionism.

Yet among the *philosophical* interpreters and adherents of Brouwer's doctrine one finds "anthropological-existential" influences [1]) which stress the subjective significance in comparison with the objective one. More radical steps in this direction, long before the rise of existentialism, were made by the signific school.

Significs starts from a critical examination of language and of methods of expression and denotation in general and is thus somewhat related to modern semiotics, especially to its pragmatic branch as conceived by certain Polish and American schools and by the Vienna Circle [2]). Significs, however, has in mind not the ordinary analysis of "meaning" but rather a theory of "psychic associations underlying human acts of language". The stress is laid on the perceptions and emotions connected with the terms or symbols. A characteristic feature of this rather psychological attitude is its conception of language as an activity by which man tries to influence the behavior and the vital power of others [3]). This applies not only to "acts of volition which demand obedience" but as well to mere "indicative" acts of communication, including the symbolic language of mathematics; according to Mannoury not even in the latter, the social significance of which should not be overlooked, is the subjective, persuasive, emotional character of language altogether lacking [4]). A mathematical formula would, then, not have a meaning *per se* but according to the purpose for which it is used. Thus Mannoury's mentioning of mathematics and mystics in one context is not accidental. The emotional and psychological moment, and this alone, is apt to explain the choice of principles (axioms) underlying mathematics and is still more perceptible in mathematical models of the external world as constructed in theoretical physics. Yet even Mannoury admits that the assertions of mathematics are to

1) Especially in O. Becker 27. Also in Borel's and Lusin's mathematical writings such influence is felt, quite independently of phenomenology.

2) Cf., for instance, Hahn-Neurath-Carnap 29, or Morris 38.

3) Here both a congeniality and a contrast with Brouwer's attitude (especially in 29), as described in § 2, become manifest. According to significs, not only the exposition of mathematics but even mathematics itself (mathematical thought) has a partly linguistic character.

4) It was mainly this attitude, related to a psychological foundation of logic (as in the schools of Heymans and Mach), which repelled most mathematicians. Cf. the ironical remarks in Jourdain 18 (p. 88) about evaluating the product 6·9 by working out, from the answers obtained in an examination among school-boys, the average to six decimal places; or about evolutionary ethics expecting to discover *what is good* by inquiring what cannibals have thought good.

a higher degree independent of sentiments of like and dislike than those of any other science. In this light one might interpret Kant's well-known maxim, measuring the scientific character of a doctrine by the applicability of mathematics to it.

§ 2. THE CONSTRUCTIVE CHARACTER OF MATHEMATICS. MATHEMATICS AND LANGUAGE

The fundamental thesis of intuitionism in almost all its variants says that *existence in mathematics coincides with constructibility*. While in intuitionism stress is laid on the very *character of mathematics* as involved by this thesis, in other trends at least the restrictions imposed on the field of admissible *mathematical procedures* derive from the thesis.

Mathematics is, according to Brouwer, not a *theory*, a system of rules and statements, but a certain fundamental part of *human activity*, a method of dealing with human experience, consisting primarily in the concentration of attention to a single one among our perceptions and in distinguishing this one from all others.

In Brouwer's first systematic exposition of his foundational views (his Ph.D. Thesis of 1907) it is argued that in mathematics (and all other activities of the intellect) the discrete and continuous aspects occur side by side as inseparable complements and that neither can be founded on the other [1]).

In Brouwer's words: "The primordial intuition of mathematics and every intellectual activity is the substratum of all observations of change when divested of all quality; a unity of continuity and discreteness". Later, after he introduced spreads, he abandoned this view and showed how to construct the continuum by means of choice sequences.

As the primitive act of intellectual construction in general, intuitionism conceives "the splitting-up of moments of life into qualitatively different parts which, separated only by time, can be reunited". Parallel to this general conception, in a remarkable similarity to a well-known idea of Plato's, the primitive act of *mathematical construction* is maintained to be "the process of stripping this splitting-up of any emotional content until the intuition of abstract bi-unity remains" [2]). Accordingly, mathematics in its entirety consists of mental processes that can be built up by an unlimited sequence of

1) Brouwer 07, p. 8. Cf. the clarification of many obscure formulations of Brouwer (particularly in the programmatic writings 07 and 29) in Van Dantzig 47 and Heyting 55.

2) Brouwer 12.

steps repeating such primitive mathematical acts indefinitely – a definition which is more comprehensive than any other and includes logical processes as well as many procedures of natural sciences. No other science – certainly not logic or philosophy – can, then, serve as a basis for mathematics. According to this conception, the use of mathematical procedures is not the exclusive privilege of *science*, but is to be found also, spontaneously and often subconsciously, in everyday thinking.

Obviously, this intuitionistic "definition" of mathematics does not supply us with an exact determination of what procedures, among those found in traditional mathematics, can be regarded as "constructive" and therefore legitimate within Brouwerian mathematics. Yet the apparently dogmatic character of the above definition is mitigated by certain remarks (above, p. 213) about the necessity and the sufficiency of intuitionistic restrictions.

The emphasis laid on the construction of mathematical entities and even the *identification between existence and constructibility in mathematics* is by no means a novelty. Definitions in mathematics usually raise the question whether an object satisfying the definition can be constructed (effective examples). Also the classical problems of Greek geometry – notably duplication of the cube, trisection of the angle, squaring of the circle – are problems of construction, in this case, of construction with ruler and compasses. Hilbert who was fought by intuitionists as the protagonist of their opponents, after having first solved a fundamental problem of invariants by existential methods [1]), found it worthwhile to look also for constructive solutions of related problems which in principle were included in the existential procedure.

In § 4 of Chapter II some procedures of a purely existential character were exhibited. Such procedures are found chiefly in analysis, geometry and set theory, but they are not altogether missing even in arithmetic and algebra. Their common feature is that certain mathematical objects (numbers, correspondences, functions, sets etc.) are shown to exist not on account of their derivation from simpler objects by a step by step construction but by means of an argumentation which appeals to logical compulsoriness. Mostly this is done in the way of an alternative between mutually exclusive possibilities in view of the principle of excluded middle (see below § 3), or by showing that the non-existence of an object with the desired quality would involve a

1) This first proof of the existence of a finite set of invariants (1890) was called by Gordan "theological" because it made essential use of an existential reasoning ("there must be an invariant of the lowest possible degree") from which an actual invariant of the kind in question cannot be constructively derived.

contradiction (whereas the nature of the contradiction does not yield a method of constructing a suitable object). A well-known instance, which appears at the very beginnings of analysis, is the theorem that every infinite bounded set of points has at least one accumulation point, proven by iterated bisection. During the 19th century demonstrations of this type were not only quite current but even specially appreciated for their acute and bold character. On the other hand, they have the disadvantage of not granting an insight into the nature of the object secured; if, for instance, the object is a number, a proof of the existence of a number with the desired property need not enable us to estimate, say, the magnitude of the number.

Intuitionism and most other intuitionistic trends, admitting construction alone as a legitimation for existence in mathematics, deny such procedures any binding power and do not accept the existence of the objects concerned as long as they are not secured in a constructive way. In particular, intuitionists maintain that identifying mathematical existence with non-contradiction would mean degrading mathematics to a mere game [1]). Only Poincaré, though in other directions taking a most radical attitude, regarded freedom from contradiction as a sufficient legitimation for existence (as does the formalistic school) and even accepted the principle of choice.

This principle and the well-ordering theorem resting on it presumably constitute the most characteristic instance of a purely existential mathematical statement [2]). In fact, the latter theorem, which asserts the existence of procedures (or sets) that well-order a given set without showing how to obtain such a procedure, is the prototype of an intuitionistically meaningless statement. The uselessness of the well-ordering theorem for ascertaining, say, the place of the power of the continuum in the series of alephs is interpreted as a mere symptom, to be expected in advance, of the theorem's voidness of meaning.

Far beyond this extreme instance all existential propositions are viewed by intuitionists with suspicion. The content of such propositions is not clear in so far as no actual effective procedure is involved in finding the object, the existence of which has been asserted. In many practical instances, however, a construction of the object can be abstracted from the proof of the existential statement. The constructive content of a statement "there exists

1) The phenomenological-existentialist brand of philosophy (cf. O. Becker 27, for instance p. 442) considered intuitionistic mathematics to be a science that "discovers actual phenomena which are comprehensible to original and adequate intuition and can be existentially interpreted".

2) Cf., however, Suetuna 51–53. (The concept of 'set' used there is somewhat obscure.)

an object *a* with the property *P*" consists of a *construction* of an object *a* and a *proof* that *a* has the property *P*. In many instances, however, the abstract has been *derived* from a statement proper; for instance, from the statement '2 is an even prime' we can derive the abstract 'there exists an even prime'.

Brouwer has compared the situation to that of coming across a document which says that a treasure is hidden somewhere. This is not a proper statement as long as the spot where the treasure is hidden has not been specified; yet the document may cause a treasure-hunter to obtain a statement of the form '*here* a treasure is hidden' by stimulating him to dig and find the treasure. From this we may proceed to the abstract which, however, has its only justification in the underlying singular statement.

In these and similar cases it is fairly obvious which statements and procedures are *not* constructive. But one can hardly maintain that the positive meaning of "constructiveness" is sufficiently clear with respect to procedures employed up to now or to be employed in the future in what has always been understood as "mathematics" – save for mathematical induction (§ 5) 1). As a matter of fact, several attempts and suggestions of defining 'construction' were made, chiefly from outside intuitionism 2). These suggestions only seem to indicate that the concept of construction has a *relative* and not an absolute character; there are higher (stronger) and weaker degrees of the concept 3) and this even applies to the domain of arithmetic 4). Thus the question loses its dogmatic character and, analogous to the axiomatic procedure, assumes the form: what parts of mathematics, or of a certain mathematical branch, can be obtained from a given starting-point by means of such-and-such "constructive" methods? In fact, an opposite (dogmatic) attitude should have induced the Greeks to infer from their conception of

1) For the meaning of 'construction' within the Paris school of intuitionism, cf. §5.

2) Among the earlier papers, before Kleene's book of 1952, we should mention Menger 28 (p. 225), 30, 31, Rosser 36, Hermes 37; particularly Kleene 43 (No. 16), 45, Post 44, D. Nelson 47 and 49. Rasiowa 54 gives a topological explanation.

3) See Mostowski 59; cf. also Heyting 58 and Dienes 52 who distinguishes between different degrees of rigor, the first of which coincides with intuitionism.

4) For instance, with respect to a sequence of integers there is a difference in constructiveness between the existence, for every n, of a rule enabling us to calculate the nth term of the sequence (or the first n terms) by means of a finite number of steps, and the existence of a rule which enables us to calculate, by means of a finite number of steps, the nth term (or the first n terms) for every n. The former possesses a weaker degree of constructiveness than the latter. The intuitionistic interpretation of logical connectives bears on the above considerations.

The indicated difference in degree of constructivness actually figures in everyday mathematics, cf. Rogers 67, p. 68, Th. XIV.

geometrical construction that, in general, there exists no angle equal to the third of a given angle.

Brouwer impetuously opposed this opinion and maintained that there is an *absolute* concept of constructiveness and that this concept settles what is comprehended in mathematics and what parts should be excluded as belonging to pseudoscience. Indeed quite a number of constructive procedures have been described and a certain "mathematical attitude of mind" may be suggested by general hints. But any list of constructive mathematical principles is not only incomplete but also *necessarily remains incompleteable* [1]); for, within the limits of the general notion given above for "mathematical" procedures, we must not limit the liberty of creative mind to further extend its constructive faculties. Hence we can never predict what special ways of construction might be needed for reaching a particular goal. The situation may be compared to a mountaineering expedition for which you may fairly well give a list of "admitted" alpinistic procedures and prohibit others, for instance driving nails into rocks, while you can never foresee all devices that may become life-saving, and therefore admitted, at a certain stage of an ascension.

The conception according to which mathematics is the *mental activity* described above rather than the oral or written *expression* of such activity has a decisive influence on *the relation between mathematics and language* [2]). Apparently the process of thinking, i.e. of mental creation, is not intrinsically connected with a linguistic expression; only for the exchange of thought (the communication of ideas) do we need spoken language or its written equivalent. In this context the weakness of mathematical language in comparison with mental construction is stressed, for any language is, says Brouwer, vague and prone to misunderstanding, even symbolic language (since mathematical and logical symbols rest on ordinary language for their interpretation). Hence mathematical language is ambiguous and defective; mathematical thought, while strict and uniform in itself, becomes subject to obscurity and error when transferred from one person to another by means of speaking or writing. It would therefore be a fundamental mistake to analyze mathematical language instead of mathematical thought. This pointed distinction between construction in mind and its expression in language is contrary not only to logicism (Chapter III) and metamathematical method (Chapter V) but also to the view, uninfluenced by mathematical considerations, of leading

1) In a certain sense, one might consider this attitude (and the maintained impossibility of including mathematics within a formal system) to be confirmed by Gödel's incompleteness theorem (Chapter V, §7).

2) See, in particular, Brouwer 29, 49, 54.

philosophers from Plato through Leibniz to W. Von Humboldt and E. Cassirer who assert that all abstract thinking is dependent on language [1]).

This attitude not only applies to mathematical theorems (and definitions) but still more so to mathematical *proof. The construction itself constitutes the proof* and one should drop the usual idea as though the demonstration were intended to convince the reader of the soundness of an argumentation by basing it, step after step, on recognized principles. Since construction is an activity it can not be communicated adequately; instead one uses the fragile crutch of words or symbols. There is neither exactness nor safety, says Brouwer [2]), in the "transfer of will" by means of language, and this also applies to the transfer of will expressed by the construction of a mathematical system. Hence *for mathematics there exists no safe language* which excludes misunderstanding in talk and prevents mistakes of memory; this defect is regarded as support of the contention directed against formalism that the exactness of mathematics should be found not "on paper" but "in the mind of man".

Brouwer in his later papers fully accepted the mathematical consequences of the above views. The fact that language is a deficient vehicle for the purpose of transferring mathematics naturally fits into a solipsist philosophy [3]), which appears to be the most suitable background for the procedures involving the mental activities of "a mathematician" (the "creative subject"). It must be stressed that the intuitionistic thesis about mathematics and language is inacceptable to the majority of mathematicians for the very reason that it makes mathematics a private affair rather than an organized intersubjective phenomenon.

Many outstanding mathematicians from Sylvester to Poincaré have compared mathematics to music. Intuitionism adds a further and more penetrating analogy; as a composer may teach a beginner how to compose a symphony – not by merely teaching harmonics but by describing how *he* had done it – so a mathematician would initiate a student in the constructive mystery of mathematical production, while the demonstration has a secondary value only.

The intuitionistic conception of language directly influences the problem of the *antinomies,* which are considered mere combinations of

1) Cf. Schlick 26. On the other hand several philosophers (for instance Greenwood 30), sometimes with a psychological argumentation, accept the distinction between mathematics and its exposition and even extend it.

2) Brouwer 29, p. 157.

3) Cf. Heyting 56, VIII, Kreisel 65, p. 119.

words, void of meaning and without constructive content. In a constructivistic conception of mathematics, based on intuitionistic principles, no meaningful formulation of the antinomies is possible. Hence for intuitionists there is, properly speaking, no problem of antinomies. It is difficult duly to appreciate the intuitionistic attitude towards mathematical language without plunging into Brouwer's general opinions on human civilization. The following brief description may serve to present the reader with an outline of Brouwer's views. [1])

The mind of an individual experiences sensations. The individual identifies certain sensations and starts to recognize iterative sequences of sensations with the property that if one of these sensations occurs the others are expected to occur also, in a specific order. Such sequences are called *causal sequences.* The individual will try to use his knowledge of causal sequences to obtain certain desired sensations by producing a sensation that precedes the desired sensation in a previously experienced, causal sequence. This shift from end to means is called "cunning act" by Brouwer. Certain complexes of sensations are independent of the order in time, and their dependence on the individual is small or nil. These complexes are called *things,* e.g. external objects, human beings. The whole of things is called the *external world* of the individual. The relation of the individual with other individuals (which are again sensation complexes, i.e. things) is described by identification of causal sequences, observed by the individual, of itself and of other individuals. This identification justifies the term "acts of other individuals". It is observed by the individual that causal acts (i.e. cunning acts based on knowledge of causal sequences) of itself and other individuals are highly dependent. Hence the need for cooperative causal acts arises. This is where scientific thinking, as an economical way to deal with large groups of these causal acts, is introduced. Scientific thinking as such is based on *mathematics.* The genesis of mathematics takes place at the creation of two-ities. Brouwer construes the two-ity from a *move of time*, which is a concept defined with respect to the individual. Namely: a move of time takes place when one sensation gives way to another. Both sensations are retained in their proper order and constitute a two-ity. The individual abstracts all quality of this two-ity and uses it as the basic ingredient for iterative processes. These iterative procedures can create predeterminately or more or less freely infinite proceeding sequences of mathematical entities previously produced.

The place of language in Brouwer's conception is that of a device for the

1) These views are expounded in Brouwer 29, 49. Cf. Mannoury 34 (see above, p. 218), Heyting 67.

transmission of will of the individuals that make up society. With respect to mathematics Brouwer considers language, including logic, as a phenomenon accompanying the wordless mathematical construction processes of the individual. As a consequence Brouwer maintains that logic does not precede mathematics but, on the contrary, is preceded by mathematics.

§ 3. THE PRINCIPLE OF THE EXCLUDED MIDDLE

In the following sections we shall survey the consequences, both for logic and mathematics, of the fundamental thesis described in § 2. §§ 3 and 4 deal with the consequences for logic, §§ 5 and 6 with mathematics. While in the present section an informal exposition is given, including references to the extensive literature on the subject, § 4 deals with the intuitionistic *logical calculus*.

From the first [1]) Brouwer raised the question which among the principles of Aristotelian logic could be retained from the intuitionistic point of view. His answer is, the principle of contradiction but not the principle of the excluded middle (*tertium non datur*) in its general sense. This constantly repeated claim as to the *invalidity of the tertium non datur in mathematics* and in infinite domains in general quickly became the centre of prolonged dispute between adherents of intuitionism and its opponents. (The Paris school and other intuitionistic trends did not accept the attitude of the Dutch school towards the excluded middle.) To perceive the great importance and the wide application of the principle of the excluded middle in mathematics let us just recall that it is the fundament of *indirect proof*. The heated discussions concerning the validity of the *tertium non datur* may erroneously have created the impression that after all the intuitionistic revolution boils down to founding mathematics on just another logic. Actually intuitionists consider the *tertium non datur* as a symptom of the disease of classical mathematics, and they do not intend to cure symptoms. Anyhow, the *tertium non datur* turned out to provide the Dutch school with a pointed slogan.

1) Brouwer 08. Cf. Brouwer 54, where he says "the long belief in the universal validity of the principle of the excluded third" . . . is considered "as a phenomenon of the history of civilization of the same kind as the old-time belief in the rationality of π . . .".

As a historical detail we may point out that in his Ph.D. thesis Brouwer defended the priority of mathematics over logic, but at the same time accepted the *tertium non datur* (Brouwer 07, p.132). In 1908 he reconsidered the *tertium non datur* (at the instigation of Mannoury (?), see Van Dantzig 57) and questioned its validity. As a matter of fact, the use of the principle of the excluded middle for *finite* domains was never questioned [1]). One should, however, be careful not to apply the principle in that case rashly. The statement "Fermat's conjecture holds or does not hold for the first 10,000 natural exponents" is only apparently an instance of the *tertium non datur* for a finite domain, since in the expanded version there occurs an unrestricted number quantifier. Brouwer's formulation reads: For every assertion of possibility of a construction of a bounded finite character in a finite mathematical domain the principle of the excluded middle holds [2]).

Note that the notion "finite" is used in its strong, intuitionistic sense. That is, a set is *finite* if there is a constructive one-one correspondence between the set and an initial segment of the natural number-sequence. Consider the set consisting of all natural numbers n with the property that the nth decimal of π is not preceded by a string of 9 sevens if a string of 9 sevens occurs in the expansion of π, otherwise let the set consist of the number 1 only. This set would be recognized as being finite by traditional mathematics. An intuitionist, however, has no means to establish the aforementioned correspondence, hence for him the finiteness of the set is an open problem.

Since the principle of the excluded third does not present difficulties in the essentially finite case, let us consider statements involving infinitely many objects.

We will examine a number of simple examples. Consider the following expressions, where n ranges over the positive integers.

1) $2^n + 1$ is a prime number;

2) the nth digit after the point in the decimal expansion of $\pi = 3.14159...$ is 7;

1) It must be pointed out, however, that generally a certain amount of idealization is involved. Instead of actually carrying out tests, etc. one convinces oneself that it is possible to do so.

2) Cf. Brouwer 52–53, p.141.

3) every geographical map containing n regions (countries) can be colored [1]) by means of at most four colors;

4) $2^n + 1$ being a prime number, there exists a greater prime number of the form $2^m + 1$ $(m > n)$ [2]);

5) n and $n + 2$ forming a pair of twin primes (i.e. a pair of prime numbers the difference between which is 2), there exists a larger pair of twin primes;

6) there exists a triplet (x, y, z) of natural numbers (a "Fermat triplet") such that

$$x^{n+2} + y^{n+2} = z^{n+2} .$$ [3])

We add the following statement which is related to 2) yet contains no free variable:

7) There exists an integer n such that the nth digit in the decimal expansion of π and its six successors (the $(n + 1)$st, $(n + 2)$nd, ..., $(n + 6)$th digit) all equal 7.

By universal quantification over n we proceed from 3) to the statement:

8) Every geographical map can be colored by means of at most four colors.

By giving n a definite value we may also turn 1)–6) into statements and then raise for each of 1) to 8) the question whether the statement is true or false. That is to say, we will examine a number of particular instances of the general principle "every meaningful statement p is either true or false", i.e. $p \vee \neg p$ (*tertium non datur*).

Let us consider our examples and weigh the grounds for accepting or refuting the principle in each separate case.

In the cases 1) and 2), it may at the present stage of research prove difficult, for large values of n, to reach the actual decision whether the statement is true or false; the same applies to 3) when $n > 37$ (for $n \leqslant 37$ it is demonstrably true). Yet *in principle* the decision can be reached; for a given n, only a finite number of tests is necessary to ascertain whether n does, or does not, have the property in question. For this very reason we may *anticipate* the result; n either does or does not have the property, i.e. the statement is either true or false. Therefore a theorem proved on either of the assumpt-

1) On condition that different colors are given to any two adjacent regions, i.e. to regions that have in common a portion of their boundaries (and not only isolated points). It is assumed that the map is drawn on a plane or on the surface of a sphere.

2) Mersenne numbers, i.e. primes of the form $2^n - 1$, yield examples similar to 1) and 4).

3) The exponent is written in the form $n + 2$ because, as is well known, there exist (infinitely many) solutions (x, y, z) of the equation $x^2 + y^2 = z^2$.

ions that $2^{131072} + 1$ is prime and that it is composite would be true also in the eyes of Brouwer, though we do not know which assumption is true.

The same argument, which relies upon the possibility of testing by finitely many steps, allows us to state, for example, with respect to the population of London at a certain moment that either each inhabitant of London is at most 99 years old or that there exists at least one such inhabitant whose age is 100 years or more. [1]). Thus, according to the Dutch school, *the principle of the excluded middle* is just a statement among millions of others, having no claim whatsoever to universal validity. In those cases where the principle holds a mathematical proof can be provided. In particular one can prove the validity of the principle in the case of a finite domain by referring to a finite examination procedure, be it an indiscriminating search procedure or an economic algorithm.

On the other hand, there is not the slightest argument in support of such a conclusion when an infinite aggregate is concerned. One should beware, says Weyl [2]), of the idea that, after an infinite aggregate had been defined, we may now proceed as if its members were spread before our eyes and we can check them one by one to find out whether there is among them an element of a certain kind; though this idea is perfectly legitimate for finite aggregates, it makes no sense with respect to an infinite aggregate.

Obviously such is the situation in the instances 4)–8) given above. The instances 4) and 5) resemble each other. It is well known that, if m is divisible by an odd integer >1, $2^m + 1$ is a composite number; among the other values of m, i.e. among the powers of 2, $m = 2^4$ is the largest integer for which $2^m + 1$ is known to be prime (as is the case for $m = 2^0, 2^1\ 2^2\ 2^3$), $m = 2^{1945}$ the largest one for which $2^m + 1$ is known at present to be composite, while between these two values of m over forty have been successfully tested (among them 2^α with $\alpha = 5$ to $\alpha = 12$ but not $2^{17} = 131072$). At present, then, it is not only unknown whether the set of primes of the form $2^n + 1$ is finite or infinite but even whether it contains just six or at least seven members; that is to say, whether $2^{16} + 1$ is the greatest prime of this form or not. 5) is also unsolved for the time being: we do not know whether there is a *last* pair of twin primes (which seems improbable) or whether there are infinitely many such pairs.

In the case of 6) – contrary to 1)–3) – not even for a given value of n

1) One should keep in mind that this type of example has slippery aspects, e.g. the finiteness of the set under consideration is often debatable.

2) Weyl 21, p. 41. Cf. the distinction made by philosophers (e.g., Husserl) between "individual" and "specific" generality; see Carnap 37, § 15.

need a decision be obtainable by a finite number of steps, for to achieve it one would have to test infinitely many triplets (x, y, z). In fact, no Fermat triplet is known and it seems probable that none exists; but while, for infinitely many values of n, the non-existence of triplets has been proved there remains another infinite set of integers n for which we do not know the answer.

Finally, 7) is somewhat similar to 4) and 5) and is related to 2) though not quite in the same way as 4) is related to 1). For a given n one can in principle decide not only whether the nth decimal of π, a_n, equals 7 but also whether $a_n = a_{n+1} = \ldots = a_{n+6} = 7$ or not. Yet 7) requires an existential quantification over n.

Likewise, 8), on account of n ranging over the infinite set of natural numbers, presents a case which has neither been proved nor refuted at the moment.

The above examples may have falsely created the impression that in the case of an infinite domain there is no hope for intuitionists to establish the validity of statements involving unrestricted quantification. That this is not the case is illustrated by the following example: For every prime number p there exists a larger prime number q. Indeed we have known since Euclid that the set of primes is infinite, but we know more: there is an effective finite procedure for calculating "the next prime". So the above statement is intuitionistically valid because of the existence of an effective procedure which provides for each p the required q.

We have discussed the examples 1)–8) at length in order to make Brouwer's attitude towards the principle of the excluded middle conspicuous. Take e.g. 4) with $n = 16$. In order to show the validity of the *tertium non datur* we have to show that *either* (a) there exists a prime of the form $2^m + 1$ with $m > 16$ *or* (b) there is no prime of the form $2^m + 1$ with $m > 16$. The latter is equivalent to (b′) all numbers of the form $2^m + 1$, with $m > 16$, are composite. However, neither (a) nor (b′) have been established until now. So the validity of this instance of the excluded third is open. This rather negative situation has sometimes been called *a third one.* It has a temporary character, for tomorrow we may discover a suitable m to validate (a), or succeed in finding a general proof for (b′). In intuitionistic literature one often encounters examples of the kind mentioned. Preferable to such special problems is a general instance of undecidability in the mathematical branch under consideration [1]); then the "third case" is justified by the assumption that there is no general decision method.

1) For branches that contain the notion of choice sequence (or real number) there are stronger means to refute the principle of the excluded middle (see § 6).

The name "third one" was doubtless an unfortunate choice, as it led some philosophers to believe that intuitionists use a three-valued logic [1]). As a matter of fact, the third case is of quite a different nature than the two others. Apart from technical reasons [2]), a three-valued interpretation conflicts with the basic meaning of the logical connectives (see § 4).

Brouwer's rejection of the principle of the excluded middle basically rests on his interpretation of the logical connectives. Let us examine a simple instance of the *principle of excluded middle* from both Platonist and Brouwerian points of view. Let p be the statement $\exists x A(x)$, where x ranges over the set of natural numbers. We first assume Platonist principles. The truth of $\exists x A(x) \vee \neg \exists x A(x)$ follows from the fact that the set of natural numbers is a completed whole and hence admits of inspection of all instances $A(0), A(1), A(2),\ldots$ simultaneously (e.g. by a Platonist supermind). Clearly, either for some n a valid instance $A(n)$ is encountered or for no $n\, A(n)$ is valid; the latter is equivalent to $\neg \exists A(x)$, so $\exists x A(x) \vee \neg \exists x A(x)$ holds. Note that this argument is an immediate generalisation of the inspection by cases for a finite domain, in fact it comes down to an infinite search procedure, considered as completed.

Now let us consider $\exists x A(x) \vee \neg \exists x A(x)$ from an intuitionistic point of view. In order to establish the validity of $\exists x A(x)$ one has to provide a construction of a natural number k and a proof of $A(k)$. The validity of $\neg \exists x A(x)$ is established by showing that it cannot be the case that $\exists x A(x)$. But if for no individual n $A(n)$ holds, then for all n $A(n)$ does not hold and vice versa. I.e. $\neg \exists x A(x)$ and $\forall x \neg A(x)$ are intuitionistically equivalent. So for the validity of the second part of the disjunction a uniform proof-schema $\pi(x)$ is required, such that $\pi(x)$ provides for each individual n a proof $\pi(n)$ of $\neg A(n)$, i.e. $\pi(n)$ proves that $A(n)$ is false. Now we sum up the validity of the principle of the excluded middle for the statement $\exists x A(x)$ as follows: Either there is construction of a natural number k and a proof of $A(k)$, or there is a uniform proof that shows the falsity of $A(n)$ for each natural number n. It is not clear at all that this last statement holds (indeed for variables ranging over reals we can show it to be contradictory), hence the intuitionist has no intuitive grounds for accepting the principle of the excluded middle. There are classes of statements p such that $p \vee \neg p$ holds, e.g. quantifierfree formulae of arithmetic. In general, however, each instance must be examined. Brouwer gives the following directions with respect to the principle of the ex-

1) Cf. Barzin-Errera 27, Heyting 33.

2) According to Gödel 33a, no finite-valued interpretation is adequate for intuitionistic propositional logic. Cf. Schmidt 60, p. 369.

cluded middle [1]): The rejection of thoughtless application of the principle of the excluded middle and the recognition of the facts that 1) the investigation of the grounds of justification and domain of validity of the principle constitute an essential object of foundational research in mathematics, 2) the domain of validity in intuitive (contentual) mathematics comprises finite systems only.

Whenever the range of the free variable that occurs in a general statement is *finite* – more precisely, when it contains only a finite number of objects each of which can be exhibited individually – the Dutch school admits that negating the statement produces an existential statement. For instance, given a finite group, negating its commutativity means that there is at least one pair of members x_0, y_0 such that $x_0 y_0 \neq y_0 x_0$. But then the existential statement obtained by negation has a constructive character since the general statement is but the *conjunction of a finite number of decidable statements*. In the example given above "all inhabitants of London are at most 99 years old" the meaning is: a_1 is at most 99 and so is a_2 ... and so is a_n, where the set $\{a_1, a_2, \ldots, a_n\}$ represents the population of London. Hence negating the statement produces the disjunction: a_1 is at least 100 or a_2 ... or a_n (in the non-exclusive meaning of 'or'). By this procedure at least one counter-instance can be constructed and the supposedly existential statement becomes intuitionistically legitimate. If, however, the variable "x" of the general statement $\forall x A(x)$ ranges over an infinite domain, there is no analogue of the replacement of a universal quantification by a finite conjunction. Hence we must look for the meaning of $\neg\forall x A(x)$ and $\exists x \neg A(x)$. $\neg\forall x A(x)$ has the intuitive interpretation (i) it is impossible that there is a proof of $\forall x A(x)$, $\exists x \neg A(x)$ has the intuitive interpretation (ii) an n can be computed such that $A(n)$ has no proof. It is fairly plausible that from a weak property like (i) no computation as required in (ii) can be abstracted. (ii) gives more information than (i)! We will return in § 4 to this problem.

Hence the use of the principle of the exluded middle in natural science would depend on the conditions of the finiteness and the atomistic structure of the universe [2]).

To avoid misunderstandings of the present point the importance of which has been stressed time and again by the Dutch school, let it be clear that the

1) Brouwer 28.

2) Brouwer 24. Yet he expressly adds the remark that, as far as mathematics is applied in natural science, the fulfilment of these conditions referring to nature does not release us from the intuitionistic purification of the mathematical procedures inasmuch as these refer to infinite and continuous domains.

legitimacy of negating general statements and of affirming existential ones, hence of applying the *tertium non datur*, does not depend so much on the *finiteness* of the domain as on the *definite limitation* of the domain's extension. This means the exclusion not only of choice sequences (§ 5), but also of certain domains of ordinary life and processes of nature; say, the domain of all plants.

An example of a similar kind is obtained as follows. Certainly the statement 'for all x, x has eight different greatgrandparents' is false when 'x' ranges over all men; it is equally false when 'x' ranges over the descendants of a certain person who still has procreative descendants. Yet the existence among those descendants of a person that has less than eight greatgrandparents cannot be stated either. This example is related to the question, broached allegedly by Aristotle, about the validity of the principle of the excluded middle for future events [1]); in particular for events subject to the freedom of will, or to indeterminism in nature.

It is true that the "ordinary" mathematician or philosopher as well as "ordinary common sense" is inclined to oppose the intuitionistic attitude towards the principle of the excluded middle. No matter whether a decision can be reached now or will be reached at any future time, the "objective" state of affairs – so they would maintain – necessarily causes the statement to be either true or false. Applied to the example 4) of p. 229, either there exist only six primes of the form $2^n + 1$ or there exist at least seven. Such argumentation, answers Brouwer, relies upon a double misunderstanding. First upon the idea that mathematics deals with external facts or with Platonic ideas existing independently of the mathematician's activity; yet just this activity and nothing else creates mathematics, as follows from what was said in § 2. Even if one believes in some "objective truth" its character is metaphysical and mathematical proofs cannot be based upon metaphysical arguments. Secondly, the assertion of the *tertium non datur* is, consciously or not, influenced by an unjustified generalization to infinite domains of a procedure which is legitimate for finite domains only; the process of perusing an infinite domain can never be finished, hence its "result" must not be anticipated, not even in principle.

Take the case of a theorem that is proved under either of the assumptions that the statement p is true (six primes only) and that p is false (at least seven primes); not even in this case may you accept, according to Brouwer, the theorem as settled, as long as the statement is not either demonstrated by a

1) For these questions, in particular that regarding future events, see for instance, Schlick 31 (p. 158), A. Becker 36, Toms 41.

general method – that is to say, by comprehending the infinite domain through a finite or intuitionistically legitimate procedure (such as a characteristic property, mathematical induction, etc.) – or refuted by a counter-instance. Hence the indirect method of proving is inadmissible in general. That is to say, indirect proofs generally prove negative facts. The usual procedure is to depart from $\neg A$ and derive a contradiction; in ordinary mathematics one then concludes A, in intuitionistic mathematics one only concludes $\neg\neg A$. A simple mathematical example (due to Heyting) may elucidate the distinction between A and $\neg\neg A$. Define $a_n = 3$ if up to the nth decimal of π no string of 7 consecutive sevens has occurred, $a_n = 0$ else. The sum $\sum_{n=1}^{\infty} a_n 10^{-n}$ is a real number a. We show that it is impossible for a not to be irrational. Suppose a is irrational, then it is impossible that a is of the form 0,333...3, so no string of 7 sevens occurs in π. Hence $a = 1/3$ but this contradicts the supposition that a is irrational. Therefore a is not irrational. The question whether a is rational cannot be answered, since we cannot compute p and q (integers) such that $a = p/q$ as long as we have no definite knowledge concerning the status of the mentioned string of sevens. Summing up, we know that $\neg\neg$ (a is rational), but we have no evidence to assert that a is rational. Counterexamples of this kind are customary in intuitionism; they are called *weak counterexamples* because they are based on insufficient knowledge *at this moment*. Sometimes one can show that a certain statement P is absurd, that is prove $\neg P$; then one calls P a *strong counterexample.* The status of propositions proved by classical means, such as the *tertium non datur,* will be discussed in §4. Some authors have tried to refute the rejection of the principle of the excluded middle by tracing out a would-be contradiction. Such attempts could not but fail; for by *omitting* one of the principles or rules of a logical system one obtains a *restriction* of its operational field and of its consequences and not an extension which might involve contradiction. Against certain misunderstandings it should be stressed that Brouwer *does not intend to replace the principle of the excluded middle by its negation*, i.e. to introduce a third case proper, opening the way for a principle of *quartum non datur*[1]). The "third" case is one for which nothing positive can be stated, which excludes its coordination with the two other cases.

1) Whereas Brouwer's attitude has to some extent revived the discussions on three-valued logic, it should be stressed that the admission of many-valued logics does not imply Aristotelian logic to be contradictory – just as the consistency of non-Euclidean geometries does not mean that Euclidean geometry contains a contradiction.

Accordingly, as confirmed by a symbolic exposition of intuitionistic logic (§ 4 or Kleene 52), the (predicate) logic of Brouwer is not in *opposition* to classical logic but constitutes a *part* of it. This torso, in consequence of its renouncing the use of the *tertium non datur*, loses a good deal of the simplicity and lucidity of traditional logic. On the other hand, even for those who oppose the Dutch school it is interesting to investigate the remaining part of classical logic and to examine how far one can proceed by its means; thus one obtains a domain apt to show the independence of the principle of the excluded middle.

Regarding the principle of contradiction, Brouwer points out that according to his attitude this principle is not actually used when one proves an impossibility by indicating a contradiction. Virtually this only means the failure of a mathematical construction which should satisfy certain conditions [1]).

Throughout this section Brouwer's attitude towards the principle of the excluded middle has been illustrated by examples referring to integers. Of course it applies all the more to *sequences* or *sets of integers*. Instead of the question "Does there exist in a given set of integers an integer with a certain property?" we then ask "Does there exist in a given set of sequences of integers (for instance, in a given set of decimals) a sequence with a certain property (meaningful for sequences of integers; for instance, periodicity)?". The criticism of intuitionists regarding a question of the latter kind does go further; for here not only do they reject the use of the *tertium non datur*, but the very *notion* of an arbitrary sequence of integers, as denoting something finished and definite, is declared illegitimate. Such a sequence is considered to be a "growing" object only and not a "finished" one (see § 5). Therefore it is in general problematic to answer the question of its having such-and-such properties. In § 5 we will examine the peculiarities of the notion of sequence.

The abandonment of the principle of the excluded middle, besides its significance for logic (see § 4), has a psychological effect extending across mathematics as a whole, namely on *the conviction that every mathematical problem can in principle be solved*. To be sure, in another sense this conviction has been shaken by Gödel's incompleteness theorem (Chapter V). But then the undecidability asserted in this theorem refers to a definite basis from which the attempt to solve the problem is being made. Intuitionism, however, rejects that conviction as a quite groundless belief for any basis and emphasizes this rejection by pointing out that one cannot make a distinction be-

1) Brouwer 07 (p. 127) and 08.

tween constructively solving a problem and the "intrinsic" or "objective" truth about the problem, as it were, independently of its solution – the latter being meaningless, a metaphysical speculation drawn from Platonism.

It is quite surprising – and scholars of the present generation are apt to forget it – how thoroughly mathematical outlook and trust have changed within a few decades. In 1900, at the Second International Congress of Mathematicians in Paris, Hilbert opened his historic lecture on (unsolved) mathematical problems [1]) with the proud words which then expressed a common belief of mathematicians: all of you certainly share the conviction that each definite mathematical problem necessarily admits of a strict settlement, and you are aware of the continuous call addressing you "behold the problem, seek its solution; you can find it by pure thought". (The term 'solution' of course includes negative solutions as for the classical construction problems of geometry (unsolved by the Greeks), and independence proofs as for the axiom of parallels, or for the continuum hypothesis – see Ch. II.) [2])

The certainty of a solution seems to distinguish mathematics from other (inductive) sciences where the ghost of everlasting failure and of a final *ignorabimus* troubles the mind of the scholar. It was this certainty which induced mathematicians to continue, through many centuries, their attempts to solve problems like that of Fermat's last theorem.

The conviction of the solvability of all mathematical problems was not based upon properly logico-mathematical arguments but mainly upon actual scientific experience and the reflection that the concepts of mathematics, hence also its problems, originate from the sphere of human thought and (internal) intuition only – in contrast with other, especially natural, sciences where external experience is essential. Accordingly, human reason ought to be capable of solving the problems put by itself, and it should consider it a point of honor to do so. Yet in addition to such emotional impulses the belief in the logical principle of the excluded middle has been, and still is, a powerful motive for the conviction of solvability; both were simply identified by Brouwer [3]).

1) Hilbert 00.

2) It is a common belief among mathematicians that the process of thinking that leads to solving a problem is always *finite.* The formalists, with their finitistic metatheory, claimed that proofs are finite objects, a fact acknowledged by most mathematicians. Notable exceptions are Brouwer (see Brouwer 27, footnote 8 and Kreisel-Newman 69) and Zermelo 35.

3) For this complex of questions on a lower level cf. Hessenberg 06 (Ch. XXII), P. Lévy 26, 26a, 27, Wavre 26. On a higher level the questions reappear in connection with the problems of completeness, incompleteness and consistency within deductive systems; see Chapter V.

Before the first quarter of the 20th century had elapsed, the belief expressed and glorified by Hilbert, which had been a strong stimulus giving scholars confidence in a final success, was breaking down and was even explicitly declared by prominent mathematicians to be an unsubstantiated prejudice. The rejection of the *tertium non datur* was not the only factor, but certainly an influential one in this surprising development; intuitionism considers, with respect to a problem *P*, various possibilities: (1) a positive solution by means of a proof, (2) a negative solution by a constructive counter-example, (3) a negative solution by a proof of the impossibility of *P* (e.g. by a reductio ad absurdum), (4) the lack of either a positive or negative solution. A solution of *P* by means of classical mathematics may very well leave the intuitionist in case (4), because of the use of non-intuitionistic reasoning.

There is no doubt that, far beyond intuitionistic circles, the conviction of universal solvability has been shaken emotionally. [1])

Recently the close connection between the problem of solvability and the theory of certain algorithms (in particular of recursive functions, see Chapter V) on the one hand, and intuitionistic logic (§4) on the other, has been stressed in a number of important papers.

§4. MATHEMATICS AND LOGIC. LOGICAL CALCULUS

In § 2 and § 3 we have dwelt on mathematics, logic and the relation between the two. From our outline of the fundamental principles of intuitionism the intuitionistic thesis concerning the status of logic can be made clear. Mathematics consists of a mixed collection of procedures and objects ranging in nature from strictly constructive to freely generated. Logic, on the other hand, is the theory of forms which express thought, hence the theory of mathematical *exposition* which arises subsequent to mathematical construction and constitutes an abstraction from mathematics. Thus logic is chiefly degraded to a "phenomenon of language"; Brouwer's attitude towards the relation between mathematics and language, as explained on p. 226ff, completely determined his conception of logic. In his doctoral thesis Brouwer

1) It is characteristic (though methodically conforming to the line of metamathematics) that Hilbert himself later (in 25; cf. also 18, pp. 412 ff) formulated the problem in the sense that general solvability should be *consistent*, i.e. non-contradictory. It is true that in later essays (especially in 31) he proceeds from 'non-contradictory' to 'true'. Yet these researches, in spite of the title of the paper 31, belong to metamathematics (cf. Chapter V).

writes: It is evident that in the language, accompanying mathematics, the order of words is determined by laws; it is, however, erroneous to consider those laws as fundamental in the creation of mathematics. The laws of logic become the laws of symbolization of thinking, hence they may be applied only as far as they are compatible with the intuitive basis and the constructive development of mathematics. In particular, the principle of the excluded middle is applicable in finite domains and is in them both a consequence and an anticipation of a certain property of constructions within such domains (p. 228).

This order of precedence, conceiving logic as succeeding mathematics, is well suited to the character of arithmetic – which is significant since arithmetic is the principal basis of mathematics according to intuitionism. In fact arithmetical statements are proved more naturally by construction than by logical deduction from more general laws.

Yet it has remained difficult to grasp from Brouwer's explanation what precisely is meant by *intuitionistic logic* as abstracted from mathematics and still more difficult to comprehend the principles leading to his conception of the relation between mathematics on the one side, language and logic on the other. A good deal of these difficulties derive from the informal and rather rhetorical character of Brouwer's explanations which he justifies, and even insists upon, in view of the informal, nonsymbolical, dynamic character of construction itself.

Therefore it was a decisive step towards clearing up the nature of the controversy – the most decisive step since the establishment of intuitionism in 1907 – when in 1930 Brouwer's pupil Heyting undertook *to present the main contents of intuitionistic logic and mathematics in a symbolic form* of essentially the usual kind, subject to a few modifications and additions evolving from certain peculiarities of the new logic [1]).

To be sure, the Dutch school does not regard Heyting's system as an

1) Heyting 30 and 30a; cf. Heyting 30b, 34, 46, 48, 55, 56; also Freudenthal 35 and 36a, McKinsey-Tarski 48, Lorenzen 50, 62, Kleene 52, Myhill 67. In particular, see Kreisel 62, 65, Beth 59, Kleene-Vesley 65, Kripke 65 and their bibliographies. McKinsey 39 proved the independence of Heyting's primitive symbols (for the propositional calculus).

Fitch 49 extended Heyting's calculus to *modal* concepts such as necessity and possibility. Topological interpretations of Heyting's propositional calculus were given (independently) in the comprehensive essays Stone 37 and Tarski 38a; cf. Beth 59 (Ch. 15), Rasiowa-Sikorski 63.

A useful modification of Heyting's propositional calculus (and of Johansson's minimal calculus), fit for a splitting-up into subsystems of axioms, was given in Schröter 57. Cf. Wajsberg 38, Schröter 56a. Also see Prawitz 65.

orthodox codification [1]). Such a sceptical attitude results from the fundamental impossibility of ever exhausting the totality of processes that may be considered legitimate and is connected with a more fundamental argument: an exposition in the language of symbolic logic with its static character is, in principle, inadequate to describe the dynamic and never closed domain of mathematical activity and can, therefore, only give a hint at the operational field of admitted constructions. Hence it would be an illusion to think that a system of formulae and inference rules could completely describe intuitionistic mathematics.

With such restrictions, however, Heyting's system has been accepted by the Dutch school. Hereby an enormous progress has been obtained for the purpose of comparing traditional (Aristotelian) logic and classical mathematics with their intuitionistic counterparts.

Even before Heyting's papers Glivenko had ingeniously proved [2]) two important results, viz. (1) whenever a proposition p is provable classically, the absurdity of the absurdity [3]) (hence the non-contradictoriness) of p is provable intuitionistically, (2) whenever the *absurdity of p* is provable classically it is also provable intuitionistically. Formal intuitionistic logic has developed considerably since Heyting laid down the first formalisations. In particular the emergence of semantics with a plausible foundational motivation, as provided by Beth, Kripke, Kreisel and others [4]), lent considerable impetus to intuitionistic logic. It is, however, neither possible nor appropriate to exhibit here intuitionistic logic or mathematics in its present state, as it would conflict with the scope of this book (whose main subject is set theory). For an extensive treatment of intuitionistic mathematics and logic (both intuitive and formalized) the reader is referred to the following texts: Heyting 56, Kleene 52, Kleene-Vesley 65, Kreisel 65, Troelstra 69. Instead of trying to keep up with present developments we will concentrate on the basic features of logic and mathematics. We begin with an exposition of intuitionistic logic and its divergencies from the classical one. Heyting constructed a formal system for intuitionistic propositional logic and predicate logic, using a standard formalisation. Let us first consider the propositional calculus. Its language contains the connectives $\wedge$, $\vee$, $\rightarrow$, and $\neg$, and the set of axioms consists of

1) Cf. Heyting's own criticism, for instance in 54 and 56.

2) Glivenko 29. See the exposition in Kleene 52, pp. 492 f.

3) The term 'absurd' used by Brouwer for intuitionistic negation signifies, in accordance with its meaning in Dutch, 'contrary to reason' in view of a proof (and not 'nonsensical' or 'ridiculous').

4) Beth 56a, Kripke 65, Kreisel 65, Grzegorczyk 64, Goodman 68, 70.

intuitionistically valid (in the intuitive sense!) propositions. Heyting's system actually is a subsystem of classical propositional logic, obtained by replacing the axiom of the excluded third (i.e. $p \vee \neg p$) by a weaker statement which states that from p and $\neg p$ every statement q can be deduced (i.e. $p \to (\neg p \to q)$). Johansson [1]) even rejected the latter proposition, thus arriving at his *minimal calculus*. It has been shown in various ways that in Heyting's system $p \vee \neg p$ is not derivable, hence his calculus is a proper subsystem of the classical calculus. This also shows that Heyting's system is not complete, i.e. the addition of an underivable proposition to the axioms does not produce an inconsistent system.

The crucial factor in ascertaining the validity, according to intuitionistic standards, of the axioms is the availability of a standard semantics comparable to the two-valued semantics for classical logic. Unfortunately intuitionism, with its character of perpetual development, on principle defies such an ultimate characterization of its proof procedures and linguistic description. We will proceed with the description of an interpretation of the logical connectives, which is motivated by the fundamental principles of intuitionistic mathematics [2]). The following interpretation, which was implicit in Brouwer's writings, was formulated by Heyting [3]). Kreisel extended these ideas to the extent of a theory of constructions [4]). Recalling the nature of mathematics, according to Brouwer's philosophy, we recognize that the truth of a statement concerning a state of mathematical affairs can only be established by a construction. In simple cases, like $5 + 2 = 3 + 4$, it is fairly clear what kind of construction is needed to establish the truth. In general, rather complicated complexes of constructions, constructions applied to constructions etc. may be required. It is quite in accordance with the intuitionistic principles to state that proofs are constructions [5]). Now let us interpret the meaning of the logical connectives in terms of proofs (i.e. constructions). First we formulate the interpretation in intuitive terms:

A proof of $A \vee B$ is obtained by providing a proof of A or a proof of B (in the sense that we effectively know which of the two cases applies).

1) Johansson 36.

2) An early interpretation that takes account of the intuitionistic principles was given by Kolmogoroff 32. He interpreted intuitionistic logic as a calculus of problems. In many respects it resembles Heyting's interpretation.

3) Heyting 56 Ch. VII, Heyting 30b.

4) Kreisel 62, 65, Goodman 68, 70.

5) The fundamental role of the notion of construction is strikingly analogous to the role of the notion of set in classical mathematics.

A proof of $A \wedge B$ is obtained by providing a proof of A and a proof of B.

A proof of $A \to B$ is obtained by providing (i) a construction which converts any proof of A into a proof of B and (ii) a proof of (i).

A proof of $\neg A$ is obtained by providing a proof of $A \to \bot$, where $\bot$ is a simple false statement like $0 = 1$.

A proof of $\exists x A(x)$ is obtained by providing an element e and a proof a such that a proves $A(e)$.

A proof of $\forall x A(x)$ is obtained by providing (i) a construction a such that, for any element e, $a(e)$ proves $A(e)$ and (ii) a proof of (i).

A formula is said to *hold* or to be *true* if it has a proof in the above sense.

This intuitive form has been given a precise formulation by Kreisel and Goodman. For the purpose of heuristic guidance the refinement in the interpretations of $\to$ and $\forall$ can be dropped. That is to say, it generally suffices to convince ourself of the existence of a construction that converts proofs of A into proofs of B, in the case of implication, and of a construction that for each e produces a proof of $A(e)$, in the case of general quantification. For a deeper analysis of the theory of constructions, we refer the reader to the expositions in the literature. One fundamental point, that was stressed by Kreisel, should, however, be mentioned. The notion "a is a proof of A" is decidable, i.e. the intuitionist recognizes a proof when he sees one! This certainly is a quite reasonable idealization if one departs from the solipsist point of view. The interpretation hence accomplishes a reduction of the meaning of the logical connectives to ordinary, decidable two-valued logic.

The interpretation of the implication is of particular interest, as the classical procedure of definition by cases does not apply. The basic assumption of two-valued semantics is the possession of truth-values by all statements (*wahrheitsdefinit* [1])), whereas intuitionists reject this assumption on principle. The following example [2]) shows that an implication can be true although the truth of its components is undecided. Let A be: "there is sequence of 9 consecutive nines in the decimal expansion of π", and B: "there is a sequence of 8 consecutive nines in the decimal expansion of π". Clearly $A \to B$ is true, as every proof of A can be trivially converted into a proof of B (and the proof of this fact is likewise trivial). Guided by the above interpretation the reader will be able to check the validity of a reasonable supply of formulas. E.g. consider $A \to (B \to A)$; this statement holds if we have a construction for converting any proof of A into a proof of $B \to A$, i.e. into a construction converting a proof of B into a proof of A. Now, given a proof a of A the (constant)

1) Lorenzen 62.
2) Cf. Heyting 36.

construction c that converts any proof b of B into a satisfies the last requirement, so the construction we look for is the one that has for argument a the value c (in λ-symbolism: $\lambda a \cdot (\lambda b \cdot a)$). The interpretation of the quantifiers is illuminating because in the case of the existential quantifier it brings out the constructive character, and in the case of the universal quantifier it shows that (independent of the intended domain) there is no commitment to infinite sets as completed objects. That is to say, $\forall x A(x)$ holds if there is a proof-schema a such that, for any object p that is presented, $a(p)$ is a proof of $A(p)$. One should note the uniform character of the proofs of the various formulas $A(p)$. This is in accordance with the basic principles: if we have evidence *now* that, for all possible p's, $A(p)$ has a proof, then the evidence must provide us with a method for constructing all possible proofs required in the future. Such a method is exactly the proof-schema we mentioned above.

The constructive character of the existential quantifier is in complete accordance with Brouwer's view that an existence statement is nothing but an abbreviation of a statement of the form: "I have a construction such that ...". As indicated, the interpretation of the negation is but a special case of the interpretation of the implication.

Returning to Heyting's formalization of intuitionistic logic we remark that the axioms are all valid under Heyting's interpretation [1]). As Heyting's system is a subsystem of classical logic, the consistency of the latter system entails that of the former system. There is a remarkable relation between intuitionistic and classical logic, as pointed out by Glivenko and Gödel [2]). Let $\vdash$ ($\vdash_i$) stand for "derivable in classical (intuitionistic) logic". Then we have, for propositional logic, $\vdash A$ if and only if $\vdash_i \neg\neg A$, in particular $\vdash \neg A$ iff $\vdash_i \neg\neg\neg A$. But $\vdash_i \neg\neg\neg A \leftrightarrow \neg A$, so $\vdash \neg A$ iff $\vdash_i \neg A$. This result sheds light on the special character of negative propositions.

For predicate calculus a more restricted version holds. Gödel formulated a translation of the formulas of arithmetic by the following inductive procedure:

(1) For atomic formulas A, $A' = A$,

(2) $(A \wedge B)' = A' \wedge B'$,

(3) $(A \vee B)' = \neg(\neg A' \wedge \neg B')$,

(4) $(A \to B)' = \neg(A' \wedge \neg B')$,

(5) $(\neg A)' = \neg A'$,

1) For a detailed account see e.g. Goodman 68, 70.
2) Glivenko 29; Gödel 32, 33; Kleene 52, § 81.

(6) $(\forall x A(x))' = \forall x A'(x)$,

(7) $(\exists x A(x))' = \neg \forall x \neg A'(x)$.

Note that $\vee$, $\rightarrow$, and $\exists$ have been eliminated and that the meaning has been weakened by the translation.

For this translation we have $\vdash A$ iff $\vdash_i A'$. This result can be summarized as: intuitionistic arithmetic is only apparently weaker than classical arithmetic since the latter can be reproduced (according to the translation) within the intuitionistic system. It follows that intuitionistic arithmetic is consistent if and only if classical arithmetic is.

The translation preserves the classical validity (meaning), but certainly not the intuitionistic validity (meaning), so the fragment of arithmetic, defined by the translation, cannot be accepted by intuitionists as a substitute of arithmetic proper.

Heyting's formalization was given in the form of a Hilbert-type system. Other formalizations were provided by Gentzen 34a, Beth 56a, 59, Spector 62, and others.

A large number of formal properties of intuitionistic logic have been discovered in various ways. Some of them we mention below.

a) Intuitionistic propositional logic was shown to be decidable [1]). Although this is a very pleasing result no special intuitionistic significance is attached to it.

b) Intuitionistic predicate logic and arithmetic have the disjunction property and the existential property [2]), which (formulated for arithmetic) read as follows:

If $\vdash_i A \vee B$, then $\vdash_i A$ or $\vdash_i B$;

if $\vdash_i \exists x A(x)$, then $\vdash_i A(n)$, for some numeral n.

This result is very gratifying from the intuitionistic point of view as it shows that the formal systems reflect the strong properties of disjunction and existential quantification. The properties, however, can get lost if one takes intuitionistic logic and adds axioms in a careless way [3]).

c) Intuitionistic predicate calculus is semantically complete with respect to

1) Jaśkowski 36; Gentzen 34; McKinsey and Tarski 46, Rieger 49, Kleene 52, § 80.

2) Gödel 32, Gentzen 34, Harrop 56, Kleene 62, Joan R. Moschovakis 67, Rasiowa-Sikorski 63.

3) For example, if one adds the *tertium non datur*; see Kreisel-Putnam 57, Kleene 62.

a number of semantics [1]). Since all these semantics freely use a non-intuitionistic meta-language, the result is not acceptable to intuitionists. Kreisel [2]) has pointed out that intuitionistically acceptable completeness proofs meet with serious obstacles. In particular, assuming Church's Thesis one can show the incompleteness of intuitionistic predicate logic.

d) In contrast to classical logic, the intuitionistic monadic predicate calculus is undecidable [3]).

e) It has been shown by McKinsey that the connectives of intuitionistic propositional calculus are independent.

Since Heyting wrote down the first formal systems of intuitionistic logic in 1930, several others have been proposed, each having certain advantages for the particular goals of the respective authors [4]). The system listed below is taken from Kleene's *Introduction to Metamathematics*.

Axiom-schemata for propositional logic.

1. $A \to (B \to A)$
2. $(A \to B) \to [(A \to (B \to C)) \to (A \to C)]$
3. $A \to (B \to (A \wedge B))$
4. $(A \wedge B) \to A$
5. $(A \wedge B) \to B$
6. $A \to (A \vee B)$
7. $B \to (A \vee B)$
8. $(A \to C) \to [(B \to C) \to (A \vee B) \to C)]$
9. $(A \to B) \to [(A \to \neg B) \to \neg A]$
10. $\neg A \to (A \to B)$.

Rule of inference. If A and $A \to B$ then B.

Axiom-schemata for predicate logic.

11. $\forall x A(x) \to A(t)$
12. $A(t) \to \exists x A(x)$.

Rules of inference. If $C \to A(x)$, then $C \to \forall x A(x)$
If $A(x) \to C$, then $\exists x A(x) \to C$.

In the last two rules x does not occur free in C, and in 11 and 12 t is a term which is free for x in $A(x)$ [5]).

1) Rasiowa-Sikorski 63; Beth 56a, Kripke 65.

2) Kreisel 62a, 70.

3) Maslov et al. 65; Kripke, unpublished.

4) Heyting 30, Gentzen 34, Beth 56a, Rasiowa-Sikorski 63, Schütte 68, Kleene 52, Spector 62.

5) I.e. no variable in t is bound in $A(x)$ after substitution of t for x.

One obtains the axiom-system for classical logic by replacing 10 with 10°: $A \vee \neg A$.

The axiom $\neg A \to (A \to B)$ can be interpreted as: from a contradiction any statement may be concluded (*ex falso sequitur quodlibet*). This axiom has met with criticism from those who believe that it embodies an illegitimate strengthening of the notion of implication. I. Johansson [1]) developed *minimal logic,* obtained from the above system by dropping 10. In order to understand the problem, let us consider 10*: $(\neg A \wedge A) \to B$, which is equivalent to 10. The interpretation of 10* reads: there is a construction c such that $c(a)$ is a proof of B if a is a proof of $\neg A \wedge A$. However, it is easy to verify that $\neg A \wedge A$ has no proof. Since the proof-interpretation presupposes decidability of the proof relation, ordinary two-valued logic can be applied. As a consequence, for any construction c we have that "a is not a proof of $\neg A \wedge A$ or $c(a)$ is a proof of B", which is correct.

Among the many semantics that have been provided for intuitionistic logic there is a closely related group which has a rather good intuitionistic motivation, although the metalanguage is non-intuitionistic in each case. The first of these semantics was introduced by Beth, in connection with his semantic tableaux. Later versions were presented by Kripke and Grzegorczyk [2]). As Kripke's methods present certain advantages we will sketch here the *Kripke-semantics.* Imagine an (idealized) mathematician, who is pursuing mathematical research and let his activity proceed in stages. The set of all possible relevant stages is partially ordered by the relation "after"; note that in general each stage of research can be continued "in several directions". Now the meaning of the logical connectives is explained in terms of the research and its stages. It is assumed that the mathematician at each stage establishes atomic facts and from these deduces more complicated statements by certain rules. So far the intuitive picture; to define a Kripke model we consider a partially ordered set S (with order $\leqslant$) and an assignment I of sets of atomic formulas to elements of S such that $I(\alpha) \subseteq I(\beta)$ if $\beta \leqslant \alpha$ (intuitively $I(\alpha)$ is the set of formulas established at stage α, the condition therefore states that the mathematician never forgets). The assignment I is extended to all formulas by the convention that

$$A \vee B \in I(\alpha) \quad \text{iff} \quad A \in I(\alpha) \text{ or } B \in I(\alpha)$$

$$A \wedge B \in I(\alpha) \quad \text{iff} \quad A \in I(\alpha) \text{ and } B \in I(\alpha)$$

[1]) Johansson 36.

[2]) Beth 56a, 59, Kripke 65, Grzegorczyk 64; see also Schütte 68.

$$A \to B \in I(\alpha) \quad \text{iff} \quad \text{for all } \beta \text{ such that } \beta \leqslant \alpha, \; A \notin I(\beta) \text{ or } B \in I(\beta)$$

$$\neg A \in I(\alpha) \quad \text{iff} \quad \text{for all } \beta \text{ such that } \beta \leqslant \alpha, \; A \notin I(\beta)\,.$$

We refrain from interpreting the quantifiers; the reader can consult the existing literature. The intuitive meaning of $\vee$ and $\wedge$ is clear. $A \to B$ holds at stage α if at any later stage we know that B holds as soon as A holds, likewise $\neg A$ holds at stage α if A holds at no later stage. These interpretations fit very well in the intuitionistic ideas. A formula A holds (is true, valid) in a Kripke model if it holds at all stages, i.e. if $A \in I(\alpha)$ for all α. A formula A is true if it holds in all Kripke models. The Kripke semantics is complete with respect to Heyting's predicate calculus, i.e. truth in the sense of Kripke semantics and provability are equivalent. Therefore the Kripke models prove to be very suitable for independence results, etc.

One example will suffice: Let $S = \{0,1\}$ and $0 \leqslant 1$. Define I by $I(1) = \emptyset$, $I(0) = \{A\}$. Now $\neg A \notin I(1)$, because $A \in I(0)$, so neither A nor $\neg A$ hold at stage 1, therefore $A \vee \neg A$ does not hold at stage 1. The *tertium non datur* is not true in the Kripke semantics, hence it is not derivable in intuitionistic logic.

No discussion of intuitionistic logic and semantics is complete that does not mention *Kleene's realizability notion.* This notion, which can be viewed as a constructive semantics, is the result of an application of recursion theory. Kleene [1]) defined the notion x *realizes* A, where x is the Gödel-number of a partially recursive function. Let us not go into details and loosely interpret x *realizes* A as "x effectively validates A". It is interesting to check the case of implication in Kleene's inductive definition: x *realizes* $A \to B$ if for every y such that y *realizes* A we have $\{x\}(y)$ *realizes* B [2]).

In our loose interpretation this becomes: "If y effectively validates A, then $\{x\}(y)$ effectively validates B". As in the proof interpretation there is a uniform procedure to find the "effectively validating" number. Kleene and others used the realizability concept for many proof-theoretic purposes [3]). Of a similar nature is the functional interpretation introduced by Gödel [4]).

It must be remarked that intuitionistic predicate logic is not complete for either Kleene's realizability or Gödel's (*Dialectica*) interpretation (Gödel 58).

1) Kleene 45, 52, §82. Nelson 47.

2) $\{x\}$ denotes the partial recursive function with Gödel number x; it is assumed that the function is defined for y.

3) Kleene 62, Nelson 47, Kreisel-Troelstra 70.

4) Gödel 58, Kreisel 59, Spector 62, Howard 68, see also Mostowski 66, p. 90 ff.

As a matter of fact, both strengthen the intended meaning of the logical connectives.

A list of the most conspicuous formulas that fail to be intuitionistically valid is displayed below.

(1) $\nvdash_i \neg\neg A \to A$

(2) $\nvdash_i \neg(A \wedge B) \to \neg A \vee \neg B$

(3) $\nvdash_i (A \to B) \vee (B \to A)$

(4) $\nvdash_i ((A \to B) \to A) \to A$

(5) $\nvdash_i (A \to B) \to (\neg A \vee B)$.

Particularly important is the failure of the following rules on negating quantified statements.

(6) $\nvdash_i \neg \forall x A(x) \to \exists x \neg A(x)$

(7) $\nvdash_i \neg \forall x \neg A(x) \to \exists x A(x)$

(8) $\nvdash_i \neg \exists x \neg A(x) \to \forall x A(x)$

(9) $\nvdash_i \neg\neg \exists x A(x) \to \exists x \neg\neg A(x)$

(10) $\nvdash_i \forall x \neg\neg A(x) \to \neg\neg \forall x A(x)$

(11) $\nvdash_i (A \to \exists x B)) \to \exists x (A \to B(x))$ (x not free in A)

(12) $\nvdash_i \forall x (A \vee B(x)) \to (A \vee \forall x B(x))$ (x not free in A)

(13) $\nvdash_i (\forall x A(x) \to B) \to \forall x (A(x) \to B)$ (x not free in B).

Using the construction interpretation one can make the invalidity of these formulas plausible. E.g. take (6); from the fact that we have a construction that converts every proof of $\forall x A(x)$ to a proof of a contradiction it is hard to see how we can construct an element e such that there is a proof of $\neg A(e)$, let alone that we have a uniform construction for the transformation of the proofs. In general, however, an ordinary mathematical example will be of just as much, or more, assistance, A counterexample to (6) is obtained by taking $x = 0 \vee x \neq 0$ for $A(x)$, where x ranges over real numbers. It has been shown that $\neg \forall x (x = 0 \vee x \neq 0)$ holds (see §5). On the other hand $\exists x \neg (x = 0 \vee x \neq 0)$ implies $\exists x\, (x \neq 0 \wedge \neg x \neq 0)$, which is contradictory. So (6) does not hold!

The study of formal systems (and their interpretations) for analysis has been initiated by Kleene. Many results are now available in the literature [1]).

Noting how much of the standard machinery of classical logic has been abolished by intuitionists, he who is familiar with the lucidity and simplicity of the classical logical calculus will not readily submit to the complications involved by the mentioned restrictions. It must be remarked, however, that a number of mathematicians never felt at home in the paradise created by Cantor; in particular, constructive and numerical mathematics side with Brouwer [2]) rather than with Hilbert, E.g. the intermediate value theorem of Weierstrass [3]) is not much help if one wants to compute a zero of a function; one has to employ some approximation procedure, i.e. an intuitionistically significant procedure!

A reconciliation between classical and intuitionistic mathematics was suggested by Brouwer in 1928. Another attempt in this direction was made by Van Dantzig along the lines of a weak embedding of classical logic in intuitionistic logic as given by Gödel [4]). Van Dantzig considered the *stable* part of intuitionistic mathematics, where a formula A is called *stable* if $(A \leftrightarrow \neg\neg A)$, with connectives such as in Gödel's translation (see p. 243). This procedure works well in certain theories; in the theory of real numbers, however, it breaks down. Van Dantzig also developed a part of intuitionistic mathematics, involving only positive statements, which he called *affirmative mathematics.*

Serious objections against the concept of negation were raised by G.F.C. Griss. [5]) Griss argued that our knowledge and insight in properties of mathematical objects is solely based on constructions that are actually – or at least can be – performed. Hence a notion, based on the impossibility of a construction, cannot be clear. As a consequence, negation does not deserve a place in mathematics. Griss calls intuitionistic mathematics, trimmed according to his views, *negationless mathematics.* In spite of the negative name Griss' program is strictly positive, more so than Van Dantzig's affirmative mathematics.

Although some work has been done within negationless mathematics, Griss's program has generally been considered a curiosity as far as actual everyday mathematics is concerned.

1) Kleene-Vesley 65, Kreisel-Troelstra 70. Scott 68, 70.

2) See, for example, Bishop 67, although Brouwer is not spared criticism there.

3) If a continuous function f satisfies $f(0) < 0$ and $f(1) > 0$, then there exists a number a such that $0 < a < 1$ and $f(a) = 0$.

4) Van Dantzig 47, written in 1942, cf. Gödel 32, Kleene 52, §81.

5) Griss 44–51, Vredenduin 54, Heyting 56, § 8,2.

Brouwer staunchly maintained his views on the admissibility of the notion of negation and its use in mathematics [1]). We shall return to his views in § 5. Viewed in retrospect, Brouwer's arguments seem to outweigh Griss's objections.

A satisfactory presentation and a criticism of the program of Griss was given by Gilmore [2]). By introducing a certain deductive theory H formalized within intuitionistic logic, a subtheory of which contains the theory of the relation # (§ 6), Gilmore was able to express the consequences of Griss's criticism which would discard many of the predicates and statements of H because they do not satisfy the condition of 'positive nonnullity'. (This condition rejects the empty set (null-class) and any statement which is not true. More precisely, a predicate or statement p is admitted only after the proof of $\exists xp$, where $\exists x$ stands for a row of existential quantifiers (empty when p is a statement], namely one for each free variable of p.) Yet, in addition to the predicates and statements admitted by Griss, Gilmore exhibits a further class of predicates and statements whose introduction does not contradict Griss's principles, namely those which are equivalent to contradiction within H; e.g., $x_1 \# x_1$.

Now Griss's main innovation, viz. the rejection of negation, can be introduced by defining negation as the implication of a predicate which plays the role of the null-predicate (provided such a predicate exists in the system). Also the limitation in the use of disjunction, conjunction, implication, and quantification can be mastered. The effect of Griss's criticism of the intuitionistic predicate calculus can be expressed, according to Gilmore, by adding a certain axiom for every "atomic" predicate and statement.

To be sure, Griss would not have accepted the deductive theory H on which Gilmore's results depend – similar to the orthodox intuitionistic attitude towards Heyting's system (p. 240).

Gilmore's achievement is, on the one hand, a step forward within Griss's program of constructing a system of negationless mathemetics but, on the other hand, also a *criticism* of Griss's ideas inasmuch as his 'nullity', the rejection of which is Griss's cornerstone, proves to be a relative and not an absolute concept. While seemingly Van Dantzig's "affirmative mathematics" and Griss's "negationless mathematics" are in agreement, since both reject negation and demand that theorems expressed by means of negation either be translated into a positive form or, when this proves impossible, be dropped – there is the difference that in Griss's system of mathematics negation can be *defined*

1) Brouwer 48.
2) Gilmore 53.

and the system can thus be converted to a system like Heyting's. This does not hold for Van Dantzig's system which rejects disjunction, the essential tool of that conversion

The latest development in the constructive foundations of mathematics is the so-called ultra-intuitionism, which was discussed by A.S. Esenin-Volpin in 1959 [1]). Esenin-Volpin's criticism cuts deeper than any of the earlier attempts to reform mathematics, it does not even spare Brouwer. The ultra-intuitionistic criticism is directed against the current interpretation of the notion "finite". Indeed objections of a similar kind have been raised earlier [2]) but they were not employed before as a point of departure for systematic foundational work.

Esenin-Volpin considers natural sequences which are discrete procedures, that start with an initial event (say 0) and such that with each event a there exists a successor event a'. The following conditions are imposed: $\neg 0 = a'$ and $a' = b' \rightarrow a = b$. The natural numbers form such a natural sequence. According to Esenin-Volpin it is highly questionable whether for instance 10^{12} is a natural number, or put otherwise, whether the natural sequence of counting will ever lead up to 10^{12}. This attitude entails the rejection of the full principle of complete induction, the closure of the natural number sequence under the sum-, product- and exponentiation-operation and the following consequence of the rule of modus ponens: if $\vdash A$ and $\vdash A \rightarrow B$, then $\vdash B$. Thinking of lengths of proofs, to conclude $\vdash B$ from $\vdash A$ and $\vdash A \rightarrow B$ one needs extra proof steps. Likewise the rule "if $\vdash A$ and $\vdash B$, then $\vdash A \wedge B$" is questionable. In particular Esenin-Volpin exploits this phenomenon in the case where A and $\neg A$ have been proved, but no proof of $A \wedge \neg A$ is available.

He defines a theory T to be contradictory if $T \vdash A \wedge \neg A$, so the derivability of A and $\neg A$ does not yet make a theory contradictory. Esenin-Volpin develops from theses ideas a sketch of a consistency proof for ZF. The program of ultra-intuitionism has not yet been carried out in detail. Both proof theory and model theory are still lacking. In its present form the program presents an attitude considerably weaker than the intuitionistic one. Considering the stress on finiteness it would perhaps be proper to classify the program as finitistic in the sense of Hilbert.

1) Esenin-Volpin 61, 70.
2) E.g., Van Dantzig, 56.

§5. THE PRIMORDIAL INTUITION OF INTEGER. CHOICE SEQUENCES AND BROUWER'S CONCEPT OF SET [1])

The mathematician, tired of dogmatic principles underlying the intuitionistic program and of the painful restrictions imposed on classical procedures of definition and proof, will want to learn how mathematics can retain its *infinite character* without which it would loose much of its interest. It seems that Weyl, who had first created his own peculiar semi-intuitionistic system and then accepted the attitude of Brouwer, providing it with additional arguments, was the originator of the slogan: "mathematics is the science of infinity".

The nucleus of positive intuitionistic principles, common to all intuitionistic trends from Kronecker to the present day, is the "primordial intuition" (*Urintuition*) of *positive integer* or of the construction by *mathematical induction*. Induction is also the prototype of those constructions with which mathematical activity is identified (pp. 226ff). This, of course, does not mean that Brouwer would accept the infinite sequence of positive integers as a mathematical object or as a legitimate idea at all – as Weyl's semi-intuitionism still did; it is the *law* for the construction of integers and not their would-be aggregate which is accepted. As Poincaré says, "*Quand je parle de tous les nombres entiers, je veux dire: Tous les nombres entiers qu'on a inventés, et tous ceux que l'on pourra inventer un jour .... et c'est ce "que l'on pourra" qui est l'infini.*" It therefore makes no sense to ask whether there exists an integer with a given property – except for the cases where an integer with the property can be named, or else it can be proved that no such integer exists.

We shall not enter into a philosophical analysis of "primordial intuition". O. Becker (originally from Heidegger's existentialist school) who devoted two books chiefly to an analysis of intuitionism maintains [2]) that one should not conceive it as a "sensorial" or "empirical" intuition but as the kind of immediate certainty with which we are given the fundamental facts of logic, arithmetic, and combinatorics. This intuition is peculiar to mathematics and can be reduced to the indefinitely repeated bisection of the unit, a process whose

1) Most subjects of §§ 5 and 6 are treated in more detail (and partly with closer adherence to the Dutch school) in the book Heyting 56. (Cf. the penetrating review by S. Kuroda 56.) Moreover, as far as § 6 and the second half of § 5 are concerned, a more modern viewpoint is taken in the book Kleene-Vesley 65.

2) O. Becker 27, pp. 446 ff (cf. p. 463) in the *Jahrbuch* pagination. Becker even ventured to base transfinite ordinals upon such intuition.

fundamental significance has already been pointed out by Plato. As will be shown presently, Brouwer utilizes the primordial intuition for defining choice sequences and sets, thus extending the realm of intuition from the discrete and countable domain of arithmetic to the continuous (and somehow non-denumerable) domain of analysis. From this primordial intuition of number the name *intuitionism*, which can easily mislead the reader, has been derived.

The precursor of mathematical intuitionism, Kronecker, coined the slogan: "God made the integers, everything else is the work of man." Poincaré, who adopted a similar attitude, exhibited in full detail [1]) the idea that mathematics, seemingly constituting an immense tautology consisting of *analytic* statements which are obtained by syllogistic inferences, derives its *synthetic* and creative character chiefly from a synthetic principle *a priori* in the sense of Kant, viz. from mathematical induction [2]). Weyl stressed [3]) that mathematics altogether – including "the logical forms of its expansion" – is dependent on natural number. Of course, intuitionists do not accept the view, suggested by the axiomatic attitude, that mathematical induction might be considered an ingredient of a *definition* of integers, for then one ought to show the existence of an object satisfying the definition which again would require induction.

The weight attributed to induction by intuitionists of all trends is explained by the fact that constructing through finite procedures can only yield statements of a finite character whose contents can, in principle, be verified by finitely many tests. Yet the main statements of arithmetic do not have this character but are transfinite; for instance, '$x + 1 = 1 + x$ for each positive integer x' or 'for each prime number p there exists a greater prime number in an interval dependent on p'. To be sure, there is a fundamental difference between such statements of analysis and set theory which are transfinite in a higher sense; for instance, the theorem that every bounded set of real numbers has a least upper bound, or the well-ordering theorem. The above arithmetical statements can be finitely verified in each particular case (each value of x, or of p) though the general statement, just because of its transfinite

1) Particularly in Poincaré 03, Chapter I.

2) In later years (in particular in Poincaré 08) he modified his earlier attitude of attributing to induction alone a creative character and considered it to be only the simplest among various creative principles. It seems that he even regarded the axiom of choice as one of them. This shows the fundamental difference between him and the Dutch school. On the other hand, scholars akin to this school (for instance, Van Dantzig 32) pointed out that the creation of new formalisms, on account of analogy etc., also belonged to the intuitive activity of mathematicians.

3) In his comprehensive summary 21 of the development of intuitionism, p. 70.

nature, would require an infinite sequence of verifications which cannot be accomplished. Poincaré believed that this possibility of verifying the particular cases of the intuitionistically legitimate transfinite statements is the very reason that enables mathematicians – in contrast, for instance, to philosophers – to understand each other in spite of the defects of language; whenever the situation becomes doubtful it can be put to the test by a verification which serves as the supreme arbiter. On the other hand, the transfinite statements of analysis and set theory which cannot be tested in a similar way – the classical example being the well-ordering theorem – remain obscure and eternally exposed to misunderstanding. It is remarkable that since the 20's and 30's of the present century leading representatives of the so-called formalist school, beginning with Hilbert, von Neumann, Gentzen, etc., have adopted this intuitionistic attitude and nevertheless obtained remarkable results, notwithstanding the threat of Gödel's incompleteness theorem (Chapter V).

While the semi-intuitionists, especially of the French school (including Weyl in his first period), were at least theoretically ready to put up with the blockade of analysis and geometry involved by the restriction to induction as *the* transfinite process in mathematics, Brouwer's seemingly more radical trend endeavored to create an outlet through the concepts of *choice sequence* and of *spread* and *species*.

After the concept of positive integer, the next or equally important mathematical (and scientific) concept is the *continuum* (cf. § 1). One may classify various attitudes in the foundations of mathematics according to the kind of "continuum" they admit. The classical continuum concept of the nineteenth century corresponds, on the one hand, to the Greek conception which accepts the continuum as "given" by nature or as its mathematical reflection, on the other hand to Cantor's conception of set according to which a set is given if, for any object, it is "internally settled" whether it belongs to the set or not. The deficiency of these conceptions compels us either to take an axiomatic attitude which renounces a definition of set altogether (Chapter II) or to put up otherwise with more restricted conceptions. In this respect semi-intuitionists have taken different and not always consistent attitudes [1]) which were sometimes interpreted in the sense of "extensional definiteness" [2]); at any rate, the set of all properties of integers,

1) The presumably last expression of Lebesgue's attitude (of 1938) is found in Lebesgue 41; cf. Sierpiński 41, p. 137.

2) So, in particular, Weyl's "atomistic" continuum (Weyl 18, cf. 19 and 21; cf. the philosophical analysis of these and the neo-intuitionistic attitudes by O. Becker, also Skolem 29, § 3). In Becker's work the alleged dependence of mathematics on the notion of *time* and in general the anthropologic-existentialistic arguments are closely connected

which essentially coincides with the classical continuum, is not legitimate in this sense. (Cf. the *impredicative definitions*, Chapter III.) Hence the continua of Weyl, Lebesgue, Lusin, etc. are denumerable, though they cannot be enumerated within the system.

Brouwer has introduced a fundamentally novel and important moment into this discussion. His constructivistic attitude conforms to the views mentioned: a single real (irrational, transcendental) number is conceived as a law defining it and the classical continuum cannot be attained by arithmetical operations [1]). Nevertheless, he maintains that a veritable continuum which is not denumerable can be obtained as a *medium of free development*; that is to say, besides the points which exist (are ready) on account of their definition by laws, such as e, π, etc., other points of the continuum are not ready but *develop* as so-called *choice sequences*. As it has been said, the choice sequences "free the infinite from the concept of law".

Thus both the rejection of geometry altogether (Weyl) and the distinction between a discrete-atomic analysis and a continuous geometry (Hölder, Lusin) become unnecessary for Brouwer.

The following attempt to analyze the concept of choice sequence is mainly historically motivated: to exhibit the creation of this remarkable concept by Brouwer and its later development by him and others. For a more systematic mathematical-logical treatment of this concept, as well as the concepts of spread and species (set), the reader should turn to the literature [2]). The most familiar example of a choice sequence is the one with natural number values.

with intuitionistic conceptions (especially in O. Becker 27, 29, 30); cf. the well-founded refutation in Cassirer 29.

Weyl does not pretend that his continuum reflects the classical notion; on the contrary, he despairs, for logical reasons, of the possibility of such a reflection and is forced to be satisfied with extracting discrete "atomic drops" from the continuous "pulp" by rather arbitrary principles of construction. A similar attitude is established by Borel's restriction to *nombres calculables*, i.e. real numbers which can be effectively approximated by rational numbers; however, Borel's notion is rather involved (cf. Sierpiński's example 21, p. 114, of an effectively defined *integer* which is not "calculable").

1) It is remarkable that as early as 1892, when such ideas were rather unusual, Hölder (cf. 24, pp. 193 f, 349 ff, etc.) denied this possibility. However, not being ready to abandon the mathematical branches which depend on the classical continuum, he accepted it, if not as an arithmetical-analytical construction, at least – similarly to the Greek attitude – as a kind of aprioristic schema which cannot be constructed by thinking but may serve as an object of thinking. Similar ideas appear among the French semi-intuitionists; cf. Borel 14, note IV, and Lusin 27, p. 33.

2) E.g. see Heyting 56, Kleene-Vesley 65, Kreisel 65, 68, Troelstra 69, 70, Kreisel-Troelstra 70, Myhill 67.

Such a sequence α is obtained by successive choices of natural numbers in such a way that in the course of the process restrictions on future choices may be placed. One may represent a choice sequence as a sequence of pairs $\langle a_0, R_0\rangle, \langle a_1, R_1\rangle, \langle a_2, R_2\rangle, \ldots$ where the R_i's are conditions on sequences of natural numbers. The conditions may get "narrower", i.e. if $R_i\alpha$ holds then $R_{i+s}\alpha$ holds too. An extreme case presents itself if the relation R_0 determines a unique sequence α; we then say that α is determined by a law (i.e. R_0). The class of sequences predetermined by a law is called the class of *lawlike* (or *constructive*) sequences. Another extreme is met if no restrictions at all will be placed on future choices; these sequences were termed *lawless* by Kreisel [1]). These choice sequences cannot be conceived as finished, completed objects; at every moment only an initial segment is known. It was Brouwer's brilliant idea to employ this device of choice sequence to overcome the difficulties of a constructivist's continuum. One can simply take the collection of dyadic intervals $\lambda_{k,l} = [a_k/2^l, a_k+2/2^l]$, where $\lambda k\text{-}a_k$ is a fixed enumeration of the integers, and consider choice sequence of nested intervals $\lambda_{k,l}$. The continuum is then made up of all equivalence classes of choice sequences, among which there are lawlike reals like e, $\sqrt{2}$;π, but also many others. In this way, one obtains the continuum as an unfinished, growing object. Besides choice sequences of natural numbers there are choice sequences of various mathematical objects. Knowledge about properties of a choice sequence α has to be established at a certain moment when only a finite initial segment of α is available. So if, for example, one has established that α represents a positive real number, then this is so because one of the choices is an interval $[a,b]$ with $a>0$. Automatically all choice sequences with the same initial segment, up to $[a,b]$, represent positive reals.

Accordingly a choice sequence is not the kind of object to consider isolated. The conception of the continuum as an aggregate of existing points (members), which is at the bottom of nineteenth century analysis and of Cantor's set theory, is replaced by an aggregate of parts which are partially overlapping and which are so to speak the manifestations of real numbers still to be generated. In Brouwer's notion of *spread* (German: *Menge*) the properties of choice sequences are exploited. Brouwer's original definition is rather complicated. By now some clear expositions are available [2]). Nevertheless we will

1) Kreisel 68. Note, however, that among intuitionists there is some controversy whether this concept is legitimate. See for example Brouwer 52–53 p. 142, footnote.

2) Heyting 56, Kleene-Vesley 65, Troelstra 69.

give an informal exposition here because the notion plays an important role in intuitionistic mathematics. Since we will consider natural numbers, finite sequences of natural numbers and (infinite) sequences of natural numbers, let us adopt the following notation: i,j,k,l,m,n... (if necessary with index) denote natural numbers; a,b,c,d... denote finite sequences of natural numbers, where a_i is the i-th number in the sequence; $\alpha, \beta, \ldots$ denote infinite sequences (functions); $\bar{\alpha}k$ is the sequence $\langle \alpha 0, \ldots, \alpha(k-1) \rangle$. $*$ denotes the concatenation operation, i.e. $\langle n_1 \ldots n_k \rangle * \langle n_{k+1} \ldots n_l \rangle = \langle n_1, \ldots, n_l \rangle$ [1]).

A *spread-law* is an effective procedure S that operates on finite sequences of natural numbers; we say that S accepts a if $Sa = 1$, otherwise $Sa = 0$; S is subject to the following conditions:

(i) S accepts at least one sequence of length 1.
(ii) If $Sa * \langle k \rangle = 1$ then $Sa = 1$.
(iii) If $Sa = 1$ then there exists at least one k such that $Sa * \langle k \rangle = 1$.

In plain words a spread-law determines a collection of finite sequences with at least one member, closed under predecessor and with at least one successor for each member. One can conceive this collection as a tree. The infinite sequences each initial segment of which is accepted by S constitute a *spread*. So far only spreads with number sequences have been defined. One can generalize the concept by introducing an extra mapping D which assigns mathematical objects to acceptable sequences. Thus infinite sequences of mathematical objects are obtained. The pair $\langle S, D \rangle$ is called a *dressed spread*; D is the *dressing* of S.

Examples of spreads are:

1. The spread determined by S_1, with the property
   $S_1\langle 0 \rangle = S_1\langle 1 \rangle = 1$,
   $S_1 a * \langle k \rangle = 1$ iff $S_1 a = 1$ and $k \leqslant 1$.
2. The spread determined by S_2, with the property
   $S_2 = \lambda a \cdot 1$
3. The dressed spread with
   $S_3\langle 0 \rangle = S_3\langle 1 \rangle = S_3\langle 2 \rangle = 1$
   $S_3 a * \langle k \rangle = 1$ iff $k \leqslant 2$
   and
   $D(\langle k_0, \ldots, k_n \rangle) = (k_0-1)2^{-1} + (k_1-1)2^{-2} + \ldots + (k_n-1)2^{-n-1}$.

The spread determined in the respective cases 1, 2, 3 is called the *binary fan*,

1) In the literature one mostly uses codings of sequences instead of sequences.

the *universal spread,* and the closed interval [−1, 1] [1]). The spreads of the examples 1 and 3 have the special property that for each a there are finitely many k's such that $Sa * (k) = 1$; such spreads are called *finitary spreads* or *fans*. In general, it is the dressed spreads that occur in actual mathematics such as analysis or topology [2]), but for simplicity we consider plain spreads only. From the point of view of foundations not much is lost thereby.

The notion of spread is the constructive notion of set in intuitionism; even though it is infinite and unfinishable, it has the important characteristic of a step by step approximation. Another notion introduced by Brouwer is that of a *species.* A species is (the extension of) a property of previously defined mathematical objects [3]).

Examples of species are

1. $X_1 = \{n | n > 2 \wedge \exists x > 0 \, \exists y > 0 \, \exists z > 0 \, (x^n + y^n = z^n)\}$ [4])
2. The species X_2 of all prime twins.
3. The species X_3 of all irrational reals.
4. The species X_4 of all subspecies of N (N is the species of natural numbers).

From the examples one may conclude that species can be very wild indeed. E.g. for X_1 it is unknown whether it is empty or not, for X_2 it is unknown whether it is finite or infinite.

The theory of species is still poorly developed and it may have surprises in store. For instance, the following uniformity principle is consistent with respect to intuitionistic analysis: $\forall X \exists x A(X,x) \rightarrow \exists x \forall X A(X,x)$ (due to A.S. Troelstra). One gathers from it that " the existence of a number for each species such that..." can be very strong.

For the theory of choice sequences a number of intuitionistically plausible (if not valid) principles have been put forward, which make intuitionistic analysis divergent from classical analysis instead of merely a subtheory (as in the case of arithmetic).

Brouwer's principle or the *continuity principle*: If for each choice sequence α we can determine a natural number n such that $A(\alpha, n)$ holds, then n can already be determined on the knowledge of an initial segment of α [5]).

1) Notice the analogy between the binary fan (the universal spread) and the Cantor space (the Baire space).

2) Freudenthal 36 was the first to apply the spread techniques to the study of topology without recourse to a metric. See also Troelstra 67.

3) Note the connection with the comprehension principle.

4) As usual $\{n | A(n)\}$ stands for the species (set) of all n with the property A.

5) Brouwer 27, Kleene-Vesley 65, Howard-Kreisel 66.

A formalization reads: $\forall\alpha\,\exists x A(\alpha,x) \to \forall\alpha\,\exists x\,\exists y\,\forall\beta(\bar{\alpha}y = \bar{\beta}y \to A(\beta,x))$. Kleene also introduced a continuity principle for functions (choice sequences) which claims that if to each α a β is correlated, the value of $\beta(t)$ is determined by t and an initial segment of α. Unfortunately this principle conflicts with another intuitionistic principle (Kripke's schema).

In intuitionistic analysis various forms of *choice principles* are acceptable [1]); we will discuss a simple case here. Suppose that for each natural number x a natural number y exists such that a certain relation $A(x,y)$ is fulfilled. Interpreted intuitionistically, this means that given an x the corresponding y can be computed; moreover, we know that this can be done for all natural numbers. Clearly we must then possess a computation schema that allows us, whenever a number x is presented, to compute the corresponding y. But then, in effect, we have an effective function α that computes y. Therefore the choice principle $\forall x\,\exists y A(x,y) \to \exists\alpha\,\forall x A(x,\alpha(x))$ is intuitionistically true.

Kreisel has introduced an important class of operations, the so-called Brouwer operations. This class K consists of neighbourhood functions on the universal spread (classically one can prove that K is exactly the class of all representing functions of all continuous mappings from the Baire space into a set of natural numbers). The arguments of functions of K are finite sequences of natural numbers. For convenience we define the ***x-shift*** of a function α by $\alpha_x\,(\langle n_1, ..., n_k\rangle) = \alpha(\langle x, n_1, ..., n_k\rangle)$.

Now K is inductively defined as the smallest class P such that 1) It contains all non-zero constant functions, 2) if $\alpha(0) = 0$ and for all x, $\alpha_x \in P$, then $\alpha \in P$.

Let us denote elements of K by e, f, g. One can visualize the action of e by imagining the universal spread (N^N), where e successively computes $\langle n_1\rangle$, $\langle n_1, n_2\rangle, \langle n_1, n_2, n_3\rangle, ...$ along some infinite branch. In general e will first produce zero's until it, after a finite number of steps, produces a positive number, from then on it remains constant. In effect one easily proves by induction:

$$\forall\alpha\exists x(e(\bar{\alpha}x) \neq 0)$$

and

$$\forall a\,\forall b(e(a) \neq 0 \to e(a) = e(a * b))\,.$$

The elements of K represent continuous functionals in the following natural way: $\Phi_e(\alpha) = y$ iff $\exists x(e(\bar{\alpha}x) = y + 1)$, i.e. let e work on initial segments of α and pick the first positive value minus one.

1) Cf. Kreisel-Troelstra 70, p. 263, Kleene-Vesley 65, p. 14.

Now the use of K for the study of intuitionistic analysis is embodied in the (intuitionistically plausible) postulate: *each continuous functional is a* Φ_e (for some $e \in K$). The consequences of this postulate are far-reaching. E.g. Brouwer's *bar theorem* is a consequence (actually in some sense the bar theorem is equivalent to the postulate [1])). The bar theorem was first used by Brouwer in the famous proof that every real function on a closed interval is uniformly continuous [2]). The bar theorem reads $[\forall a\, \forall b(P(a) \to P(a * b)) \wedge \forall a(\forall x\, Q(a * \langle x \rangle) \to Q(a)) \wedge \forall \alpha\, \exists x P(\bar{\alpha}x) \wedge \forall a(P(a) \to Q(a))] \to \forall a Q(a)$. As a matter of fact the bar theorem can be viewed as an induction principle for well founded sets of finite sequences [3]). From the bar theorem and Brouwer's principle one deduces the *fan theorem*: If to each α of a fan a natural number n is associated, then there exists a number k such that for all sequences α the value is already determined by an initial segment of length k. The fan theorem itself is instrumental in establishing the uniform continuity of functions defined on closed intervals. From the continuity theorem Brouwer extracted a refutation of the principle of the excluded middle [4]) as follows: let Rat(α) stand for "the real number α is rational". Suppose $\forall \alpha \in [0, 1]$ $(\mathrm{Rat}(\alpha) \vee \neg \mathrm{Rat}(\alpha))$. Then the characteristic function of Rat is fully defined on [0, 1]. But according to the above continuity theorem this characteristic function is uniformly continuous; as there is at least one rational number in [0, 1], the function is constant, with value 1. This contradicts the fact that there is at least one irrational point between 0 and 1. So we conclude $\neg \forall \alpha \in [0,1]\, (\mathrm{Rat}(\alpha) \vee \neg \mathrm{Rat}(\alpha))$.

As Brouwer created intuitionistic mathematics before the notion of recursive function was conceived, it is interesting to see what the impact of the theory of recursive functions on intuitionism is.

At many points in the intuitionistic literature effective procedures or computable functions are made use of and one may well wonder whether recursive functions would do. The question to be asked is: does Church's Thesis hold? Church's Thesis claims that every effectively computable function is (general) recursive. It is noteworthy that in contrast to classical mathematics Church's Thesis can be formulated in formalized intuitionistic mathematics. There are several ways of doing so; for example Kreisel [5]) used a system with variables for lawlike functions to obtain a formulation. One

1) Troelstra 69, pp. 40, 41.

2) For a detailed analysis see Kleene-Vesley 65, Ch. I, p. 6.

3) For the connection between the bar theorem and transfinite induction see Howard-Kreisel 66.

4) Brouwer 28.

5) Kreisel 65, 2.7.

can, however, avoid the use of special variables by going back to the meaning of the statement $\forall x \exists y A(x, y)$, where A does not contain choice parameters. To an intuitionist the assertion $\forall x \exists y A(x, y)$ means "there is an effective procedure to compute for each x the corresponding y". So a natural formulation of Church's Thesis is the following schema:

$$\forall x \exists y A(x,y) \rightarrow \exists n [\forall m \exists k T(n, m, k) \wedge \forall x A(x, \{n\}(x))] \ .$$

(T is Kleene's T-predicate, which has the heuristic meaning: "$T(n, m, k)$ iff the Turing machine with Gödel number n and input m performs the computation k".) So the thesis actually belongs to the object language and hence may be provable or refutable. So far Church's Thesis still presents an open problem. It has been shown to be consistent with various versions of intuitionistic analysis [1]). It must, however, be remarked that on the basis of the intuitionistic foundations of mathematics a strictly mechanistic characterization of effectivity does not seem probable.

Quite a different question is whether the universe of analysis, i.e. the collection of number theoretic functions, can be taken to consist of recursive (or even lawlike) functions. Here the answer must be negative, the notion of choice sequence being clearly of vital importance for intuitionistic mathematics. A formal hint in that direction is the failure of the fan theorem in the case that one admits recursive functions only [2]).

Apart from Brouwer's work there have been other attempts to develop a constructivistic theory of the continuum.

Of the "semi-intuitionistic" continua, the continuum introduced by Weyl in 1918 (and later abandoned by him in favour of Brouwer's continuum) has been examined most thoroughly, notably by Grzegorczyk [3]). The latter's formalization of Weyl's informal restriction to certain "e(lementary) d(efinable)" analytical methods starts with the arithmetic of integers; the class of e.d. relations shall be closed with regard to the logical operations of the propositional calculus and to universal quantification over an integer variable. The notion of an e.d. function is defined by means of the minimum (least integer) of an e.d. number-theoretical relation, provided a minimum exists. An e.d. functional is analogously defined over a finite number of functions and such that it assumes integral values.

By means of these notions, e.d. analogues of the classical concepts of real

1) Kreisel-Troelstra 70.

2) See Kleene-Vesley 65, p. 112.

3) Weyl 18 and 21, Grzegorczyk 54.

number and of sequences and sets of real numbers are defined. It is confirmed that the "semi-continuum" (field) of e.d. real numbers is denumerable, and an essential part of the e.d. analysis of continuous functions is developed. Differentiation and integration on families of continuous functions prove e.d. and an e.d. continuous function defined on an interval whose ends are e.d. assumes its maximum at an e.d. value. The theory of recursive functions also provided a suitable starting point for a constructive version of analysis. One can introduce a "continuum" consisting of *recursive reals* (i.e. reals with a recursive Cauchy sequence). Many interesting results have been obtained; for instance, a version of the continuity theorem [1]). The usual techniques of recursion theory provide also many counterexamples to theorems from classical analysis.

Under the influence of the mathematician A.A. Markov, a Russian school of constructive analysis came to flourish [2]). The Russian school actually employs constructive (i.e. intuitionistic) logic in its mathematical research. Markov introduced, however, one new principle that lacks intuitionistic motivation: $\neg\neg\exists x\, Ax \rightarrow \exists x\, Ax$ for primitive recursive A (*Markov's principle*).

The relation of Markov's principle to formal intuitionistic systems has been studied by a number of logicians [3]).

Thus far we have examined the connection between Brouwer's definition of set and certain fundamental concepts of the foundations of mathematics in general. But we may also raise the "practical" question, *what category of sets in classical mathematics is represented by Brouwer's sets.* Menger 28 showed that there is a close relation [4]) between the sets in Brouwer's sense and the abstract analytic sets [5]); however, the *tertium non datur* is required to show this relation. To clarify this connection we start with the concept of *ramified set* [6]).

Let D be a class of objects in which an operation is defined such that, to any (finite or denumerable) sequence of objects from D, an object corresponds which need not

1) See Y.N. Moschovakis 64, Klaua 61, Mazur 63.

2) The activity of the school covers a wealth of subjects. For information the reader is referred to the *Mathematical Reviews* or the *Zentralblatt.* As examples we just mention here Markov 58, Šanin 62.

3) Kleene-Vesley 65, Kreisel 62a, Troelstra 71.

4) Heyting criticizes Menger for not appropriately considering choice sequences.

5) The concept of analytic set was first introduced by M. Souslin in 1917; therefore some authors, for instance Hausdorff, use the term 'Souslin's sets'. The first expositions of the theory of analytic point sets and also of abstract analytic sets are found in Hausdorff 22, § § 19 and 32, and in Lusin 27 (cf. 30).

6) Menger 28, p. 213. A very interesting historical outline is given in Sierpiński 63.

belong to D. If $d_1, d_2, ..., d_k, ...$ are objects of D given in this succession, we denote the result of the operation by $\psi(d_1, d_2, ..., d_k, ...)$.

Furthermore, to any *finite* sequence of positive integers $(n_1, n_2, ..., n_k)$, let there correspond an object of D which shall be denoted by $d_{n_1 n_2 ... n_k}$.

For a given infinite sequence of integers $\nu = (n_1, n_2, ..., n_k, ...)$ we consider the sequence $(d_{n_1}, d_{n_1 n_2}, ..., d_{n_1 n_2 ... n_k}, ...)$ and denote $\psi(d_{n_1}, d_{n_1 n_2}, ..., d_{n_1 n_2 ... n_k}, ...)$ by d_ν. Each of the objects $d_{n_1}, d_{n_1 n_2}, ..., d_{n_1 n_2 ... n_k}$ is called a *constituent* of d_ν. Then the set of all objects d_ν, when ν runs over all infinite sequences of positive integers, is called a *ramified set*; more precisely, the ramified set obtained from the class of objects $\{d_{n_1 n_2 ... n_k}\}$ by the operation ψ, where $k, n_1, n_2, ...$ are any positive integers. $\{d_{n_1 n_2 ... n_k}\}$ shall be called the class *producing* the ramified set.

If the objects of D are sets and ψ is the operation of intersection $\cap$, the ramified set is the set of all intersections d_ν, ν denoting all infinite sequences of positive integers. The union of all sets d_ν *is called the analytic set* produced by the class of sets $\{d_{n_1 n_2 \cdots n_k}\}$. (The same analytic set may be produced by different classes of sets.)

One easily sees that there is a close connection between this concept and Brouwer's concept of set (spread). By essentially using the principle of the excluded middle, Menger showed that *the sets in the sense of Brouwer's definition coincide with those ramified sets for which every object* $\neq O$ (the null-object) *of the producing class is a constituent of at least one* d_ν. Thus, by means of the *tertium non datur,* every ramified set proves to be a set in Brouwer's sense.

The main importance of these results lies in the fact that analytic sets have been extensively studied and are, incidentally, considered by many (especially French) mathematicians, who are remote from intuitionism, to be the only "definable" sets. On the other hand, the results have not much significance from the intuitionistic point of view because they depend on the principle of the excluded middle.

In the years following 1948, Brouwer, in a number of papers, further exploited the characteristic properties of intuitionistic mathematics. He based the proofs of several theorems on the activity of the creative mathematician. These theorems (e.g.: "apartness is strictly stronger than inequality") had been considered extremely plausible, but until then no proof had been given.

Lately, Brouwer's papers on essentially negative properties, contradictoriness of parts of classical mathematics, etc. have led to a more systematic study of the idealized mathematician and his manifestations. Notably Kreisel, Kripke, and Myhill have considered the subject [1]). It is important as it allows a closer analysis of Brouwer's historical arguments and because it has important consequences for the existence of functions.

1) Kreisel 67, Myhill 67.

Brouwer explicitly introduced a subjective element in mathematics by reference to the activity of a creating subject [1]). Kreisel has laid down a few fundamental principles which involve a new primitive notion $\vdash_n$, which is interpreted as "the creative subject has at stage n evidence for ...". The underlying idea is that the creative subject's activity proceeds in stages. The following principles were proposed [2]).

(i) $\vdash_n A \vee \neg \vdash_n A$

(ii) $(\vdash_n A) \rightarrow A$

(iii) $(\vdash_n A \wedge m > n) \rightarrow \vdash_m A$

(iv) $A \rightarrow \neg\neg \exists n \vdash_n A$

(iva) $A \rightarrow \exists n \vdash_n A$.

(i) asserts that the notion "$\vdash_n$" is decidable, (ii) asserts that A holds as soon as the creative subject has evidence for A, (iii) states that the creative subject does not forget, and finally (iv) states that if A is true, it is absurd that the creative subject will never find evidence for it. Actually a stronger form (iva) is plausible on the solipsist basis: if A holds then the creative subject must already have evidence for A. The new notion is still highly problematic, careless handling can easily produce contradictions [3]). The introduction of a new primitive notion can be avoided, while at the same time preserving the advantages of the method by *Kripke's schema.*

Suppose one defines a function α by

$$\alpha(n) = \begin{cases} 0 & \text{if } \neg \vdash_n A \\ 1 & \text{if } \vdash_n A \end{cases} \quad \text{for a given } A;$$

then

(w) $$\exists\alpha[\{\forall x\,(\alpha x = 0) \leftrightarrow \neg A\} \wedge \{\exists x\,(\alpha x \neq 0) \rightarrow A\}]$$

holds if one uses (iv). By applying (iva) one shows that

(s) $$\exists\alpha[\exists x\,(\alpha x \neq 0) \leftrightarrow A] .$$

1) Brouwer 49.
2) Kreisel 67, pp. 158–161.
3) Cf. Troelstra 69, p. 105.

(w) is called Kripke's schema; we will call (s) Kripke's strong schema. One notes here the analogy with the comprehension schema. It turns out that Kripke's schema is powerful enough to formalize most of Brouwer's historical arguments [1]). Instead of pursuing Brouwer's arguments, let us look closer at the functions provided by Kripke's schema. If the formula A does not contain choice parameters, we conclude from the decidability of the notion $\vdash_n$ that the function α is lawlike (effective). Myhill called these functions *empirical*. The empirical functions embody a considerable expansion of the known lawlike functions. The empirical functions have many unexpected properties, for instance it is provable that the class of empirical lawlike functions is not enumerable by a lawlike function. This shows that Church's Thesis does not hold for that class, a result that is not surprising if one realizes that the Thesis was formulated for "mechanically" computable functions. However, there is no doubt that Kripke's schema is intuitionistically well motivated. In effect Myhill, in his formalization of intuitionistic analysis [2]), includes it among his axioms.

§ 6. MATHEMATICS AS TRIMMED ACCORDING TO THE INTUITIONISTIC ATTITUDE

After the preceding sections it is clear that the historical field of mathematics has to undergo serious restrictions in order to conform to the principles imposed by the intuitionists. In this final section we shall give a survey of the effect produced by these restrictions and of what can be saved; mostly we shall refer for details (in a positive or negative direction) to the intuitionistic literature, elaborating the arguments in a few characteristic cases only. In general we limit ourselves to Brouwer's intuitionism, while restrictions eliminating negation (pp. 249ff) shall be disregarded altogether.

Arithmetic and algebra in the strict sense are less affected than other branches. In those parts of algebra where only discrete species appear so that for any two members it can be decided whether they are equal or not, constructive methods should be applied which produce a result by a finite number of steps, as Kronecker had already done in his treatment of certain arithmetical problems, rejecting the (simpler) "classical" methods. The finite extensions of the field of rationals (or of a field with p^n members) belong to this part of algebra. But far-reaching modifications prove necessary wherever

1) Hull 69.
2) Myhill 68.

the distinction between $\neq 0$ and $\# 0$ (see below) is unavoidable; for instance in the field of real numbers. Hence the very definition of a field becomes complicated, for not every non-zero element has an inverse. In the field of all real numbers, for instance, a^{-1} only exists if $a \# 0$, i.e. if a rational number can be named which lies between 0 and a [1]). The apartness relation # was introduced as a positive analogue of $\neq$. Heyting characterized it by the properties

1) $\neg a \# b \leftrightarrow a = b$

2) $a \# b \rightarrow \forall c\, (a \# c \vee b \# c)$.

From 1) and 2) one immediately concludes $a \# b \leftrightarrow b \# a$ and $a \# b \rightarrow a \neq b$. A somewhat surprising feature is the identity $\neg\neg a = b \leftrightarrow a = b$; this follows from 1) by taking double negations. The presence of an apartness relation is of the greatest importance for the building of a positive theory. In the theory of reals, $a \# b$ is defined by $\exists k(|a-b| > 1/k)$.

The theory of *groups* with an apartness relation is not greatly different from the ordinary theory, but the properties of *rings* and *fields* become rather involved. In a ring without divisors of zero or in a field one cannot, from $ab = 0$ and $a \neq 0$, conclude $b = 0$; if this conclusion is generally valid the ring is called 'regular', and if it moreover contains for every $a \# 0$ an inverse a^{-1}, a 'field'. According to this definition, the rationals as well as the real and the complex numbers constitute fields. An intuitionistic treatment of the fundamentals of algebra was given by Heyting [2]). Other constructive approaches to algebra led to the so-called computable algebra [3]), based on recursion theory.

In *analysis* the difficulties are much more fundamental, and the restrictions involved prove rather catastrophic. With the introduction of real numbers by the device of choice sequences, the comparison of numbers according to magnitude becomes impossible in general [4]); in other words, given real numbers a and b, the trichotomy $a \gtreqless b$ ceases to be valid. A characteristic example is the following. It is unknown whether in the decimal expansion of π a sequence of (at least) seven 7's will occur; if so, let h denote the place after the decimal point (i.e. the index of the respective digit in $\pi =$

1) For this and the following see Heyting 34 (or 55), 56, 41.

2) Heyting 41, 56.

3) Fröhlich-Sheperdson 56, Rabin 60, Lambert 68, Eršov 68, Mal'cev 61. For earlier constructive treatments see Kronecker 1882, 1883, Van der Waerden 30, Vandiver 34–35.

4) See also Brouwer 50, in addition to his earlier papers. For a thorough treatment, see Kleene-Vesley 65, Ch. III and IV, and Kreisel-Troelstra 70.

$3.a_1 a_2 \ldots a_h \ldots$) where for the first time such a sequence starts. Now we define a decimal ρ, which begins with 0.777 ..., as follows: if h exists as defined, replace the digit 7 at the hth place after the point in ρ by 6 if h is odd and by 8 if h is even, hence *all* digits of ρ are 7 "if no h exists". The real number ρ is effectively defined since, simultaneously with the expansion of π, the successive digits of ρ can be calculated. But without the *tertium non datur* it cannot be asserted that one of the cases $\rho = \frac{7}{9}$, $\rho < \frac{7}{9}$ or $\rho > \frac{7}{9}$ holds true, or even that either $\rho = \frac{7}{9}$ or $\rho \# \frac{7}{9}$. This does not contradict the fact that the rational numbers are comparable; for ρ cannot, according to our present knowledge, be considered rational – though after a possible decision regarding the integer h (non-existent, odd, even), ρ proves to be a rational in each case. Brouwer, in his later papers, proved an even more amazing fact: $\neg \forall a(a \neq 0 \rightarrow a \# 0)$, where a ranges over reals [1]).

The incomparability of real numbers fatally affects many classical proofs in analysis. While in a few cases certain intuitionistically invalid proofs have been replaced by constructive ones, for the majority this has not been achieved nor is there any prospect of achieving it; still worse, in some cases the negation of the formula under consideration can be proved.

An elementary example may illustrate the situation. For the so-called *fundamental theorem of algebra* dozens of proofs have been given since the end of the 18th century. Many of the proofs use various theorems from the theory of (real or complex) functions, sometimes also topological facts, while others (such as the second proof of Gauss, simplified by Gordan) are completely arithmetical except for the use of the elementary analytic theorem by which every algebraic equation with real coefficients and an *odd* degree has a real root. Proofs of the first kind had already been rejected before the turn of the 19th century by intuitionists such as Kronecker and Mertens.

Yet the "arithmetical" proofs, too, contain ingredients which are not acceptable to intuitionists; in fact, in 1924 new proofs of the fundamental theorem were independently published by scholars from three different countries [3]) who maintained that no sound proof had been given before. An essential shortcoming of most "arithmetical" proofs can be described as follows. In order to get rid of multiple roots of the given equation one has to ascertain whether its discriminant equals 0. For instance, the equation $x^3 - \frac{49}{27}x + 2\rho^3 = 0$, where ρ is the real number defined above, has the discriminant $D =$

1) Brouwer 49, Heyting 56, § 8.1.2, Hull 69.

2) Brouwer 48; cf. Brouwer 49 and Van Dantzig 49.

3) Brouwer-de Loor 24, Skolem 24, Weyl 24; Rosenbloom 45. Cf. Van der Corput 46, Rice 54. Fundamentally, this argumentation, too, goes back to Kronecker (1882; cf. Fine 14, Vandiver 34–35 [*Annals of Math.*]).

$-108(\frac{7}{9})^6-\rho^6)$. In this case D is intuitionistically neither equal to nor apart from 0. Generally speaking, the impossibility of basing the proof on the alternative that the discriminant is either 0 or a positive or a negative number, makes the proof illusory. Therefore a proof is required (and can be given) which enables us to calculate the roots of the equation numerically with any desired precision, starting with an approximation of the coefficients by rational numbers; for the discriminant of an algebraic equation with *rational* coefficients is also intuitionistically either equal to or apart from 0. As a matter of fact, the constructive proof brings its own reward: as a by-product one obtains that that root depends continuously on the coefficients (in a certain range).

In the foundations of analysis, the theorem of Bolzano-Weierstrass [1]), the convergence of a bounded monotone sequence of real numbers, the existence of a zero for a continuous real function $f(x)$ with $f(a) < 0$ and $f(b) > 0$ [2]), the theory of Dedekind cuts, and the existence of the least upper bound for a bounded set (species) of real numbers and of a maximum for a continuous real function in a closed interval are among the victims of the intuitionistic criticism; these theorems prove either false or meaningless. In the case of Bolzano-Weierstrass, for instance, the obstacle is the alternative between mutually exclusive cases which uses the *tertium non datur*. It is also required for the comparison of two Dedekind cuts $\kappa = (K_1 | K_2)$ and $\lambda = (L_1 | L_2)$ in the form: either all rationals of L_1 belong to K_1 or else there exists in L_2 a rational which is not contained in K_1.

Regarding convergence and the theory of infinite series, Brouwer claimed [3]) that the notion of convergence ought to be split up into different notions, and Belinfante in a series of ingenious papers [4]) showed that Brouwer's distinction between "positive" and "negative" convergence leads to two quite different theories only the first of which is similar to the classical theory. Absolute convergence and summability are also dealt with by Belinfante. Interest in these traditional areas of calculus has waned [5]). The collection of intuitionistic real functions is severely hamstrung, compared to the classical counterpart, by Brouwer's theorem that every function, which is

1) For the use of the *tertium non datur* in this theorem, cf. Billing 49 and Rice 54.

2) Note that in the premise $f(a)$ and $f(b)$ are supposed to be apart from 0.

3) Brouwer 25.

4) Belinfante 29, 30, 38, 38a. See also Dijkman 46, 52, 62 and Van Rootselaar 52. In particular, Dijkman modifies the concept of "negative convergence" so as to preserve the classical theorems about the convergence of the sum and the product of convergent series.

5) However, quite recently W. Gielen has proved the equivalence of absolute and unconditional convergence using Brouwer's principle for numbers (unpublished).

defined on a closed interval, is uniformly continuous [1]). This is a very deep theorem, based on fundamental intuitionistic principles. One might think that a function f with the following definition

$$f(x) = \begin{cases} 0 & \text{if } x = 0 \\ \dfrac{|x|}{x} & \text{else} \end{cases}$$

provides a counterexample. This f, however, is not everywhere defined in $[-1,1]$, namely not for those arguments x for which it is unknown whether $x = 0$ or $x \neq 0$.

One has to keep in mind that the often deplored impoverishment of mathematics by the intuitionistic purge is but one side of the story. Certainly, quite a number of theorems are rejected, but also some quite strong theorems are added. The mentioned continuity theorem is one of these. For other examples the reader is referred to Heyting's monograph 56.

M.J. Belinfante has developed large parts of the theory of complex analytic functions. He proved some of the fundamental theorems, including the integration of the logarithmic derivative and Picard's theorems [2]). The area as such does not seem to present intuitionistically challenging problems that could not be formulated in the theory of real functions.

Generally, in arithmetic and still more in analysis even those results which remain meaningful and true in the eyes of intuitionists mostly require new proofs which are very complicated; one of the reasons is the *impossibility of using indirect proofs* (see above p. 235) [3]). In mathematical practice, intuitionists presuppose knowledge of the "illusory" classical proofs in order to motivate their criticism and to make it clear why new demonstrations are required.

There exists no autonomous *geometry* for intuitionism, no more than for Weyl's semi-intuitionism – except for the very concept of continuum which is saved by the choice sequences. The notions of real number and continuum precede the notion of space. In the familiar cases spaces are constructed as manifolds over the continuum. Of course, one can apply the axiomatic

1) Brouwer 24a, 27, Heyting 56 (p. 46), Kleene-Vesley 65 (§ 15).

2) Belinfante 31. Cf. Goodstein 51 Ch. VI, where besides functions of a complex variable, measure theory and constructive topics of topology are also treated from a viewpoint not much different from Brouwer's.

3) Cf. the historical exposition Goodstein 48.

method (based on intuitionistic reasoning (logic)), but one has to keep in mind that it is not autonomous. Axiomatic theories are nothing but convenient vehicles economizing intuitive reasoning. So the meaning of a theorem T being derived from an axiom system Σ is that, whenever we have constructed a mathematical system that satisfies Σ, we automatically know that T is satisfied, too. In the particular case of geometry, it is by analytic geometry that we provide meaning to the axiomatic frame [1]).

In a somewhat paradoxical way the crisis of the *tertium non datur* influences geometry; Desargues's theorem, for instance, after having been proven on the one hand for triangles situated in the same plane and on the other hand for triangles in distant planes, should not be considered to be proven altogether, because "in general" a decision between the two cases cannot be reached [2]). In topology, especially in combinatorial topology and Euclidean spaces, Brouwer provided the basic notions and laid down the basis for later work; he also reformulated part of his earlier work in topology (e.g. the Jordan curve theorem, 25a, the notion of dimension, 26, the fixed-point theorem, 52). In 1926 Brouwer investigated the problem which spaces are intuitionistically meaningful using methods based on a metric. Freudenthal, in 1936, succeeded in constructing topological spaces in an intrinsically topological way, independent of metric notions. Since then Troelstra has considerably extended Freudenthal's methods and results [3]). A fruitful notion introduced by Brouwer [4]) is that of a *catalogued or located set* (species). In general one cannot expect a subset B of A to be decidable [5]), i.e. $\forall x \in A\,(x \in B \vee x \notin B)$; as a matter of fact it is a corollary of the continuity theorem that the only decidable subsets of $[0,1]$ are the empty set and $[0,1]$ [6]). The closest we can get to a notion of a "well-behaved" set S in a metric space is to require that for every point p its distance from S can be determined. Sets with that property are called catalogued. Catalogued sets are basic in branches of intuitionistic mathematics that involve topology [7]). In measure theory and functional analysis the early work of Brouwer was con-

1) See Heyting 28a, 59, Van Dalen 63. Applications of the axiomatic method are also made in Algebra (Heyting 41), Hilbert space (Heyting 53), Topology (Freudenthal 36, Troelstra 68).

2) Heyting 28a, p. 511.

3) Troelstra 66, 68.

4) Brouwer 19, p. 13.

5) *Removable or detachable,* in intuitionistic literature.

6) The "Unzerlegbarkeit" of the continuum, cf. Heyting 56, p. 46.

7) Outside intuitionism the concept has been employed by Bishop 67. See further Van Dalen 68.

tinued by Heyting and Van Rootselaar [1]). Heyting considered the intuitionistic theory of Hilbert spaces and proved the Riesz-Fischer theorem. Van Rootselaar generalized the notion of the Brouwer integral. Recently, Ashwinikumar and Gibson [2]) have further extended the work of Heyting and Van Rootselaar. The most revolutionary changes, naturally, are found at the very roots of intuitionism, where such notions as choice sequence, spread, and species are introduced. These concepts were explained in the preceding section. The construction of the continuum as a spread, compared to various other constructive conceptions of the continuum, is certainly a major achievement of Brouwer. Moreover, facts like the continuity theorem and the indecomposability of the continuum confirm that the continuum has the desired intuitive properties. Brouwer [3]) called spreads and elements of spreads (i.e. choice sequences) mathematical entities. Starting with mathematical entities he built a hierarchy of species: a property of mathematical entities is a species of the first order, a species of the second order is a property of mathematical entities and of species of the first order, etc. Thus "species" means approximately what is meant by "set" in the constructive stage of classical set theory. As the elements of a species S must already be defined (independent of S), impredicative definitions are excluded. Many notions of set theory are split up into several non-equivalent notions in the theory of species. E.g. the identity relation between sets has two intuitionistic counterparts:

(i) the identity relation: $\forall a \in A \, \exists b \in B \, (a=b) \wedge \forall b \in B \, \exists a \in A \, (a=b)$,
(ii) the congruence relation: $\neg \exists a \in A \, \forall b \in B \, (a \neq b) \wedge \neg \exists b \in B \, \forall a \in A \, (a \neq b)$.

This splitting of classical notions is a general phenomenon [4]). The theories of cardinals, ordered species, and ordinals (well-ordered species) are considerably more complicated then their counterparts in classical mathematics. A few examples may suffice.

J.J. de Iongh considered the following notions of finiteness: (i) S is finite if S is in one-one correspondence with an initial segment of the natural number sequence, (ii) S is quasi-finite if it is the image of a finite species, (iii) S is pseudo-finite if S is a subspecies of a finite species, (iv) S is bounded in number by n if S does not contain a subspecies of n elements, and some refine-

1) Brouwer 23, Heyting 51, 53 (cf. 56, Ch. VI), Van Rootselaar 54.
2) Ashwinikumar 66, Gibson 67.
3) Brouwer 18–19, 25–27.
4) Brouwer 24b, Dijkman 52, cf. Heyting 56, 7.3.2.

ments. Mixtures like pseudo-quasi are also permitted. De Iongh's results are contained in Troelstra 67, where countability predicates are also considered.

It turns out that most of the finiteness notions are distinct; for example quasi-pseudo-finite implies pseudo-quasi-finite, but the converse is not true. With respect to the theorem of Cantor-Bernstein, recently two results have been obtained. Troelstra [1]) showed that if one allows empirical functions, the Cantor-Bernstein theorem holds for species of natural numbers. On the other hand, VanDalen [2]) proved the Cantor-Bernstein theorem for fans to be contradictory. Let us call two species S and T equivalent (*gleichmächtig*) if there exists a one-one mapping of S onto T, and let $S \lneq_1 T$ stand for "there is a one-one mapping f: $S \to T$, but no one-one mapping g of T onto S". Now consider the continuum C and the species N of natural numbers. The existence of a one-one mapping of N into C is clear. Suppose now that there is a one-one mapping f: $S \to N$; according to the continuity theorem (p. 260), f must be continuous and hence constant. This contradicts the supposition that f is one-one. We conclude now that $N < C$ [3]). Note that no diagonal argument is used! Essentially the same continuity arguments show that $N \lneq_1 B \lneq_1 C$, where B is the binary fan. Apparently there is an abundance of distinct cardinalities, however, all of which are, from a classical point of view, $\leqslant 2^{\aleph_0}$. There is no evidence whatsoever for higher cardinalities. Brouwer, in his thesis, denied them any mathematical content. Should one for example want to consider C^C, then all one obtains is the species of all continuous real functions. So the diagonal procedure is not applicable. From the classical point of view this does not represent an increase in cardinality.

The status of the axiom of choice in intuitionistic mathematics is rather remarkable. Under certain circumstances there are arguments in its favour and under different circumstances the axiom is refutable. Consider a statement of the form $\forall x \exists y A(x,y)$, where x and y range over natural numbers. This statement holds intuitionistically if, for each x that is presented, a corresponding y can actually be calculated and, moreover, there must be a uniform procedure for the calculation. This uniform procedure provides us with a choice function, so the axiom of choice holds in this case [4]). The axiom does not hold in general. Consider for example the statement: "for each real number α there exists a natural number n such that $\alpha < n$". The statement

1) Troelstra 69, p. 104.
2) Van Dalen 67.
3) See Brouwer 25–27, p. 253.
4) Cf. Kleene-Vesley 65, p. 17, 2,2.

evidently holds, but there is no choice function, because if there were one, it would have to be continuous and hence constant. Note that the dependence of n on α is intensional (i.e. n depends on the representation of α) while the notion of function is extensional.

The intuitionistic theory of order necessarily diverges from the usual one, as the trichotomy $a \gtreqless b$ fails in a number of important cases. Instead the following formulae may hold: (i) $\neg(a<b) \wedge \neg(b<a) \rightarrow a=b$, (ii) $a<b \rightarrow \forall c$ $(a<c \vee c<b)$. For a survey of the theories of order we refer to Heyting's monograph. Brouwer used his method of the creative subject to establish some strong facts, e.g. $\neg \forall a \ [a \neq 0 \rightarrow (\neg a>0 \vee \neg a<0)]$ [1]) in the theory of reals. Kleene extensively analyzes a number of Brouwer's results in the framework of the formal system of Kleene-Vesley [2]).

The theory of well-ordered species is based on the following constructive generation principles:

(i) the ordered union of a finite number of well-ordered species is a well-ordered species,

(ii) the ordered union of a countable sequence of well-ordered species is a well-ordered species.

The generation process starts with the singleton. Note that this approach to well-ordering goes back to the early papers of Cantor.

The well-ordered species share many properties with their classical counterparts, such as "every well-ordered species has a first element", "every descending sequence $a_1 \, a_2 \, a_3 \ldots$ in a well-ordered species has a last element (i.e. is finite)", and the principle of transfinite induction. However, the property that is mostly used to characterize well-ordered sets, i.e. "every non-empty subset of a well-ordered set contains a least element" does not hold for well-ordered species. Even the weaker statement, "every inhabited [3]) sub-species contains a least element", fails.

It must be noted that all well-ordered species are countable. Considerable parts of the theory of well-ordering have been developed by Brouwer [4]). A relation between the class K (p. 259) of neighbourhood functions and the class of well-ordered species was noticed by Troelstra [5]); as a matter of fact he showed that each is explicitly definable in the other.

1) See Heyting 56, p. 118.

2) Kleene-Vesley 65, Ch. IV.

3) S is inhabited if $\exists x \, (x \in S)$, to be distinguished from "non-empty" (i.e. $\neg \forall x$ $(x \notin S)$).

4) Brouwer 25–27 III.

5) Troelstra 69, § 14.5.

The notion of well-ordering played a key role in the bar theorem [1]). The fundamental importance of well-ordering appears from Brouwer's insistence on the legitimacy of infinite proofs (as well-ordered species of "elementary conclusions") [2]). Intuitionistic set theory, in the form of a theory of species and a theory of choice sequences and spreads, has certainly developed since its initiation. Although the progress along the traditional lines, pointed by Brouwer, is modest, the study of formal aspects has come to flourish by the work of mathematicians such as Kleene, Kreisel, Myhill, Troelstra, and Vesley [3]). In view of the severe restrictions that intuitionism imposes on mathematics it is not surprising that only a handful of mathematicians have been willing to accept the intuitionistic principles as far as the daily practice of mathematics is concerned. On the other hand, the study of formal properties of intuitionistic logic and mathematics has enjoyed popularity ever since Heyting's formalization in the thirties. The school of Hilbert and some other trends show already for some time full understanding for the basic attitude of intuitionism, as is apparent from the work of Clifford Spector and others. Moreover, there is the remarkable stimulating influence of intuitionism which, mainly in connection with recursion theory (see Chapter V), has suggested a number of improvements in classical analysis [4]) and produced a wealth of new (classically significant) problems and results in proof theory and model theory, notably through the work of Kleene and Kreisel.

In conclusion one can say that the fierce disputes between formalists and intuitionists belong to the past. Although both sides stick to their fundamental principles, a mutual appreciation has developed, which has already begun to bear fruit [5]).

1) See Brouwer 27, 54.

2) Brouwer 27, footnote 8, cf. Kreisel-Newman 69.

3) *Loc. cit.*

4) See for instance Grzegorczyk 59, Bishop 67.

5) The conference on intuitionism and proof theory (Buffalo, 1968) convincingly bears witness to that, see Kino-Myhill-Vesley 70.

CHAPTER V

METAMATHEMATICAL AND SEMANTICAL APPROACHES

§ 1. THE HILBERT PROGRAM

So far, we have discussed three main approaches to the problem of rebuilding the foundations of set theory, the "naive" Cantorian conception of which was so badly shaken by the antinomies. The *Brouwerians* believed that this conception was wholly wrong from the beginning. They accused it of misunderstanding the nature of mathematics and of unjustifiedly transferring to the realm of infinity methods of reasoning that are valid only in the realm of the finite. By regaining the right perspective, mathematics could be constructed on a basis whose intuitive soundness could not be doubted. The antinomies were only the symptoms of a disease by which mathematics was infected. Once this disease was cured, one need worry no longer about the symptoms. All *Russellians* thought that our naiveness consisted in taking for granted that every grammatically correct indicative sentence expresses something which either is or is not the case, and some – among them Russell himself – believed, in addition, that through some carelessness a certain type of viciously circular concept formation had been allowed to enter logico-mathematical thinking. By restricting the language – and proscribing the dangerous types of concept formation – the known antinomies could be made to disappear. Their faith in the consistency of the resulting, somewhat mutilated, systems was less strong than that of the Brouwerians, since certain intuitively not too well founded devices had to be used in order to restore at least part of the lost strength and maneuverability. *Zermelians,* finally, thought that our blunder consisted in naively assuming that to every condition there must correspond a certain entity, namely the set of all those objects that satisfy this condition. By suitable restriction of the axiom of comprehension, in which this assumption is formulated, they tried to construct systems which were free of the known antinomies yet strong enough to allow for the reconstruction of a sufficient part of classical mathematics. Their faith in the consistency of the resulting systems was based on nothing more – but also on nothing less – than the fact that the usual ways of deriving the known antinomies could not be reproduced.

The adherents of the axiomatic approach to the foundations of set theory, and to a somewhat lesser degree the adherents of the type-theoretical approach, were badly in need of a *proof* of the consistency of their systems. The classical method of providing such a proof, viz. the exhibition of a *model* taken from a theory whose consistency was not in doubt – in the tradition of Beltrami who in 1868 proved that certain non-Euclidean geometries were consistent (relative to Euclidean geometry, as we would say now) by constructing a model for them within Euclidean geometry, or of Hilbert who in 1899 constructed a model for Euclidean geometry within real number theory, thereby proving its consistency (relative to real number theory) – could not be applied: finite models were obviously not suitable, and no conceptual framework within which an infinite model could be constructed might be regarded as safe, in view of the antinomies. A different method was needed. It was Hilbert who supplied it, dimly in 1904, but with increasing precision and pregnancy from 1917 [1]): it must be shown that the standard mathematical proof procedures are strong enough to derive *all* of classical mathematics – including *all* of Cantor's set theory – from suitable axioms but not so strong as to derive a contradiction. That classical mathematics was basically sound was an act of faith with Hilbert which was never shaken by the antinomies.

Hilbert intended to carry his program through in two steps: first, all of mathematics – as a matter of fact, he was thinking mainly of arithmetic, analysis and set theory, – had to be *formalized* [2]), i.e. a *formal system,* or *formalism*, had to be constructed from whose axioms should be derived at least the beginnings of mathematics to a degree corresponding to that achieved, say in *Principia Mathematica*, with the help of a certain definite set of rules of inference. The system was to be formal in the sense that only the kind and order of symbols, upon whose sequences the rules of inference were to operate, was to be taken into account but not, for instance, their "meaning". Such a system could be sufficiently mastered by a minimum of intuition, the so-called "global intuition" required to decide whether two symbol occurrences are occurrences of the same symbol or of different symbols, a kind of intuition that requires no intellectual powers at all and can be built into suitably constructed machines. Whether a series of symbol sequences is or is not a proof of its last sequence can, in such a system, be

1) Cf., e.g., Hilbert 22, p. 160.
2) *Ibid.,* p. 174.

mechanically checked by using essentially nothing more than a purely mechanical operation of matching. This complete formalization goes far beyond the kind of formalization provided in *formal axiomatics.* There one relies on an "understood" logic and is satisfied with a complete listing of all specific primitive terms and an enumeration of all assumptions formulated in these terms necessary for the derivation of a certain body of theorems and takes pain that the intended meaning of the specific, extra-logical terms of the theory under treatment should never enter into the derivations unless explicity expressed by the axioms. (Incidentally, it was Hilbert himself who brought the *formal axiomatic method* to its perfection.) Here, the intended meaning of *all* terms is to be disregarded, including that of the logical terms. If one wants to infer the fact that ϕ is true from the fact that ϕ *and* ψ is true, then this can no longer be done by implicit, or even explicit, reliance on the meaning of 'and' but this inference has to be made on the basis of suitable axioms and rules exclusively. Hilbert's kind of formalization is based on what is sometimes called the *logistic method*. Hilbert was, of course, in a position to rely on existing axiomatizations of certain parts of mathematics, such as Peano's axiomatization of number theory and Zermelo's axiomatization of set theory, as well as on existing formalizations of logic, provided by Frege and Russell-Whitehead, though especially the system of *Principia Mathematica* was not quite up to the standards necessary for his purpose. The various systems of formalized arithmetic, developed by Hilbert and his school [1]) over many years of hard work, started indeed from an almost mechanical superposition of Peano's axiom system on the *Principia Mathematica* first-order predicate calculus, though later developments went far beyond this stage and exhibited great originality.

In the second step, Hilbert planned to show that the application of the rules of inference to the axioms could never lead to contradiction, or rather to *formal inconsistency* in one of the senses of this term (with which we shall deal later, in § 4, at length), e.g. that there could exist no valid formal proof the endformula of which would be '1 = 2'. The argumentation by which this impossibility metatheorem was to be established had to be of such an elementary character that its soundness could not possibly be doubted. All those kinds of argumentation which the intuitionists found objectionable *within* mathematics, such as the use of the *tertium non datur* for infinite sets or the inference from the falseness of a universal statement to the truth of a certain existential statement, not to mention impredicative concept forma-

1) For a careful and and very detailed description of these systems, see Hilbert-Bernays 34–39.

tions and the use of the axiom of choice, were to be banned from that metatheory in which the proof procedures of mathematics were to be investigated and which Hilbert called therefore *metamathematics* or *proof theory*. As a matter of fact, he even went beyond the intuitionistic strictures when he insisted that only *finitary* arguments were to be allowed in proof theory. But just as the intuitionists never committed themselves to a complete specification of what proof procedures were admissible – such a commitment would have militated against their basic stand – so Hilbert never produced a univocal statement as to what procedures were regarded by him as finitary. The nearest to such a specification can perhaps be found in the following quotation from Herbrand, that highly gifted French Hilbertian whose early death put an end to great hopes: [1])

> We understand by an intuitionistic [i.e. finitary] argument an argument that fulfils the following conditions: one always deals with a finite and determined number of objects and functions only; these are well defined, their definition allowing the univocal calculation of their values; one never affirms the existence of an object without indicating how to construct it; one never deals with the set of all the objects *x* of an infinite totality; and when one says that an argument (or a theorem) holds for all these *x*, this means that for every particular *x* it is possible to repeat the general argument in question which should then be treated as only a prototype of these particular arguments.

It should be stressed that the task of metamathematics was not only to show the consistency of mathematics proper by safeguarding it against the antinomies; among other things, it was meant to protect mathematics from the restrictions that other schools, especially those of intuitionistic provenience, were trying to impose upon it [2]).

Had Hilbert succeeded in carrying out his original program, this might have been the end of foundational research for most mathematicians as it would have been for Hilbert: this kind of research for him was a not too pleasant duty that he felt obliged to perform but which distracted him from other more attractive occupations. True, a minority would still have claimed then, as they do now, that consistency of a formalism as such is by no means a sufficient condition for the material truth of any of its interpretations. Even if a formal system in which, say, an axiom of choice is contained, is

1) Herbrand 32, p. 3, footnote 3.
2) See Hilbert 22, p. 174.

demonstrably consistent, *selection functions* – so this minority would argue – just *don't exist*. To which Hilbert – in this respect being, strangely enough, in the good company of Poincaré – would have replied that mathematical existence is nothing but the consistency of the system [1]). But we shall leave the discussion of mathematical existence to the last section.

Fortunately – if we may be allowed to be for a moment somewhat light-hearted on such a serious matter – neither Hilbert nor any of his brilliant followers and associates did succeed in accomplishing this program, not because of any lack in ingenuity but – just because it could not be done [2]). However, during the pursuit of this, as we now know by hindsight, Utopian aim – as it so often happened in the history of mathematics – an enormous wealth of new theories, concepts, and techniques was developed which have already proved to be extremely interesting and fruitful and promise to become even more so in the future. The Gödel theorems of 1931, from which the inaccomplishability of the original Hilbert program can be deduced (see § 6), did shatter certain illusions, to be sure, but they have also been hailed – rightly, we believe – as belonging to the greatest achievements of abstract human thinking in recent times. Hilbert was wrong when he attempted to belittle the crisis into which mathematics had be thrown by the antinomies, and his belief in the essential decidability of all mathematical problems turned out to be unjustified, at least *if* decidability is understood as decidability in one specific formal system. No unique and universally accepted way of reconstructing mathematics exists or is in view, and in this sense the foundational crisis is still in force. But many a scientist wishes that his field were in as "critical" a state as mathematics, and few are the mathematicians who are really depressed by the existing uncertainties in the foundations. *Dealing with these foundations has, surprisingly enough, turned out to be not only a job that had to be undertaken for reasons of intellectual sincerity or philosophical meticulousness but something that was infinitely rewarding, exciting, and fruitful.*

We shall not try to present the rise and decline of the Hilbert program in all its historical details, interesting as such a presentation would doubtlessly be [3]). We shall rather introduce first the concepts in which its present status can be understood and then describe in outline the main results of the

1) For some recent discussions of the relationship between consistency and mathematical existence, see Bernays 50 and Beth 56.

2) For a detailed account of the present status of the Hilbert program, see Kreisel 64.

3) For one such presentation, see Heyting 55, pp. 36 ff; for another, older one, see Bernays 35 a.

theories that were developed during its pursuit. At the end of the chapter, and thereby of the book, we shall deal with some of the philosophical problems connected with the foundations of set theory.

§ 2. FORMAL SYSTEMS, LOGISTIC SYSTEMS AND FORMALIZED THEORIES

Let us start with a general description of the structure of a certain important class of formal systems and then illustrate it by a partial description of a certain number-theoretic formalism as well as by a complete description of a set-theoretical formalism such as might arise from a complete formalization of the set theory ZF presented in the first section of Chapter II.

We have no intention to describe here the structure of all possible formal systems. This was not even quite done by Carnap who dedicated more than 120 pages of his masterwork, *The Logical Syntax of Language,* to General Syntax, i.e. to the theory of formal systems in general [1]).

The following remark as to terminology is appropriate: for any of the technical terms to be used in this and the following sections, there exist many, occasionally very many, synonyms and near-synonyms. On the other hand, many of these terms are homographs and carry different meanings, sometimes with the same author in different or even in the same publications. The reader would only have become bewildered had we tried to list all the synonyms in every case. Therefore, we shall use in general only one or two terms for each concept. When comparing our statements with those found in the literature, great care will therefore have to be taken in determining what corresponds to what. As an illustration only let us notice that 'formal system' has the following synonyms: *formalism, calculus, formal calculus, uninterpreted calculus, abstract calculus, syntactical system, formal language, formal logic, codificate,* and many others.

A formal system [2]) is determined by the following five sets:

(1) A set of *primitive symbols,* the (*primitive*) *vocabulary,* divided into various kinds, such as *variables, constants, and auxiliary symbols.* If the vocabulary is finite, it can be given simply in form of one or more lists. If it is infinite, and in practice also if it is very large, its membership is established in the metalanguage by inductive means, using syntactical variables. In any case, being a primitive symbol has to be an effective notion, enabling us to determine in a finite number of steps whether or not a given symbol is primitive. The primitive symbols are to be regarded as indivisible whatever their external look and whatever the way of their specification.

1) Carnap 37, pp. 153–275; see especially p. 167.

2) Curry's views on the nature of formal systems, expressed in Curry 50, 51, 54 and 63, though idiosyncratic and occasionally somewhat obscure, should be consulted by any serious student of this problem.

In almost all formalisms the number of variables belonging to the primitive vocabulary is infinite. The following inductive specification would still make the notion of being a variable an effective one:

(a) 'x' is a variable.

(b) If ξ is a variable, then $\xi_{|}$ is a variable.

(c) The only variables are those provided by (a) and (b).

(Thus a "general" variable is of the form '$x_{|||\cdots|}$'.)

Any finite string of symbols is called an *expression* [1]). The number of expressions is therefore at least denumerably infinite even if the vocabulary is only finite.

(2) A set of *terms* as a subset of the set of expressions, determined by effective rules.

Continuing our illustration, we might have the following rules:

(d) Each variable is a term.

(e) If α [2]) is a term, then $S(\alpha)$ [3]) is a term.

(f) If α and β are terms, then $(\alpha+\beta)$ and $(\alpha\cdot\beta)$ are terms.

(g) The only terms are those provided by (d), (e), and (f).

It can easily be checked that being a term is indeed an effective notion in this formalism.

(3) A set of *formulae* as a subset of the set of expressions, determined by effective rules with the help of the notion of term, whenever this notion has been determined (which is not always the case).

We might have, for instance, the following rules:

(h) If α and β are terms, $\alpha=\beta$ is a formula.

(i) If ϕ is a formula, then $\neg(\phi)$ is a formula.

(j) If ϕ and ψ are formulae, then $(\phi)\rightarrow(\psi)$ is a formula.

(k) The only formulae are those provided by (h), (i), and (j).

Once more the effectiveness of the notion of formulahood is evident.

One might occasionally want to introduce terms and formulae by a simultaneous induction. We could have, e.g., such a rule as

(l) If ϕ is a formula and ξ is a variable, then $(\lambda\xi)\phi$ is a term (where 'λ' is a certain primitive constant).

1) In spite of our general decision not to mention divergent terminologies, a comparison with the terminology used in Church 56 is often indicated, in view of the extraordinary meticulousness with which this terminology has been prepared and explained. Church uses 'formula' instead of our 'expression' and 'well-formed formula' instead of our 'formula'.

2) Where α is a *syntactical* variable ranging over the expressions of the formalism.

3) Where 'S' is a constant of the formalism; 'S' is used autonymously – see Chapter II, p. 20 – in '$S(\alpha)$'. Had there been some point in avoiding such a usage, we could have said instead: "... then 'S' $\frown$ α (the *concatenation* of 'S' and α, i.e. the sequence consisting of 'S' followed by α) is a term".

(4) A set of *axioms* as a subset of the set of formulae. If this set is finite, the axioms can, at least in principle, be listed. If not, they may be given through *axiom-schemata* formulated in the meta-language with the help of syntactical variables. The rules have to be such as to guarantee the effectiveness of the notion of axiomhood.

Still continuing our illustration, we might have, among others, the following axiom-schema:

(m) All formulae of the form $(\alpha = \beta) \rightarrow ((\phi) \rightarrow (\psi))$ are axioms, if α and β are terms, ϕ and ψ are formulae, and ψ differs from ϕ only in containing β at a place where ϕ contains a free ocurrence of α.

It is again obvious that checking whether a formula is or is not an axiom in accordance with (m) is an effective procedure.

(5) A finite set of *rules of inference* according to which a formula is *immediately derivable* as *conclusion* from an appropriate finite set of formulae as *premises.*

Most formalisms contain a rule of inference either identical with or equivalent to the *rule of detachment* (or *modus* (*ponendo*) *ponens*):

(n) From a set of formulae consisting of two formulae of the form ϕ and $(\phi) \rightarrow (\psi)$, ψ is immediately derivable.

A finite sequence of one or more formulae is called a *derivation from the set* Γ *of premises* if each formula in the sequence is either an axiom or a member of Γ or immediately derivable from a set of formulae preceding it in the sequence. The last formula of the sequence is called *derivable from* Γ. A derivation from the empty set of premises is called a *proof* of its last formula, hence of each of its formulae. A formula is called *provable* or a *formal theorem* if there exists a proof of which it is the last formula.

It follows that the notion of being a proof is effective. But it does not follow that the notion of theoremhood is effective. There need not exist, from all we know so far, a general method that would, for a given formula, either tell us how to construct its proof or else show that no proof is possible. This does not exclude, of course, that theoremhood might not still be a demonstrably effective notion for certain formal systems. Once upon a time, some mathematicians had hoped that all of mathematics could be formalized in a formal system of the latter kind.

The rules determining the membership of the sets (1) – (3) above, i.e., the rules determining the language, are called *rules of formation,* those determining the membership of sets (4) and (5) above are called *rules of transformation.*

A formal system may be schematically regarded as an ordered quintuple of sets fulfilling certain requirements.

Terms and formulae will be called *open* [1]) if they contain at least one free occurrence of a variable, *closed* if they contain no free variables. Instead of 'closed formula' we shall usually say 'sentence'. [2])

We are now ready to present one of the many formal systems into which the set theory ZF can be formalized.

A. *Rules of formation*

1. α is a *primitive symbol* (of ZF) if and only if it is one of the following symbols:

a. *Individual variables:* 'x', '$x_{|}$', '$x_{||}$', ... (ad infinitum). For the 10 first variables following 'x' we shall use as abbreviations the following symbols, in this order: 'y', 'z', 'w', 'u', 'v', 'r', 's', 't', 'p', 'q'.

b. *Primitive predicates*: the binary predicate '$\in$'.

c. *Logical constants*:

α. Connectives: '$\neg$', '$\vee$', '$\wedge$', '$\rightarrow$', '$\leftrightarrow$'.

β. Quantifiers: the *universal quantifier* $\forall$,
the *existential quantifier* $\exists$.

γ. The equality symbol: '='.

d. *Auxiliary symbols*: '(', ')'.

3. φ is a *formula* if and only if it has one of the following forms:

a. $\xi = \eta$ or $\xi \in \eta$, where ξ and η are variables; a formula of these forms is called *atomic*.

b. $\neg(\psi)$, $(\psi) \vee (\chi)$, $(\psi) \wedge (\chi)$, $(\psi) \rightarrow (\chi)$, $(\psi) \leftrightarrow (\chi)$, where ψ and χ are formulas.

c. $(\forall\xi)(\psi)$, $(\exists\xi)(\psi)$, where ξ is a variable and ψ is a formula.

B. *Axioms and rules of inference*

4. φ is an *axiom* if and only if it has one of the following forms:

a. *Axiom schemata of the propositional calculus*:

$(\psi) \rightarrow ((\chi) \rightarrow (\psi))$, $(\psi \rightarrow \chi) \rightarrow ((\psi \rightarrow (\chi \rightarrow \rho)) \rightarrow (\psi \rightarrow \rho))$ [3]),
$(\psi \rightarrow \chi) \rightarrow ((\psi \rightarrow \neg\chi) \rightarrow \neg\psi)$, $\neg\neg\chi \rightarrow \chi$,
$\psi \rightarrow (\chi \rightarrow \psi \wedge \chi)$, $(\psi \wedge \chi) \rightarrow \psi$, $(\psi \wedge \chi) \rightarrow \chi$,
$\psi \rightarrow (\psi \vee \chi)$, $\chi \rightarrow (\psi \vee \chi)$, $(\psi \rightarrow \rho) \rightarrow ((\chi \rightarrow \rho) \rightarrow (\psi \vee \chi \rightarrow \rho))$,
$(\psi \leftrightarrow \chi) \rightarrow (\psi \rightarrow \chi)$, $(\psi \leftrightarrow \chi) \rightarrow (\chi \rightarrow \psi)$, $(\psi \rightarrow \chi) \rightarrow ((\chi \rightarrow \psi) \rightarrow (\psi \leftrightarrow \chi))$,
where ψ, χ and ρ range over all formulae [4]).

1)An interesting light is thrown on the intricacies of the terminological situation by the fact that Carnap, who until 1950 was employing the term 'open sentence' for our 'open formula', started using the term '(sentential) matrix' in a book – Carnap 50 – published that year, thereby adopting a term advocated by Quine in publications prior to 1950. Quine himself, however, in a book published in that very year – Quine 50 – decided to use 'open sentence', admittedly following Carnap's usage; see Quine 50, p. 90n.

2) We shall, however, continue to use the term 'statement' instead of 'sentence' as an informal term applying to natural languages and unformalized theories.

3) In this and the following formulae, some parentheses are omitted, under certain obvious tacit conventions.

4) This is taken from Kleene 52, p. 82.

b. *Axiom schemata of the predicate calculus*:
$\forall_\xi \psi \to \chi$, $\chi \to \exists_\xi \psi$, where ψ ranges over all formulae, ξ and η range over all variables and χ is a formula obtained from ψ by replacing in it one or more of the free occurrences of ξ which are not within the scope of a quantifier $\forall\eta$ by occurrences of η.

c. Axioms of equality: $\forall x\ (x=x)$, $\xi = \eta \to (\psi \to \chi)$, where ψ is any formula and χ is a formula obtained from ψ as in 4b.

d. *Specific axioms* (and *axiom schemata*):

I. (*Axiom of Extensionality*)
$\forall x \forall y [\forall z (z \in x \leftrightarrow z \in y) \to x = y]$.[1])

II. (*Axiom of Pairing*)
$\forall x \forall y \exists z \forall u (u \in z \leftrightarrow u = x \vee u = y)$.

III. (*Axiom of Union*)
$\forall x \exists y \forall z (z \in y \leftrightarrow \exists t (t \in x \wedge z \in t))$.

IV. (*Axiom of Power-Set*)
$\forall x \exists y \forall z (z \in y \leftrightarrow \forall u (u \in z \to u \in x))$.

V. (*Axiom-Schema of Subsets*)
$(\ )\quad \forall x \exists y \forall z (z \in y \leftrightarrow z \to x \wedge \psi)$.

where 'y' does not occur free in the formula ψ and where '( )' stands for a string of universal quantifiers binding all the free variables of ψ (except 'z').

VIc. (*Axiom of Infinity*)
$\exists z \{\exists u [\forall v (\neg v \in u) \wedge u \in z] \wedge \forall x \forall y [x \in z \wedge y \in z \to \exists t (t \in z \wedge \forall w (w \in t \leftrightarrow w \in x \vee w = y))]\}$.

VII. (*Axiom Schema of Replacement*)
$(\ )\quad \forall u \forall v \forall w (\varphi(u,v) \wedge \varphi(u,w) \to v = w) \to [\forall x \exists y \forall v (v \in y \leftrightarrow \exists u (u \in x \wedge \varphi(u,v))]$,

where $\varphi(u,v)$ is any formula, $\varphi(u,w)$ is the formula obtained from $\varphi(u,v)$ by substituting w for all free occurrences of v in $\varphi(u,v)$ and by relettering the bound occurrences of v in $\varphi(u,v)$ if necessary to avoid collision of variables, and where ( ) is as for Axiom-Schema V above.

IX* (*Axiom of Foundation*)
$\forall x \{\exists y (y \in x) \to \exists y [y \in x \wedge \neg \exists z (z \in x \wedge z \in y)]\}$.

The reader will have noticed that the symbolic versions of the specific axioms are sometimes more cumbersome than those given in Chapter II. This is due to the fact that in the formulation presented here we have completely avoided any use of defined symbols such as '0' or '∪'. This was done because we wished to dodge the problem of the status of definitions in formal systems. Let us only remark that if definitions are to be allowed as a part of a formalism (and not as metatheoretical devices), certain rules of definition have to be laid down with which the definitions will have to comply.

1) All parentheses, brackets, and braces are to be regarded as variants of the same two auxiliary symbols.

5. φ is *immediately derivable* from (the set of) the formulae ψ and χ if

a. (*Rule of detachment, modus ponens*): χ is $\psi \rightarrow \varphi$,

or else if ψ has the form $\rho \rightarrow \sigma$ and φ has the form

b. (*Rule of consequent universalization*): $\rho \rightarrow \forall \xi \sigma$, where ξ is not free in ρ, or

c. (*Rule of antecedent existentialization*): $\exists \xi \rho \rightarrow \sigma$, where ξ is not free in σ.

Rules 4a and 5a form a complete set of transformation rules for the propositional calculus. Together with rules 4b and 4c and with the rules of inference 5b and 5c they form a complete set of transformation rules for the first-order predicate calculus with equality. We now have a clearer image of the sense in which the system ZF has this calculus as its basic logical discipline.

The notion of formal system, as developed so far, is a rather narrow one, with strong requirements of effectiveness, and is probably very close to, if not identical with, the one Hilbert had originally in mind. For certain purposes, however, it is useful, and for other purposes even necessary, to investigate systems with weaker requirements of effectiveness. Whether we call such systems 'formal systems', too, perhaps with some qualifying phrase, or artificially confine one or more of the synonyms mentioned above (p. 280) for this wider conception, or coin some new terms, is of course a purely terminological matter. We shall from now on use *'logistic system'* for 'formal system with strong requirements of effectiveness' and employ qualifying phrases, where necessary, to distinguish between the different shades of formal systems.

Until recently, only formal systems with an at most denumerably infinite vocabulary were investigated. Lately, however, the revival of the algebraic approach to logic has also caused an increase of interest in systems with a non-denumerable vocubulary [1]).

Expressions are usually regarded as *finite linear* symbol strings, but occasionally expressions of infinite length are taken into consideration [2]).

Then there are formal systems in which the notion of formulahood is not effective. To mention just one example, Hilbert and Bernays introduce the iota-operator of definite description in such a way that an expression of the form $(\iota\xi)\varphi$, where ξ is a variable occurring free in φ, is regarded as a term only when the "uniqueness condition" $\exists\eta\forall\xi[\varphi\leftrightarrow(\xi=\eta)]$ is provable [3]). Since provability is, in general, not an effective notion, as we saw above, there exists no

1) The use of a non-denumerable vocabulary is now a standard practice in model theory – see e.g. A. Robinson 63 or Kreisel-Krivine 67.

2) See, e.g., Karp 64, Barwise 68 and 69.

3) Hilbert-Bernays 34–391, pp. 383 ff. Other authors, following Frege and Russell, take care that iota-expressions should always be terms; cf. Carnap 47, pp. 33 ff and Schröter 56.

generally effective method of determining whether $(\iota\xi)\phi$ is a term, hence no effective method of determining whether certain expressions containing $(\iota\xi)\phi$ as a part are formulae [1]).

With regard to axioms, we had already allowed their number to be infinite, even for a logistic system, so long as their specification within the metalanguage is such as to provide for an effective procedure of testing whether a given formula is or is not an axiom. Nevertheless, since systems with a finite number of axioms have definite theoretical advantages, the problem of the *finitizability* of a given infinite axiom system, i.e. the establishing of a finite axiom system whose set of theorems coincides with the set of theorems of the former system, is an important one that has often been discussed with respect to the various set theories (cf. below, p. 324).

Sometimes, however, one has occasion to deal with formal systems whose notion of axiomhood is only *semi-effective* in the sense that though there exists a mechanical procedure which, for any given formula, would determine after a finite number of steps that it is an axiom if it is one, there exists no mechanical procedure that would determine, after any number of steps, that a formula is not an axiom if it is not one [2]).

With regard to the rules of inference, whereas for logistic systems the number of premises in any immediate derivation is by definition always finite, there are formal systems that contain rules of inference in which a conclusion is immediately derivable from a denumerably infinite number of premises. The best-known of such rules is the following so-called *rule of infinite induction*. Assume that one is dealing with a system which has a formula $\nu(\xi)$ "asserting" that ξ "is" a natural number and which contains a symbol **n** for each natural number *n*. The rule of infinite induction states that for each formula $\varphi(\xi)$, $\forall\xi(\nu(\xi)\rightarrow\varphi(\xi))$ is immediately derivable from the infinitely many

1) The arguments brought forward in Church 56, pp. 52–53, to the effect that systems, whose rules of formation and transformation are non-effective, are not suitable for purposes of communication do not sound too convincing. Communication may be impaired by this non-effectiveness but is not destroyed. Understanding a language is not an all-or-none affair. Our quite efficient use of natural languages shows that a sufficient degree of understanding can be obtained in spite of the fact that "meaningfulness", relative to natural language, is certainly not effective.

For a discussion of calculi for which some or all of the notions of formula, axiom, and rule of inference are non-effective, see especially Carnap 37, § 45.

2) It has, however, been proved by Craig that for any formal systems with a semi-effective set of axioms there exists an equivalent formal system with an effective set of axioms; see Kleene 52a and Craig 53. The theorem as such is formulated not in the intuitive terms 'effective' and 'semi-effective' but rather in their strict counterparts '(general) recursive' and 'recursively enumerable' to be discussed below, pp. 308f.

formulae $\varphi(0), \varphi(1), \ldots$. The extreme intuitive plausibility of such a rule does not alter the fact that it transcends the framework allowed in a logistic system. We shall come back to this rule later.

Finally, occasions arise to investigate systems whose set of sentences, meant to be the counterparts of the true sentences of the intuitive theories which these systems are designed to systematize, can no longer be assumed altogether, or at least *ab initio,* to be identical with the set of sentences formally derivable from some set of axioms. Let us not stretch the term 'formal system' any further to make it cover even such systems, but let us rather use the term *'formalized theory'* for that purpose (each formal system being a formalized theory but not vice versa, just as we decided before to use the term 'logistic system' such that each logistic system is a formal system but not vice versa). Whereas a formal system is determined by its set of *provable* sentences, a formalized theory is determined by its set of what we shall call *valid* sentences. The exact extension of this term has to be defined from case to case, the only general condition which such a definition will have to fulfil being that the set of valid sentences should be *closed with respect to derivability,* i.e. that every sentence derivable from valid sentences by the rules of inference should be valid itself; this again entails that all *logical axioms* as well as the sentences derivable from them, i.e. all *logically valid* sentences, will be among the valid sentences. If our formalized theory is a *formal system* then the valid sentences are exactly the provable sentences, or theorems, of the system. The schematical representation of a formalized theory will then consist of an ordered sixtuple of sets: a set of symbols, a set of terms, a set of formulae, a set of logical axioms, a set of rules of inference, and a set of valid sentences, with various relations obtaining between, and various conditions fulfilled by, these sets. [1]).

Whether a given theory, presented initially as a formalized theory, can also be equivalently represented as a formal system, perhaps even as a logistic system, in other words, whether the set of valid formulae of this theory can be exhaustively described as the set of sentences derivable from some initial set of axioms by some rules of inference, these axioms and rules fulfilling more or less rigid requirements of effectivity, is a problem, often a very difficult one, always a decisive one. In still other words, the problem is whether a given formalized theory is *axiomatizable,* i.e. equivalent to an *axiomatically built* formal system. As a matter of fact, Hilbert and many other mathematicians and logicians had assumed that all of mathematics is

1) E.g., the set theory ZF* of footnote 3 on p. 141 is essentially presented there as a formalized theory.

axiomatizable. Anticipating results to be mentioned later, let us state, however, that many mathematical theories, including such "simple" ones as the arithmetic of natural numbers, have turned out to be nonaxiomatizable, to the amazement of them all.

§ 3. INTERPRETATIONS AND MODELS

A formalized theory is usually set up in order to formalize some intuitively given theory. Whether, and to what degree, this aim is achieved can only be determined after the formalized theory is provided with an *interpretation*, with the help of suitable *rules of interpretation*, turning thereby into an *interpreted calculus*. These rules can take many forms, but their common function is to provide each sentence of the formalized theory with a meaning such that it turns into something that is either true or false, i.e. into a statement, though no effective method need of course be provided for deciding whether it is the one or the other.

We shall sketch here one way of providing an interpretation for *first-order theories*, both because they are the most important and best studied sorts of formalized theories and for the sake of simplicity.

A full-fledged first-order theory T contains some standard set of logical constants (including a symbol of equality) and auxiliary symbols, a denumerable set of variables ranging over the same set of entities, and an at most denumerable set of extra-logical constants comprising individual constants, unary, binary, ..., n-ary predicates and operation symbols; this last set is assumed to be ordered in some sequence without repetitions. (The auxiliary symbols and the operation symbols are theoretically superfluous but are in practice often very convenient.) Such a theory is interpreted by providing the logical symbols with their usual signification, by fixing the *universe* (*of discourse*)U over which the variables range, and by assigning, through *rules of designation,* to each individual constant some member of U, to each unary predicate a certain subset of U and in general to each n-ary predicate a certain n-ary relation whose field is a subset of U, finally to each n-ary operation symbol an n-ary function from ordered n-tuples of U to members of U. Calling the sequence consisting of U and these individuals, sets, relations, and functions ordered in a way similar to that of the constants to which they are assigned, a *structure, semi-model,* or (*possible*) *realization* of T, *rules of truth* finally determine under what conditions a formula of T is *true in a given structure*, relative to a given *value-assignment* to its free variables if it contains such. (These rules determine, of course, also truth conditions for all the *sentences.*)

An interpretation of a given theory T is called *sound* if under it all the valid sentences of T become true in the structure determined by the interpretation; this structure itself – and often also the assigning function – it then called a *model* of the theory.

In the special case, when the given formalized theory is a formal system, if the axioms of this system become true under *some* sound interpretation, and its rules of inference are *truth-preserving,* i.e. leading always from true premises to true conclusions, then the theory has a sound interpretation, in other words, there exists a model of the theory.

Let us illustrate by giving a complete set of rules of interpretation for Z, i.e. ZF minus the axiom VIII of replacement which is a set theory more similar to the original theory of Zermelo.

a. If ϕ and ψ are sentences, then $\neg\phi$ is true if and only if ϕ is not true, $\phi\vee\psi$ is true if and only if either ϕ or ψ (or both) are true, etc.

b. The universe U is the set $R(\omega\cdot 2)$, where R is the function defined on p. 94; i.e., the universe contains the null-set O, its power-set **P**O, the power-set $\mathbf{P}^2$O of **P**O,..., $\mathbf{P}^n$O (for all finite n), the union $R(\omega)$ of these, $\mathbf{P}^n R(\omega)$ (for all finite n), as well as all the members of these sets.

c. '$\in$' designates the relation being-a-member-of (with its field restricted to U), i.e. the set of all the ordered pairs whose first element is a member of the second one, where both belong to U.

d. An atomic formula of the form $\xi=\eta$ (where ξ and η are variables) is true, relative to a given value-assignment to ξ and η, if and only if the same set is assigned as value to ξ and η.

e. An atomic formula of the form $\xi\in\eta$ (where ξ and η are variables) is true, relative to a given value-assignment to ξ and η, if and only if the set assigned as value to ξ is a member of the set assigned as value to η.

f. A formula of the form $\forall\xi\phi$ (where ϕ is a formula and ξ a variable) is true, relative to a given value-assignment to all the free variables of ϕ, if and only if ϕ is true, relative to any value-assignment that differs from the given one at most by the value assigned to ξ.

g. A formula of the form $\exists\xi\phi$ (where ϕ is a formula and ξ a variable) is true, relative to a given value-assignment to all the free variables of ϕ, if and only if ϕ is true, relative to at least one value-assignment that differs from the given one at most by the value assigned to ξ [1]).

Let us now introduce a few semantical terms which we shall need in the sequel. An open formula will be called *valid* [2]) in a given structure if it is true in this structure relative to every value-assignment to its free variables, *satisfiable* if true relative to at least one such value-assignment. A sentence

1) In order to see that the interpretation provided by these rules is sound, we have to verify that all axioms of Z are true in the structure $\langle$ *U, is -a-member-of* $\rangle$. This is easy.

2) The concept of a formula valid in a given structure should not be confused with the concept of a sentence valid in a formalized theory.

true in every structure is *logically true*, and an open formula valid in every structure – *logically valid.* A sentence true in every structure in which all the sentences of a given set of sentences are true is called a *logical consequence* of this set.

When a sentence ϕ is derivable from a set of sentences Γ, it is also a logical consequence of Γ, but the converse does not generally hold. It does hold, however, with regard to first-order theories, according to the extended Gödel completeness theorem (p. 296 below). For such theories we have then also that a formula is logically true if and only if it is logically provable, in other words, that the proof procedures of the underlying logic are complete [1]).

Whereas *sound interpretability (having a model)* is a *semantic* property of a formalized theory T, *interpretability in another formalized theory* T′ is a *syntactic* property of T, to the definition of which we turn now[2]).

Let T and T′ be first-order theories. The syntactic notion of an interpretation of T in T′ originates, like almost all interesting syntactic notions, from semantical ideas. We have in mind first an interpretation (in the informal sense) of the *language of* T as having a universe of discourse which is a subclass of the universe of discourse of T′, i.e., in the language of T one talks about the members of a subclass of the class of objects about which one talks in T′. Moreover, this subclass is to be a subclass which can be referred to in T′, and therefore it is to be a subclass given by a formula $\chi(x)$ of T′ with no free variables other than x. A statement "for all x..." of the language of T is now interpreted as the statement "for all x such that $\chi(x)$ ..." of T′, and a statement "there is an x such that ..." of the language of T is interpreted as "there is an x such that $\chi(x)$ and ...". So far we have interpreted in T′ only the universe of discourse of the language of T and we have still to interpret in T′ the extralogical symbols of the language of T. An n-ary relation symbol of this language is interpreted in T′ by an n-ary relation expressible in T′, i.e., by a formula $\rho(x_1,...,x_n)$ of T′ with no free variables other than $x_1,...,x_n$; individual constants and operation symbols of the language of T are interpreted similarly.

Therefore, an interpretation of the language of T in T′ is given syntactically by a formula $\chi(x)$ of T′ as above, and by an assignment which assigns to each n-ary relation symbol of the language of T a formula $\rho(x_1,...,x_n)$ of T′ as above, to each individual constant of T a term τ of T′ without free variables, and to each n-ary operation symbol of T a term $\tau(x_1,...,x_n)$ of T′ with no free

1) Cf. the remarks on p. 298 in regard to elementary theories.

2) We follow Tarski-Mostowski-Robinson 53, where, on p. 29, our interpretability is called *relative interpretability*.

variables other than $x_1,...,x_n$. Let φ be a formula of T. The interpretation φ^* of φ is obtained from φ as follows. We replace in φ each quantifier $\forall x$ and $\exists x$ by $\forall x(\chi(x)\rightarrow$ and $\exists x(\chi(x)\wedge$, respectively, each individual constant t by the term τ assigned to it, each term $F(s_1,...,s_n)$ by $\tau(s_1,...,s_n)$, where $\tau(x_1,...,x_n)$ is the term assigned to F and $s_1,...,s_n$ are terms (which also undergo the same changes), and we replace each atomic formula $P(s_1,...,s_n)$, where P is an n-ary relation symbol other than the equality sign, by $\rho(s_1,...,s_n)$, where $\rho(x_1,...,x_n)$ is the formula corresponding to P. Since the universe of discourse is always assumed to be non-void in logic we shall also demand that $\exists x\chi(x)$ be valid in T. Once this requirement is satisfied, one can easily show that the interpretation φ^* of every logically valid sentence φ of the language of T is a valid sentence of T′ [1]); moreover, if φ is a logical consequence of $\varphi_1,...,\varphi_n$ then $\varphi_1^*,...,\varphi_n^*$ imply φ^* in T′. If also the interpretation φ^* of every sentence φ valid in T is valid in T′ we say that we have an *interpretation of* T *in* T′, rather than just an interpretation of the language of T. If T is given by means of a set of axioms and we want to show that a given interpretation of the language of T in T′ is indeed an interpretation of T itself in T′, it is enough to show that for every *axiom* φ of T, φ^* is valid in T′; since every theorem ψ of T is a logical consequence of a finite number $\varphi_1,...,\varphi_n$ of axioms of T, also ψ^* is valid in T′. We say that T is *interpretable* in T′ if there is an interpretation of T in T′.

One of the main uses of interpretability is to obtain proofs of relative consistency. If we have an interpretation of the formal system T in a formal system T′ and T′ is consistent then so is T. To see this we notice that the interpretation $(\varphi\wedge\neg\varphi)^*$ of $\varphi\wedge\neg\varphi$ is $\varphi^*\wedge\neg\varphi^*$; if T were inconsistent then $\varphi\wedge\neg\varphi$ would be a theorem of T, and therefore its interpretation $\varphi^*\wedge\neg\varphi^*$ would be a theorem of T′, which is impossible if T′ is consistent. This is the way in which it was proved, e.g., that if the set theory ZF is consistent then so is ZFC (p.60). That was done by constructing an interpretation of ZFC in ZF. The universe of discourse of ZFC was interpreted as the class of all constructible sets, i.e., the formula $\chi(x)$ was taken to be a formula which asserts that x is constructible, and the membership relation of ZFC was interpreted as membership. This is indeed an interpretation of ZFC in ZF since the interpretations of all the axioms of ZFC are theorems of ZF [2]).

1) The requirement that $\exists x\chi(x)$ be valid is indeed necessary, since the interpretation of the logically valid sentence $\exists x(x=x)$ is $\exists x(\chi(x)\wedge x=x)$ which is obviously equivalent to $\exists x\chi(x)$.

2) For some proofs of relative consistency, and for some other applications, more general notions of interpretability are needed. This is the case, e.g., with respect to the proof of the consistency of classical number theory relative to the intuitionistic one – see p. 243.

We saw that if we have an interpretation of T in T′ then we can prove the metamathematical statement that if T′ is consistent then so is T. To what extent can we trust such a proof of relative consistency? That depends on the means used in that proof. If we were to use in that proof all the means of a theory which is as strong as T itself, then such a proof would not be very convincing; e.g., the knowledge that we can prove in T itself the metamathematical statement that if T′ is consistent then so is T [1]) yields no information on the consistency of T, since if T happens to be inconsistent then every statement is provable in T, and in particular the statement that T is consistent. However, it turns out that the proofs of relative consistency carried out by means of interpretations are, as a rule, such that they use only strictly finitary means and their formal counterparts can be carried out in primitive recursive arithmetic [2]). The proof of the relative consistency shows actually that, under some very natural assumptions, we have a simple procedure such that once we are given a formal proof of a contradiction in T the procedure yields a formal proof of a contradiction in T′. In the case of ZFC and ZF mentioned above, and in many other cases, one can prove that the length of the proof of the contradiction in T′ and the number of steps needed to obtain this proof from the proof of the contradiction in T are of the order of magnitude of the length of the latter proof.

As we saw, the notion of an interpretation of T in T′ is a syntactical notion; however, it is also related to important semantical facts. Given an interpretation of T in T′ and a structure $\mathfrak{A}$ which is a *model* (or sound interpretation) of T′, then the interpretation of T in T′ determines a model $\mathfrak{B}$ of T as follows. The universe of $\mathfrak{B}$ is the set of all members of $\mathfrak{A}$ which satisfy $\chi(x)$ in $\mathfrak{A}$ (i.e., the members u of $\mathfrak{A}$ such that the value-assignment which assigns the object u to the variable x makes $\chi(x)$ true in $\mathfrak{A}$). The n-ary relation of $\mathfrak{B}$ denoted by the relation symbol P is the set of all n-tuples of members of $\mathfrak{A}$ which satisfy $\rho(x_1,...,x_n)$ in $\mathfrak{A}$, where $\rho(x_1,...,x_n)$ is the formula of T′ assigned to P by the interpretation of T in T′. We similarly obtain the distinguished individuals of $\mathfrak{B}$ (denoted by the individual constants) and the operations of $\mathfrak{B}$.

Since, as we saw, an interpretation of T in T′ directly determines a model of T for every model of T′, an interpretation of T in T′ is often referred to as a "model of T in T′." It is not uncommon among mathematicians who set up an interpretation of one

1) Whether this can be proved directly if T "speaks about" expressions of its language, or indirectly via a Gödel-numbering (see § 6 below) if T contains arithmetic.

2) See Goodstein 57 for primitive recursive arithmetic. The notion of a finitary proof is explained on p. 278; the proofs in primitive recursive arithmetic are finitary even according to the most exacting standards.

theory in another one to have in mind only the model-theoretic construction we mentioned and not the syntactical notion of an interpretation which yields the finitary relative consistency proof.

§4. CONSISTENCY, COMPLETENESS, CATEGORICALNESS, AND INDEPENDENCE

A formalized theory is interesting only if it is free from contradiction. It gains in interest if it answers all pertinent questions. And one might wish that it should uniquely characterize its subject matter.

These formulations are of course very vague. They form nevertheless the starting-point for the rigorous definitions of the many metatheoretical concepts to which we shall turn now. Whereas at the beginning it was thought that the rigorous concepts answering the three above-mentioned desirable properties of theories would turn out to be relatively simple and unambiguous, at least in the case of axiomatically built theories, we have come to realize that there exists a whole battery of terms [1]) explicating the various aspects of these properties. We shall not try to be exhaustive, especially since some of these terms refer to exceedingly complex situations whose description would require many pages.

A formalized theory (formal system) T is called formally *consistent* if not every sentence of T is valid (provable), otherwise formally *inconsistent*. If T is a first-order theory (but not only then), its formal consistency implies that there is no sentence ϕ of T such that both ϕ and $\neg\phi$ are valid (provable) in T. A class of formulae of a given formal system is called *consistent as to derivability* (*as to consequences*) [2]) if not every formula of this system is derivable (is a logical consequence) from this class. It follows that the class of extra-logical axioms of a consistent formal system is consistent as to derivability.

One can easily deduce from these definitions, together with those of the preceding section, that a formal system is soundly interpretable (has a model) if and only if its class of extra-logical axioms is consistent as to consequences. On the other hand, it is clear that if a class of formulae is consistent as to

1) Many of these terms were used before in this book informally or semiformally, sometimes with forward references. We are now paying our debts.

2) These are Church's terms; see Church 56, p. 327. Hermes-Scholz 52, pp. 31–32, use 'syntactically consistent' and 'semantically consistent' in approximately the same sense.

consequences, it is consistent as to derivability. It follows that if a formal system has a model, it is formally consistent. For first-order theories, the converse of this last statement holds too, since in these theories the concepts of logical consequence and derivability are entirely equivalent.

One way, then, of proving the formal consistency of a theory is to show in its metatheory that it possesses a model. Such a consistency proof, however, can be regarded as *absolute* only if the metatheory is unimpeachable. Otherwise, it carries conviction only *relative* to the degree in which we are convinced of the consistency of the metatheory. For certain theories, such as number theory, analysis, and set theory, it looks hopeless to find a suitable metatheory that would not be at least as suspect as these theories themselves, in fact, that would not contain counterparts of these very theories themselves [1]). It was this hopelessness that made Hilbert invent a different, namely the syntactical method of providing consistency proofs.

Another way of proving the formal consistency of a theory T is by establishing a normal truth definition for it in some other theory T′. One says, following Tarski [2]), that a formal theory T′ possesses a *truth-definition* for another formal theory T if there exists in T′ a predicate, say '*Tr*', such that all sentences (of T′) obtainable from the expression

$$Tr(x) \text{ if and only if } p$$

by substituting for '*x*' a name (or some other designation) of any sentence of T and for '*p*' a translation of this sentence into T′ are provable in T′. The truth-definition is called *normal*, following Wang [3]), if in addition the formal counterpart of $\forall x[x$ is a theorem of $\mathrm{T} \rightarrow Tr(x)]$ is provable in T′.

The possibility of establishing a truth definition in T′ for T does not necessarily mean that T′ is stronger than T [4]) – though T′ must not be identical with T, or a subtheory of T, in view of Tarski's truth theorem (cf. below, p. 312) – but this feature does not by itself guarantee the possibility of establishing the consistency of T in T′, contrary to what was generally believed for a time [5]). Only the possession of a *normal* truth definition does

1) Cf. Kleene 52, p. 131.

2) Tarski 36. (56, VIII, pp. 187 ff).

3) Wang 52b; this important paper contains a very thorough investigation of the whole problem of the relationship between truth-definitions and consistency proofs.

4) *Ibid.*, pp. 269 and 272.

5) This belief originates with Tarski 36 (56, VIII, p. 236) who created the method of proving consistency through the establishment of a truth definition. Wang 50 and Mostowski 51 showed that this belief was not quite justified, as the metatheory T′ might not be powerful enough to contain a certain strong form of the principle of mathematical induction.

guarantee the consistency, but then T′ must be definitely stronger than T, which once again destroys the epistemological value of the consistency proof.

Formal systems with the natural numbers as individuals and containing a notation, primitive or defined, for each such number – we may then as well assume that they contain the ordinary numerals '0', '1', etc. – are called *ω-consistent* [1]) if there exists no formula ϕ free in ξ such that $\phi(0)$ (i.e. the formula resulting from ϕ by substituting everywhere '0' for ξ), $\phi(1)$, $\phi(2)$, ..., and $\neg(\forall\xi)\phi$ are theorems, otherwise *ω-inconsistent.* ω-consistency clearly implies formal consistency but the converse does not hold.

A formally consistent but ω-inconsistent system would not be intuitively satisfactory. A mathematical system might well be demonstrably consistent but, being ω-inconsistent, might still contain certain theorems that would be regarded as intuitively false.

A formalized theory T is called *formally complete* if there is no proper consistent extension of T with the same vocabulary, otherwise *formally incomplete.* If T is a first-order theory (but not only then), its completeness implies that, for every sentence ϕ of T, either ϕ or $\neg\phi$ is valid in T. A class of formulae of a given formal system T is called *complete as to derivability* (*as to consequences*) if from it, for every sentence ϕ of this system, either ϕ or $\neg\phi$ is derivable from (is a logical consequence of) this class. It follows that the class of extra-logical axioms of a formally complete first-order theory is complete as to derivability, hence also complete as to consequences.

In addition to the absolute notions of formal consistency and completeness, there are a host of relative notions of consistency and completeness applying to formal systems, where the relativization concerns either some property of the provable formulae or some intended interpretation. If in a formal system all sentences that are true under all possible interpretations, or under all intended interpretations, or under all interpretations of a certain kind, or under some specific interpretation, are provable, then this system is said to be *complete with respect to truth under all interpretations,* etc. [2]). Of special interest are the two concepts of completeness with respect to truth under *all* interpretations and under *all intended* interpretations, either of which – not properly distinguished until the investigations of Henkin in 1947 [3]) – might have been referred to when one was speaking about (*semantic*) *completeness of a formal system* (without qualification).

1) The term was introduced in Gödel 31. It was later generalized in many different directions.

2) Cf. Kleene 52, p. 131; Wang 53, p. 440.

3) Cf. Henkin 50; the idea as such probably originates with Skolem.

The best known and most important completeness theorem is the *Gödel completeness theorem.*[1]) It asserts that the first-order predicate calculus is complete with respect to all interpretations of this calculus and even with respect to all interpretations of this calculus in denumerable structures, i.e., every sentence of first-order predicate calculus which is true in every (denumerable) structure is provable in first-order predicate calculus. Applying the Gödel completeness theorem to a sentence $\neg\phi$ we find out that if $\neg\phi$ is not provable then it is false in some structure. In other words, if no contradiction can be derived from ϕ then ϕ is true in some structure. The *extended Gödel completeness theorem* asserts that if ϕ is a set of sentences consistent as to derivability then Φ has a model (sound interpretation), i.e., Φ is consistent as to consequences. As a consequence of this theorem we know that in the first-order predicate calculus a sentence ψ is a logical consequence of Φ just in case that ψ is derivable from Φ[2]).

The situation is different with regard to higher-order predicate calculi. The second-order predicate calculus, for instance, differs from the first-order calculus in that it has quantifiable predicate variables. Its rules of interpretation must provide the predicate variables with suitable ranges of values. In the intended (principal) interpretations, the range of each n-ary predicate variable is the set of *all* n-ary relations over the universe U, i.e. the set of *all* sets of ordered n-tuples of members of U. If the ranges are arbitrary subsets of these sets, we get a larger class of interpretations among which those that are non-intended and sound are called the *secondary* interpretations. It follows from Gödel's incompleteness theorem (below, p. 310) that the second-order predicate calculus is not complete with respect to truth under all principal interpretations, but Henkin [3]) succeeded in proving that it is complete with respect to truth under all sound interpretations, principal and secondary [4]), as is the calculus of order ω (the simple theory of types).

Models of second and higher order calculi, which correspond to the principal interpretations, are called *standard* by Henkin, those corresponding to any sound interpretation are called *general,* general models corresponding to secondary interpretations are called *non-standard*. We have then, in these terms, that all sentences true in every general model of the second-order calculus are provable in it but that this is not the case for all the sentences

1) Gödel 30.

2) A proof of the extended Gödel completeness theorem and a dicussion of its consequences is contained in every intermediate or advanced textbook of mathematical logic; e.g., Mendelson 64, A. Robinson 63, or Shoenfield 67.

3) Henkin 50.

4) See Church 56, § 54.

which are true in every standard model (of which there are of course more)[1]).

Among the many relative consistency notions, let us mention only *consistency with respect to satisfiability under some interpretation with a non-empty universe,* known simply as (*semantical*) *consistency*. A formal system is then *semantically consistent* if its set of provable formulae is satisfiable under some interpretation with a non-empty universe, in short, if it has a model. It follows from the extended Gödel completeness theorem that if a first-order theory is formally consistent it has a model. Since the converse of this is rather obvious, we have the important result that for such theories semantical and formal consistency coincide, hence that the in general non-finitary notion of semantical consistency is replaceable, for such theories, by the finitary notion of formal consistency. The proof Gödel himself gave of his theorem is, however, non-constructive. But Hilbert and Bernays were able to prove by finitary means a somewhat weaker completeness theorem [2]).

Formal systems fulfilling the conditions mentioned above in connection with the definition of ω-consistency are called *ω-complete* if, for every formula ϕ free in ξ, $\forall\xi\phi$ is a theorem whenever $\phi(0)$, $\phi(1)$, $\phi(2)$,... are theorems, otherwise *ω-incomplete* [3]).

A formal system is called *categorical (or monomorphic)* if it has a model and all its models are isomorphic with each other, i.e. if there exists a one-to-one correspondence between the universes of any two models such that all relations and operations are preserved [4]).

It can easily be proved that every categorical set of axioms is complete as to consequences. In fact, were it not so, there would exist a sentence ϕ in the

1) For the whole issue of intended interpretations and standard models, see the interesting paper Kemeny 56, but the reader should beware of divergencies in the terminology.

2) See Hilbert-Bernays 34–39II, pp. 252–253; Kleene 52, pp. 395–397.

3) The concept of ω-completeness – for a larger class of formal systems – was introduced in Tarski 33.

A good semi-formal discussion of various notions of completeness is to be found in Copi 54. But the reader should notice that Copi uses 'model' as a synonym for 'non-empty universe'. Cf. also De Sua 56.

4) For a rigorous definition, see, e.g., Church 56, pp. 327–330 or Tarski 56, X, p. 390. The first formulations of the idea of categoricalness are due to Huntington and Veblen (the latter of which was the first to use the *term,* following a suggestion by the philosopher John Dewey), though it was known in essence already to Dedekind and Cantor; cf. Fraenkel 28, pp. 341 ff. A comparison of the statements made there – more than forty years ago – with those contained in the present book, with regard to categoricalness of arithmetic and other aspects of axiomatics, should be quite revealing.

theory under discussion that would be true in some models of the axiom set and false in others; therefore not all models would be isomorphic.

On the other hand, there exist sets of axioms which are complete as to consequences, or even as to derivability, and non-categorical. Of this kind is, for instance, the elementary theory of succession in the natural numbers, i.e. the first-order predicate calculus with the Peano axioms but without the recursive "definitions" of addition and multiplication as its extra-logical axioms [1]).

For some time it was thought that the notion of (absolute) categoricalness would provide the basis for a significant dichotomy of *all* kinds of theories, separating *one-model theories* (see below, p. 343) such as ordinary arithmetic from *many-model theories* such as group theory. However, when it was proved that no consistent first-order theory which possesses an infinite model is categorical, if only for the simple reason that each such theory possesses models of arbitrary infinite power [2]), absolute categoricalness lost its significance with regard to such theories. Weaker notions of *relative categoricalness* were therefore introduced and, in the last years, extensively studied.

We wish to digress here and say a few words concerning arithmetic. By *Peano*'s (or *elementary*) *number theory* we mean the first-order theory with equality whose extralogical symbols are 0 , $'$ (for the successor operation), $+$ and $\cdot$, and whose axioms are

(a) $a' \neq 0$.

(b) $x' = y' \rightarrow x = y$.

(c) $\varphi(0) \wedge \forall x(\varphi(x) \rightarrow \varphi(x')) \rightarrow \forall x \varphi(x)$, for every formula $\varphi(x)$ of that language (this is the axiom-schema of induction).

(d) $x + 0 = 0 \wedge x + y' = (x + y)'$.

(e) $x \cdot 0 = 0 \wedge x \cdot y' = x \cdot y + x$.

By *second-order number theory* (or *arithmetic*) we mean the second-order theory with the same extralogical symbols and with the axioms (a), (b), (d), (e) and the following strong induction axiom:

(c′) $p(0) \wedge \forall x(p(x) \rightarrow p(x')) \rightarrow \forall x p(x)$, where p is a predicate variable.

By *n-th order number theory*, for $n > 2$, we mean the *n*-th order theory with the same extralogical symbols and the same axioms as second-order number theory.

Peano's original axiom system, first formulated in 1889, can be symbol-

1) See, e.g., Wang 53, p. 422.

2) For theories with an at most denumerable vocabulary, this result was obtained by Tarski; see Skolem 34, p. 161. It was later proved by A. Robinson and Henkin, independently, under a less restrictive assumption. A still stronger result is given in Tarski-Vaught 57, pp. 92 ff.

ized within first-order predicate calculus *with predicate variables.* In Peano's original system the successor operation is regarded as primitive, whereas the operations of addition and multiplication are introduced by definitions, through the recursion equations in (d) and (e) above. We know now that this way of introducing addition and multiplication is legitimate only within an appropriate predicate calculus of at least second order; in the framework of first-order predicate calculus these equations are taken to be the additional axioms (d) and (e).

Peano himself acknowledges his debt to Dedekind [1]) whose characterization of the natural numbers, however, is not quite axiomatic in the modern sense. Dedekind, in his turn, recognizes a kinship to Frege's work [2]), though Frege himself tended rather to stress the differences. A definition of addition and multiplication through informally stated recursion equations was already given by Peirce in 1881 but was not known to either Frege, Dedekind or Peano. [3]).

On the other hand, since Skolem was the first to point out the limitations of the axiomatic method in characterizing within a first-order calculus the set of all true arithmetical statements [4]), the term '*Skolem('s) arithmetic*' would serve as a useful abbreviation for those systems of non-axiomatic first-order arithmetic whose notion of validity is defined as truth under the ordinary interpretation.

Skolem's arithmetic (and, a fortiori, Peano's number theory) is not categorical; it has been shown that this system possesses also models whose universe contains additional entities [5]), besides the "standard" natural numbers, i.e. the number 0 and the numbers obtained from it by applying a finite number of times the successor operation. The customary arguments for the categoricalness of the arithmetic of natural numbers do not prove the absolute categoricalness of *first-order* arithmetic. What they do prove is that

1) See, e.g., Peano 03 and Dedekind 1888.

2) He refers especially to Frege 1884.

3) These historical remarks are based on Wang 57.

4) Skolem 34.

5) See Skolem 34 and 55; the second paper contains simplifications of the well-known proofs and constructions presented in the first one as well as outlines for the construction of non-standard models for various fragments of elementary arithmetic. In Henkin 50, p. 91, the result is mentioned that every non-standard denumerable model for the Peano axioms has the order type $\omega + (^*\omega + \omega)\eta$ where $^*\omega + \omega$ is the type of the integers and η the type of the rationals, both in their natural order. Malcev 36 had already proved that elementary arithmetic has models of any given infinite cardinal.

this system is categorical *relative* to the standard natural numbers (which is rather trivial). All models of elementary arithmetic whose universe contains only these numbers are indeed isomorphic [1]).

One could try to impose this limitation to standard numbers by adding an *axiom of restriction* (cf. Chapter II, p. 113) to Peano's arithmetic, say of the form

$$\forall x[x = 0 \vee x = 0' \vee x = 0'' \vee \ldots].$$

But this expression is clearly, and unfortunately, of infinite length, hence not legitimate within the framework of a first-order theory. And there is no way of reformulating it legitimately.

The absolute non-categoricalness of the usual set theories (in which number theory can be developed) follows immediately from their incompleteness (see Gödels's incompleteness theorem, below, p. 310). It is also a direct consequence of the Löwenheim-Skolem theorem (§ 5) which guarantees the existence of models with denumerable universes, whereas the intended models of these set theories have non-denumerable universes. There arises therefore the highly interesting question whether these theories are categorical in some relativized sense, e.g. whether they are categorical relative to their natural numbers or relative to their ordinal numbers [2]).

Various notions of relative categoricalness have been developed, of which categoricalness relative to the natural numbers of the theory is just an example. More generally, one says that a formal system S is *categorical relative to its subsystem* S′ if and only if any two models of S that are extensions of two isomorphic models of S′ are isomorphic, and one says that S is categorical *relative to a set of predicates* definable in it if and and only if any two models of S in which the predicates of this set are isomorphically interpreted are isomorphic. (Categoricalness relative to the natural numbers, for instance, is categoricalness relative to the set of three predicates interpreted as denoting the property of being a natural number, of being equal to zero, and the successor relation, respectively.)

A related notion is that of *categoricalness in a given cardinal*, introduced independently by Vaught and Łoś [3]). A formal system is called *categorical in*

1) Wang 53, pp. 422 and 425.

2) *Ibid.*, p. 425 and Wang 55, pp. 69–70; cf. also Wang 53a.

3) For a detailed discussion see Shoenfield 67, § 5.6.

a given cardinal if any two of its models whose universes are of this cardinality are isomorphic. Systems which are absolutely non-categorical may well be categorical in some cardinal, in many cardinals, even in all cardinals. First-order theories (with countable vocabulary) categorical in one uncountable cardinal are categorical in every uncountable cardinal [1]).

One says that an axiom, within a formal system, is *independent as to derivability* (*as to consequences*) if it is not derivable from (not a consequence of) the set of the remaining axioms. It follows immediately that an axiom is independent as to derivability if it is independent as to consequences and that an axiom is independent as to consequences if and only if the set of the remaining axioms possesses a model in which this axiom is false [2]). Establishing the independence of an axiom through exhibition of an *independence example* is, of course, the classical method that was used in effect already by Beltrami and Klein in their investigation on the consistency of hyperbolic geometry relative to Euclidean geometry and later by Peano and Hilbert in their studies in the axiomatics of arithmetic and geometry, respectively.

It follows immediately from the definition that an axiom ϕ is independent, within a first-order theory T, if and only if the theory T′ resulting from T through replacement of ϕ with $\neg\phi$ is consistent. One says, in this case, that $\neg\phi$ is *compatible* (or *consistent*) with the remaining axioms.

A set of axioms is called *irredundant* if each of its axioms is independent, otherwise *redundant*. A generation ago, there was a great deal of discussion around the importance of irredundancy in the evaluation of a formal system. Though there are even today some axiomaticians who regard irredundancy as

1) Morley 65. This is generalized to first-order theories with uncountable vocabulary by Shelah 70.

Section 2 of Tarski 56, X (pp. 308–319) contains important remarks on various other aspects of the notion of categoricalness and its application to scientific theories in general. Cf. also the closing remarks of XIII, pp. 390 ff, dealing, among other things, with the notion of *non-ramifiability*, another explicatum of the intuitive notion of completeness, and its relation to categoricalness and formal completeness. In this essay, originally published in 1935, the axiomatically built arithmetic of natural numbers is regarded as categorical, but Tarski refers there not to A^0 but to A^ω.

The notion of *notational* (or *expressive*) *completeness with respect to a given subject matter* deserves at least to be mentioned. Its meaning should be clear. As an illustration, let us only mention that the ordinary propositional calculus, based upon '$\neg$' and '$\rightarrow$' as the only connectives, is notationally complete with respect to truth-functions, in short: is truth-functionally complete, since it can easily be shown that all truth-functions are expressible on this basis. This result, together with a proof of the formal completeness (and decidability) of the propositional calculus, was first obtained in Post 21.

2) Cf. Church 56, p. 328.

a *conditio sine qua non* for a set of sentences to serve as a set of axioms, the consensus of most authors is that irredundancy as such has at most some aesthetic and didactic value, but that independence investigations and proofs are apt to provide important insights into the structure and scope of a given theory.

A rule of inference is called *independent within a theory* T if the set of theorems of T includes the set of theorems of the theory T′, resulting from T through omission of this rule, as a proper part.

In view of the far-reaching correspondence existing within axiomatic systems between axioms and (proper) theorems on the one hand, primitive and defined terms on the other hand, investigations in the irredundancy of the set of axioms of such a system are often supplemented by studies in the *irredundancy of the primitive vocabulary*, understood as the indefinability of any primitive term (with respect to the remaining primitive symbols) [1]).

§ 5. THE SKOLEM-LÖWENHEIM THEOREM; SKOLEM'S PARADOX

One of the central theorems of logic, which is also of prime importance in applications of logic to set theory, is the Skolem-Löwenheim theorem [2]). We shall present here a version of this theorem which, though not the strongest there is, is sufficient for our purposes.

Suppose $\mathfrak{A}$ is a structure which consists of an infinite set A and a finite or denumerable number of relations $R_1, R_2, \ldots$ on it. For every infinite cardinal $\mathfrak{b}$ which is less than the cardinal $|A|$ of A there is a subset B of A of cardinality $\mathfrak{b}$ and which has the following property. Let $\mathfrak{B}$ be the structure which consists of the set B and of the relations $R_1', R_2', \ldots$ which are just the relations $R_1, R_2, \ldots$ of $\mathfrak{A}$ restricted to the set B. For every sentence φ of the first-order language which corresponds to $\mathfrak{A}$, i.e., whose extra-logical symbols are just relation-symbols which refer to the relations $R_1, R_2, \ldots$ of $\mathfrak{A}$, φ is true in $\mathfrak{B}$ if and only if φ is true in $\mathfrak{A}$. Therefore, if T is a first-order theory and $\mathfrak{A}$ is a model of T, then $\mathfrak{B}$, too, is a model of T.

By the Gödel completeness theorem, if ZF is consistent then it has a denumerable model, i.e., there is a structure $\mathfrak{A}$ which consists of a denumerable set A and a binary relation E on A such that $\mathfrak{A}$ is a model for ZF with $\in$ interpreted as the binary relation E. If a member c of A stands in the relation

1) For a recent investigation along this line, see Beth 53. Cf. also A. Robinson 56, 57, and Craig 57.

2) Tarski-Vaught 57, Cohen 66, I, § 5, Shoenfield 67, Ch. 5.

E to a member b of A we shall say, for the sake of convenience, that c is a member of b in $\mathfrak{A}$ (even though c is not necessarily a member of b in the "real world"). Accordingly, we shall speak of b as if it were the set $\{c \mid cEb\}$[1]). Similarly, we shall also denote by ω the only member x of A which satisfies in $\mathfrak{A}$ the formula "*x is the least infinite ordinal*", etc. It is a theorem of ZF that there are uncountable sets, thus the set A must have a member a such that in the structure $\mathfrak{A}$ it is true that a is not denumerable, i.e., a has uncountably many members in $\mathfrak{A}$, while all the members b of a in $\mathfrak{A}$ are members of A, which is a denumerable set. This is what is known as *Skolem's paradox* [2]).

It is easy to see that Skolem's paradox is no paradox at all. When we say that a is uncountable in $\mathfrak{A}$ we mean that there is no one-one function *f in A* which maps a on ω but this does not deny the existence of a one-one function mapping a on ω which *does not belong to A*. Let us consider also another example. Since there are $\aleph_1$ countable ordinals there are well-orderings of ω of $\aleph_1$ different order types; however, only countable many of these well-orderings can be in $\mathfrak{A}$ since A is a countable set. Thus the cardinal $\aleph_1$ of $\mathfrak{A}$ is really a countable ordinal. This phenomenon of the uncountable cardinals of $\mathfrak{A}$ being countable in the universe is called the *relativeness of the cardinals*. But not only the cardinals are relative, many other notions of set theory are relative, too. Even the notion of a finite number is relative, since there are also non-standard models $\mathfrak{A}$ of ZF which contain members x which satisfy the formula "*x is a finite number*" in $\mathfrak{A}$, but which do not correspond to any finite number [3]).

This relativeness of the cardinals was very disturbing to Skolem and von Neumann. [4]) Von Neumann claimed that this relativeness brands the notions of set theory with the mark of unreality and therefore serves as an argument in favor of intuitionism. How can one, for example, trust non-denumerable cardinals when it may turn out that the structure one is speaking about is such that all the sets in it are really finite or countable?

One should recall that in the 1920's the Gödel completeness theorem had not yet been obtained and mathematicians were not yet aware of all the

1) If one assumes the existence of a well-founded model of ZF, which is a somewhat stronger assumption than the mere consistency of ZF, one can prove that there is a denumerable model A of ZF in which, for every $b \in a$, b is indeed $\{c \mid cEa\}$ (i.e., A is transitive in the sense of p. 92 and E is the restriction of the membership relation $\in$ to A); Mostowski 49, Shepherdson 51–53 I, § 1.5, or cf. Cohen 66, II, § § 7,8.

2) Skolem 23. The Gödel completeness theorem had not yet been discovered at that time; Skolem obtained this "paradox" by means of the Skolem-Löwenheim theorem.

3) See p. 299 and footnote 5 on that page.

4) Skolem 23 and 29, von Neumann 25.

aspects of the non-categoricity of the first-order theories. Also, at that time the axiom systems used by mathematicians were not formulated as first-order theories, they were formulated within informal set theory. Since these were essentially second-order theories one had indeed categorical axiom systems for natural number theory, for real number theory and for geometry. Since such an informal set theory is not available for an axiom system of set theory (if one wants to avoid vicious circles), one has to use for an axiomatization of set theory only first-order predicate calculus, and since first-order theories are not categorical, the notions of set theory are relative. As a matter of fact, the absoluteness of the notions of number theory and geometry is to some extent deceptive. Those notions are based on set-theoretical notions, and since the latter turn out to be relative, the former are relative, too[1]).

What are the philosophical projections of the relativeness of the notions of set theory? To a Platonist the notions of set theory are not really relative. From his point of view, the existence of sets and the truth of statements about sets are not determined by an axiom system. It goes the other way – the axioms are set up so as to provide true information about the existing sets.[2]) The notions of set theory become relative only when applied to models of set theory which consist of a class A and a binary relation on it, but these models are not the real thing. To a Platonist the notions of set theory are absolute almost to the same extent to which Fermat's last theorem is regarded as absolute by most mathematicians, in spite of the fact that there are non-standard models for number theory. On the other hand, a steadfast formalist is not bothered at all by the relativeness of the notions of set theory. From his point of view it is only the statements and the proofs of set theory that really count, not the vague objects and relations which they are supposed to denote. It is only those whose opinions lie in between to whom the relativeness of the notions of set theory is philosophically significant. Those are the people who believe that set-theoretical concepts do indeed exist, however not in an absolute fashion which is independent of the axioms, but rather as a consequence of the axioms. In the time that had passed since the discovery of the relativeness of the notions of set theory, even those people who hold the last mentioned point of view have learned to live with the fact that first-order axiom systems do not determine structures uniquely, and are no longer disturbed by it, no more than, say, by Gödel's incomplete-

1) von Neumann 25. For treatments of Skolem's paradox additional to those already mentioned, see Skolem 41 (and the vigorous discussion following it), Kreisel 50, Kleene 52, p. 426, Myhill 53, and Wang 53.

2) See, e.g., Kreisel-Krivine 67, Appendix IIA, and Shoenfield 67, § 9.1.

ness theorem which asserts that one cannot derive all the number-theoretic truths from a recursive set of axioms (p. 310 below).

§ 6. DECIDABILITY AND RECURSIVENESS; ARITHMETIZATION OF SYNTAX

The last property of formalized theories to which we shall turn now is decidability. The intuitive meaning of this notion is clear enough: a formalized theory T is *decidable* if there exists an effective, uniform method – a so-called *decision method* – of determining whether a given sentence, formulated in the vocabulary of T, is valid in T, and *undecidable* otherwise. The problem whether there exists a decision method for T, in other words, whether T is decidable, is then called the *decision problem* for T. If T is a formal system, we have instead of the general decision problem *for validity* the more specific decision problem *for provability*. One occasionally discusses also decision methods for other notions, such as formulahood, axiomhood, etc.

Decidability is clearly a highly desirable property of theories. In a decidable theory, every problem that can at all be formulated in its vocabulary has an answer that can be obtained by mechanically following a fixed recipe or, to use the terminology of computing machinery, a routine. It was hoped for a time – and it was the task of the Hilbert program to realize this hope – that the formal system into which classical arithmetic and analysis could be systematized and adequately axiomatized would be decidable as to provability as well as (formally and semantically) complete and categorical. The Hilbert school believed, in addition, that all this, together of course with consistency, could be proved by finitary metamathematical methods.

The pursuit of this aim led to a host of important insights, with regard to both the various interesting general connections between the mentioned metatheoretical notions and the consistency, completeness, categoricalness, and decidability of various sub-theories of the envisaged total logico-mathematical theory. Before we proceed to mention some of the results of these insights, let us briefly describe the technique of the *arithmetization of syntax* that has allowed us to deal with all these matters by a unified method whose major tool was the *theory of recursive functions*, and at the same time enabled us to replace the somewhat vague notions of effective and semi-effective procedures by the precisely defined concepts of *(general) recursiveness* and *recursive enumerability*.

We can say, with considerable justification, that the axiomatic approach to the foundations of set theory eliminates the antinomies: the logical ones, because of the limitations which the axiom system imposed upon the existence proofs, the semantical ones, because those semantical terms like 'true', 'definable', 'denotes', etc., that occupied the central place in their derivation could not be reproduced within the framework of these systems. But whereas the first contention seems to be reasonably well established, the second one was left unjustified, hoping that the reader would agree that the mentioned semantical terms could not well be expressed on the basis of a vocabulary that included, in addition to the logical terms of the first-order predicate calculus, just the symbol of set membership.

As a matter of fact, however, this second contention is far from being obviously correct. Most set theories developed (or outlined) in previous chapters of this book are strong enough, in spite of their poor primitive notation, for having all, or at least certain large parts, of classical arithmetic and analysis developed within them. Could not a method be found of developing within them their own syntax and semantics? And would then not the semantical antinomies be once more reproducible?

This was the question that Tarski and Gödel, independently, put themselves in the late twenties. The answer at which they arrived was highly surprising and led to further unexpected and interesting results. The story has been often told, in all degrees of formality and rigor, and we shall be very brief.

For purely technical reasons of expediency, let us consider a formalization of ZF which is somewhat different from the one given above pp. 283 ff. According to the new rules of formation, there are just 10 primitive symbols:

$$=, \in, x, {}_{|}, \neg, \to, \leftrightarrow, \forall, (,).$$

(It is very easy to see that this primitive notation is sufficient. There is no need to show this in detail. Nor shall we indicate the changes in the rules of transformation required for the adaptation as they, too, would be purely mechanical.) Let the numbers 0 through 9 be assigned to these symbols, say in the given order, and let to every expression (string of symbols) which does not begin with '=' be assigned as its *Gödel number* the number expressed by the corresponding string of digits. The assignment is clearly one-to-one.

The Gödel number of '$x_{|} \in x_{|||}$' would be, for instance, 2312333, and the Axiom of Power-Set, whose unabbreviated formulation would now be

$$(\forall x)\,(\neg(\forall x_{|})\,(\neg((\forall x_{||})\,((x_{||} \in x_{|}) \leftrightarrow ((\forall x_{|||})\,((x_{|||} \in x_{||}) \to (x_{|||} \in x))))))),$$

will have

872984872398488723398823312396887233398823331233958233312999999

assigned as its Gödel number.

The set of variables is represented through this device by a certain set of integers, viz. the set $\{2, 23, 233, ...\}$; similarly for the set of atomic formulae in general, of axioms etc. The corresponding sets of integers are easily recursively defined.

The number assigned to a sequence of formulae, such as a derivation composed of the formulae $\phi_1, \phi_2, ..., \phi_n$ in order, is $\Pi_{i=1}^{n} p_i^{g_i}$, where p_i is the i-th prime number and g_i the Gödel number of ϕ_i. Thereby the syntactical notions of derivation, proof, etc. become arithmetically representable. Notice that 'ϕ is a theorem' is represented by the arithmetical sentence 'there exists a number x which is the Gödel number of a proof such that the Gödel number of ϕ is the power of the largest prime number in the decomposition of x into a product of prime number powers' [1]).

We recall that the notion of proof in a logistic system is effective in the sense that there exists a mechanical procedure of testing whether a given series of formulae is or is not a proof, but that the notion of theoremhood is in general only semi-effective in the sense that there exists in general no mechanical procedure of testing whether a given formula is a theorem, though there exists a mechanical procedure of generating all theorems, one after the other, such that if the formula under investigation is a theorem, it will, after finitely many steps, appear in this list. (But there does not in general exist a mechanical procedure that would simultaneously grind out all non-theorems one after the other).

This characterization of effectiveness and semi-effectiveness suffers from the defect that the notion of mechanical procedure which plays a decisive role in it is so far left in a state of unanalyzed vagueness. Many different, partly or wholly independent, attempts were made to explicate it and, interestingly enough, almost all of the strict notions brought forward in this attempt have proved to be equivalent [2]). This being so, we can restrict

1) The most extensive treatments of Gödel representation and the arithmetization of syntax are to be found, in addition to Gödel 31 itself, in Hilbert-Bernays 34–39, Carnap 37, Kleene 52, Ladriere 57, and Shoenfield 67. A simple account is to be found in Wilder 52.

2) More specifically, the following notions were proved to be equivalent: (1) *general recursiveness,* introduced in Herbrand 31 and Gödel 34 in generalization of the notion of *primitive recursiveness* used in Gödel 31, pp. 179–180; (2) *λ-definability,* introduced in Church 36, pp. 346 ff; (3) *computability by appropriate machines* in Turing 36 and Post 36; (4) *reckonability* in Gödel 36; (5) *'regelrecht auswertbar'* ("which can be evaluated according to rule") in Hilbert-Bernays 34–39II, pp. 392 ff; (6) *binormality* in Post 43; (7) *normal algorithm* in Markov 54 (and earlier publications dating back to 1947). The equivalence of (1) and (2) was proved in Church 36 and Kleene 36, that of (2) and (3) in Turing 37; see Kleene 52, pp. 320–321.

ourselves here to mentioning the most intuitive explication of the notion of effective procedure provided independently, and with some variation, by Turing and Post in 1936 [1]) in respect of numerical computation: there exists an effective procedure of computing the value of a function from (n-tuples of) natural numbers to natural numbers, if a machine of a certain specific design (of a somewhat idealized type, as its "memory" is assumed to be infinite or at least infinitely extensible) can be programmed to perform this computation for any arguments. For an exact description of these machines – the so-called *Turing machines* – we refer to the literature [2]). We can now define: A given n-ary number-theoretical function f is called *Turing-computable* if there exists a Turing machine that computes its value for any n-tuple for which it is defined.

The thesis that Turing-computability (or general recursiveness, or any other of the equivalent notions mentioned above, p.307, footnote 2) is an adequate explicatum for the pre-systematic notion of effective calculability is known as *Church's thesis* [3]). In view of ever recurring misunderstandings (and misleading formulations in terms of 'is identical with' instead of 'is an adequate explicatum of'), it must be stressed that a thesis of this kind, claiming that a notion rigorously defined with respect to some formalized theory adequately explicates some intuitive notion, is, by definition, not susceptible to exact proof. The evidence for the correctness of Church's thesis is nevertheless overwhelming. In addition to the fact already mentioned that almost all explications proposed so far for effective calculability have proved to be equivalent, every function acknowledged to be effectively calculable, for which general recursiveness has at all been investigated, has indeed been found to be so. Since a number-theoretical property (or relation) is *effectively decidable* if and only if its *characteristic function* – i.e. the function whose value for a given n-tuple of natural numbers as argument is 0 if this property (or relation) holds for this tuple, and 1 otherwise – is effectively calculable, we see that the notion of effective decidability of number-theoretical properties and relations, hence of number-theoretical predicates, is adequately explicated by general recursiveness. Since all syntactical (metamathematical) predicates are mapped by arithmetization into number-theoretical predicates, we realize that all syntactic questions such as whether a given formal system is decidable are mirrored into corresponding number-theoretical

1) Turing 36 and Post 36.

2) The most readable descriptions are given in Péter 51, Kleene 52, and Davis 58.

3) This thesis was first formulated in Church 36. Its best exposition and justification is in Kleene 52, pp. 300 ff.

questions such as whether a certain number-theoretical property is general recursive.

A set S of natural numbers is called *general recursive* if the property of being a member of S is general recursive, and *recursively enumerable* if it can be enumerated by a general recursive function f, i.e. if there exists a general recursive f such that the sequence of values $f(0)$, $f(1)$, $f(2)$,... enumerates (allowing repetitions) the members of S. A set of metamathematical objects (expressions, formulae, axioms, proofs, etc.) is called *general recursive* (*recursively enumerable*) if the set of corresponding Gödel numbers is so. It can be shown that a general recursive non-empty set of natural numbers is recursively enumerable and that a set of natural numbers is general recursive if and only if it and its complement are recursively enumerable. An example of a set which is recursively enumerable but not general recursive can be effectively constructed. That there exist also sets which are not even recursively enumerable follows already from a simple consideration of cardinalities: the set of all recursively enumerable sets is only denumerable [1]).

It follows from the definition of 'logistic system' that the expressions of a logistic system admit an effective Gödel numbering and that their properties and relations and the operations upon them are all *recursively representable.* More specifically, the axioms of a logistic system form a general recursive set, and the rules of inference determine general recursive derivability relations. The set of theorems of a logistic system is therefore recursively enumerable.

The original Hilbert program amounts, then, in these terms to the construction of a consistent, complete, decidable, and categorical logistic system whose set of provable sentences would coincide with the set of intuitively true mathematical statements and such that it could be shown, via arithmetization, in *recursive arithmetic,* i.e. in that part of arithmetic which contains general recursive functions, properties, and relations exclusively, that the system possesses all these desirable properties.

1) For all these results and in general for an admirable treatment of the theory of recursive number-theoretic functions, we refer to Péter 51, Kleene 52, Davis 58, and Rogers 67. During the 48 years of its existence (we are probably entitled to date its origin with Skolem 23a), this theory has developed into a full-fledged and constantly expanding new branch of mathematics. The related *theory of algorithms* has been mostly developed by A.A. Markov and other Russian authors and has so far received its authoritative presentation in Markov 54 and Curry 63.

§ 7. THE LIMITATIVE THEOREMS OF GÖDEL, TARSKI, CHURCH AND THEIR GENERALIZATIONS

That the attempt to carry out the Hilbert program, as formulated in the preceding paragraph, would run into troubles was not too surprising: any logistic system strong enough to have all of classical mathematics developed in it would certainly contain all of classical number theory, hence presumably all of its own syntax and semantics. But does this not entail that the semantic antinomies, of the Liar and Richard types, could be reestablished within such a system?

As already mentioned above, Gödel and Tarski asked themselves precisely this question. The partial answer at which Gödel arrived is

GÖDEL'S (INCOMPLETENESS) THEOREM: [1]) **Every logistic system rich enough to contain a formalization of recursive arithmetic is either ω-inconsistent or else contains an undecidable (though true) formula**, i.e. a formula that can be neither proved nor refuted with the means of this system (though it can be shown to be true with the help of additional means outside of the system); in other words, any ω-consistent system of this kind is (syntactically and semantically) incomplete.

Gödel established his theorem *constructively*; for a given logistic system fulfilling the mentioned condition, the Gödelian undecidable sentence can actually, time and patience allowing, be written down in primitive notation. The truth of this undecidable sentence follows from the fact that, suitably interpreted, it states that this very sentence itself (more precisely, the sentence with a certain Gödel number – which then turns out to be the Gödel number of the sentence by which this statement is made) is not provable (in the given system) [2]).

It follows immediately from Gödel's theorem that *Skolem's arithmetic is not axiomatizable.*

This, then, is the revenge of the Liar's ghost: the attempt to arrive at an antinomy through the construction of a sentence within formalized arithmetic that (intuitively) states that it itself is non-provable does not meet with success, but only because – contrary to expectation and hope – a sentence non-provable within such a formal system is not thereby necessarily refutable,

1) Gödel 31.

2) There are some authors who are still unable to swallow the fact that there should be a sentence which is not provable – in some object-language – but can be shown to be true – in its metalanguage. For one such discussion, see, e.g., Goddard 58; cf. also Wittgenstein 56, pp. 50–54.

there being sentences which are neither. That completeness' price is inconsistency, for logistic systems rich enough to contain recursive arithmetic, including all set theories worth their name formalized as such systems, is a result which was doubly unexpected: first, for its content, second, for the fact that it could be proved according to standards of rigor which were the highest known, higher even than those customarily used in mathematical proofs. With its philosophical implications we shall deal later on.

Since there exist many excellent descriptions of the background of Gödel's theorem, its methods, generalizations and ramifications, ranging from the monographs of Ladrière 57 – containing, among other things, an almost exhaustive bibliography up to 1955 – and Mostowski 52 – unifying Gödel's results with those obtained by Tarski to be mentioned shortly – through Kleene's exceedingly rigorous treatment 52 and the detailed and painstaking description given by Hilbert-Bernays 34–39, to the semi-formal exposition by Rosser 39 and the informal, but still responsible and highly readable accounts by Nagel and Newman 58 and Findlay 42 – not to mention the many shorter outlines of which the best is probably still Gödel's own informal introduction to his 1931 paper – we did not deem it appropriate to go here into further details. Let us only mention that though Gödel's undecidable sentences remind us strongly of the Liar, the argument by which he arrived at his discoveries is reminiscent rather of Richard's antinomy and is, like it, decisively based upon a diagonalization process [1]).

Gödel's theorem holds for ω-consistent logistic systems. How much can his result be generalized? Rosser [2]) showed that ω-consistency can be replaced by the weaker property of simple consistency. Other generalizations are due to Tarski and Mostowski [3]). They are based on the notion of (semantic) *definability* introduced by Tarski [4]), later generalized by Mostowski [5]) and finally still more generalized by Tarski himself [6]). We shall present here the latest version.

Let A be any formalized theory whose logical basis includes at least the first-order predicate calculus with equality and in which an infinite sequence of terms $\Delta_0, \Delta_1, ..., \Delta_n, ...$ containing no variables is available. A subset P of the set N of all natural numbers is said to be *definable* in A if there is a formula ϕ of A, free in some fixed variable ξ, such that $\phi(\Delta_n)$ – i.e. the sentence arising from ϕ through replacing ξ everywhere by Δ_n – is valid in A whenever $n \in P$ and $\neg\phi(\Delta_n)$ is valid whenever $n \in N$ and $n \notin P$. In a similar way, definability is defined for relations and functions.

1) See, e.g., Mostowski 52, pp. 7 ff. Wang 55b shows how each semantical antinomy can be made to generate undecidable sentences.

2) Rosser 36a.

3) See, especially, Tarski 39, Mostowski 52, pp. 97 ff, and Ladrière 57, pp. 335 ff.

4) See Tarski 56, VI (the enlarged English translation of Tarski 31).

5) See Mostowkski 52, pp. 74–75.

6) Tarski-Mostowski-Robinson 53, pp. 44 ff.

Let now $G(\phi)$ be the Gödel number of ϕ, according to some Gödel numbering which may be left here undetermined, and let E_n be that expression of A whose Gödel number is n. Let D (the *diagonal function*) be the function defined by

$$D(n) =_{\mathrm{Df}} G(E_n(\Delta_n))$$

(the value of the diagonal function for the argument n is the Gödel number of the expression obtained from the expression whose Gödel number is n by replacing everywhere the variable ξ by the term Δ_n; if E_n does not contain ξ, $E_n(\Delta_n)$ is, of course, just E_n itself), and let, finally, V be the set of Gödel numbers of all valid sentences of A. The following result can then be established:

TARSKI'S UNDEFINABILITY THEOREM: **If the theory** A (fulfilling the above-mentioned conditions) **is consistent, then the diagonal function** D **and the set** V **of all valid sentences** (of A) **are not both definable in** A.

From this theorem it follows immediately that if in a consistent theory the diagonalizing function is definable, its set of valid sentences is not definable. Since D can be shown to be general recursive, no theory strong enough to contain recursive number theory may contain a definition for its V. We obtain then, finally, as a special case of Tarski's undefinability theorem replacing, for historical reasons, 'validity' by 'truth' –

TARSKI'S TRUTH THEOREM: [1]) **The notion of truth** (the set of all true sentences, the truth-set) **of a consistent formalized system containing recursive number theory is not definable in this system** [2]).

Whereas Gödel's theorem shows that the *deductive power* of any given sufficiently rich system is intrinsically limited, Tarski's theorem discloses the limitations in the *expressive power* of such systems. Paraphrasing a colorful dictum by Quine, we may say that the ontology bitten off in such systems is larger than they are able to chew. This is not as surprising as it might look at the first moment. After all, the set of definable properties and relations is only denumerable – since the set of defining formulae is so, of course – whereas there are nondenumerably many subsets of the set of all natural numbers. What was not quite expected is that the set of the Gödel numbers of all true sentences of a given consistent logistic system should inevitably be among those sets for which no defining expression exists in this system.

Let us mention that Gödel's theorem has been shown to hold also for

1) See Tarski 56, p. 247; the footnote there tells the story. Cf. Quine 52.

2) An extremely elegant, simple and direct proof of Tarski's theorem is given in Smullyan 57.

certain formal systems which are not logistic systems in the sense that they allow transfinite rules of inference such as the rule of infinite induction [1]). In what sense systems which transcend the Gödel-Tarski limitations, so-called *non-Gödelian systems,* can be regarded as satisfactory foundational systems of mathematics is a question that deserves most careful investigation. Wang's system Σ, treated in Chapter III, § 6, is admittedly not a logistic system. Other systems proposed in this connection either contain no proper negation sign, such as certain systems constructed by Fitch and Myhill [2]), or no proper universal quantifier, such as systems by Church and Myhill [3]), or no proper sign of material implication, such as a system by Church [4]). The exact scope and efficiency of such heterodox systems still remains to be explored [5]).

Just as Gödel's theorem showed that the (semantical) notion of arithmetical, hence mathematical, truth could not be exhaustively mirrored by the (syntactical) notion of provability within one logistic system, thereby destroying the basis of simple-minded formalistic attitudes towards the foundations of mathematics, so did a corollary of this theorem show that the foremost aim of the original Hilbert program could not be achieved. This aim, we recall, consisted in proving the formal consistency of number theory with the help of only a part of the proof procedures employed in number theory itself, viz. the "finitary" ones. But however liberal we are ready to be in interpreting this equivocal term, so long as the finitary proof procedures are taken to transcend the proof procedures admitted in logistic systems only moderately – if at all – (one such moderate transcension would be the admission of a rule of infinite induction), the possibility of obtaining a finitary consistency proof of number theory is excluded by

GÖDEL'S THEOREM ON CONSISTENCY PROOFS [6]): **No sentence that could adequately be interpreted as asserting the consistency of a logistic system containing number theory is provable within this system [7]).**

1) See Rosser 37; Tarski 39; Kleene 43, p. 68; Mostowski 49.

2) Fitch 42, 44, Myhill 50, 50a. Later systems of Fitch contain negation which does not, however, obey the *tertium non datur*; see, e.g., Fitch 52.

3) Church 34 which contains an infinite hierarchy of universal quantifiers, and Myhill 50a.

4) The system of Church 34 exhibits also an infinite hierarchy of signs for material implication.

5) Some of the "heterodox" systems treated in the last sections of Chapter III are non-Gödelian, too.

6) For its proof, see, e.g., Kleene 52, pp. 209–210, or Shoenfield 67, § 8.2.

7) For a discussion concerning the conditions under which a sentence can be adequately interpreted as asserting the consistency of a logistic system, see Hilbert-Bernays 34–39 II, pp. 285–286 and Feferman 60.

The enormous efforts made by Hilbert and his school to carry out his program [1]) enabled them to prove by strictly finitary means the consistency of a very extensive subsystem of arithmetic, the remaining gap consisting in that this subsystem contained the inductive principle only in a weakened form which did not allow for its application to quantified sentences. Gödel's corollary shows that their failure to achieve the aim in its totality was not due to lack of ingenuity; on the contrary, we now know that they went about as far as they possibly could.

The question then arises whether the consistency of number theory could not be proved by the use of methods which, though in general forming only a proper part of all the methods used in number theory itself, transcend these methods in just one respect, the additional method, no longer finitary in the original sense, still being "finitary" in some extended sense that would conform to the basic philosophical attitude of the formalists. However, since this attitude has not been univocally and authoritatively formulated [2]), no unique and universally acceptable answer to this rather vague question can be expected. As a matter of fact, a few years after the publication of Gödel's theorem on consistency proofs, Gentzen [3]) was able to prove the consistency of number theory by using, as the only method trancending those used in number theory proper, *transfinite induction* up to the first ϵ-number ϵ_0. This method is finitary in a certain well-determined extended sense, or *constructive,* as we shall prefer to say, in order not to have constantly drag along those bothersome qualifications, "in a certain sense". That Gentzen proved his theorem for a formal system which differs in many essential features from the logistic systems considered so far – whereas the calculi we dealt with embody *theorem logics,* Gentzen's calculi embody another major form of logic, viz. *rule logic* [4]) – is of no importance for our purposes, since his results can be easily transferred to standard logistic system [5]). Gentzen's constructive proof of the consistency of Peano's number theory was later followed by other constructive proofs of the same result[6]).

In addition to its philosophical consequences concerning Hilbert's

1) Hilbert-Bernays 34–39, Ackermann 40, Kreisel 64.

2) Cf. above, p. 278.

3) Gentzen 36, 38, Schütte 60.

4) Hermes-Scholz 52 distinguish two main forms of logic: *Satzlogik* and *Regellogik* (*Folgerungslogik, Konsequenzenlogik*). Kleene 52, p. 441, distinguishes between Hilbert-type systems and Gentzen-type systems.

5) Cf. Gentzen 36 and Ladrière 57, pp. 208 ff.

6) Ackerman 40 (see Wang 63, Ch. XIV) uses the ϵ-substitution method, while Gödel 58 (see Shoenfield 67, § 8.3) uses functionals of finite type.

program, Gödel's theorem on consistency proofs is also very important from a mathematical point of view. It serves as a basic unprovability result from which one can obtain many other unprovability results. One example is the theorem that if the system VNB of Chapter II, § 7 is consistent then not all the instances of the axiom-schema of impredicative comprehension of QM (XII on p.138) are theorems of VNB[1]).

A third type of limitation on formal systems is provided by those theorems in which it is stated that the decision problem for validity (provability) is unsolvable for certain kinds of formal systems, i.e. that these systems are *undecidable* as to validity (provability) [2]).

Using once more the direct method of diagonalization, Church was able to prove in 1936 [3]) that Peano's number theory (as well as certain fragments of it) is undecidable. Since an axiomatized theory resulting from an undecidable axiomatic theory through omission of a finite number of axioms (but keeping the vocabulary) is clearly undecidable itself, Church was able to show [4]), by omitting all the finitely many extra-logical axioms from a suitable modified fragment of elementary arithmetic, that the first-order predicate calculus is undecidable, in other words, that the set of its theorems, though of course recursively enumerable, is not general recursive, in still other terms, that though this calculus contains a complete proof procedure it contains no complete disproof procedure. Since Rosser was able to prove in the same year [5]) that every consistent extension of Peano's number theory is undecidable, in short that this system is *essentially undecidable,* we have altogether the following two basic results:

CHURCH'S UNDECIDABILITY THEOREM: **The first-order predicate calculus is undecidable.**

1)See footnote 1 on p. 138. Additional applications of Gödel's theorem are given in the next section.

2) The reader should beware of the trap caused by the equivocal functioning of 'decidable'. A formalized theory that contains an *undecidable sentence,* i.e., a sentence which is neither provable nor refutable within the theory, is thereby *formally incomplete* but not necessary undecidable itself, and an undecidable theory need not contain an undecidable sentence. Gödel's incompleteness theorem has often been called an undecidability theorem, a usage we have avoided in order not to increase the danger of confusion.

For the distinction between 'undecidable' and 'unsolvable' as qualifying problems, and for the inappropriateness of speaking about "unsolvable problems", see the revealing discussion in Myhill 52a, pp. 167 ff; however, not all formulations given there are watertight. Cf. also Turing 54 and Rabin 58, p. 175.

3) Church 36.

4) Church 36a.

5) Rosser 36.

CHURCH-ROSSER'S UNDECIDABILITY THEOREM: **Peano's number theory is essentially undecidable.**

Investigations into the decidability and undecidability of the various logical and mathematical theories received an additional impulse through the advent of the fast electronic computers. With their help, it was hoped, many mathematical problems might become *practically* solvable if their *theoretical* solvability could somehow be guaranteed, in the simplest case through a proof of the recursive decidability of the theory in which these problems were formulated. It was probably under this impact that Tarski, for instance, finally published his *decision method for elementary algebra and geometry* [1]) which he had already obtained, in essence, in 1930. And though this method is still not practical enough for even the fastest existing computers, it is not inconceivable that through simultaneous improvements of the method and of the computers we will eventually reach the stage when open problems in these and other fields will be handled and solved by these machines.

One of the main methods of proving the decidability of logical and mathematical theories is the method of the *elimination of quantifiers*. This method consists of a procedure which allows one to eliminate quantifiers from formulae by passing to equivalent formulae with less quantifiers. Applying this procedure a sufficient number of times one finally arrives at particularly simple formulae for which is is easy to determine whether they are valid or not. [2]) This was the method used by Tarski to obtain his decision procedure for elementary algebra and geometry (real closed fields). Some of the other theories which have been proved decidable by this method are the first-order and second-order monadic predicate calculus (i.e., with only *unary* predicate constants and variables) [3]), the theory of dense ordered sets, the theory of a single equivalence relation, the theory of addition of integers, [4]) the elementary theory of abelian groups, [5]) and the theory of algebraically closed fields.

Several theories were proved to be decidable by different methods, such as

1)Tarski 51. For elaborations and variants of this method, see Seidenberg 54, Meserve 55, Schwabhäuser 56.

2) A detailed exposition of this method is given in Kreisel-Krivine 67.

3) Cf. e.g., Ackermann 54.

4) Presburger 30.

5) Szmielew 55.

the first-order theory of ordered sets, [1]) the theory of finite fields, [2]) and the monadic second-order theory of binary trees[3])

A simple argument shows that every logistic system which is complete (as to derivability) is also decidable. [4]) This lends additional incentive to proving that certain theories are complete, such as the first-order theory of complex and algebraic numbers. [5]) An interesting sufficient criterion for the formal completeness of a first-order theory is known as *Vaught's test*.[6]). According to it, such a theory is complete if it has no finite models and is categorical in some infinite power. Another notion useful for proving the completeness of first-order theories is the notion of *model completeness*[7]).

The proofs of the decidability of theories by methods other than the quantifier elimination method often yield only such procedures for deciding whether a sentence φ is valid or not in the respective theory which are nothing but a blind search for a proof of φ or of $\neg\varphi$, or for something else related to φ. If this is indeed a decision procedure then the search is bound to end sometime, but not necessarily during the lifetime of our galaxy. However, even such proofs contribute towards more direct solutions of the decision problem by encouraging mathematicians to look for such[8]).

On the other hand, many theories were proved to be undecidable. For a few theories this was done directly, through a kind of diagonal argument, as was the case with Church's theorem. For some others, use was made of the theorem that an essentially undecidable theory induces this property in every consistent theory which is interpretable in it [9]). Since Peano's number theory is essentially undecidable and interpretable in many axiomatic set theories, *all these theories* – if consistent – *are essentially undecidable*. The set theories VNB and G of Chapter II, § 7, being finitely axiomatizable, thereby exemplify the existence of essentially undecidable and finitely axiomatizable theories.

1) Läuchli-Leonard 66, and Läuchli 68.

2) Ax 68.

3) Rabin 69,70. Rabin used the decidability of this theory to prove the decidability of many other theories.

4) See, e.g., Shoenfield 67, p. 132.

5) A. Robinson 59; for additional examples see Keisler 64.

6) It is due to Łos and Vaught, cf. Vaught 54 or Shoenfield 67, p. 89. See also Henkin 55.

7) A. Robinson 63, Ch. IV.

8) For example, after Ax-Kochen 65 and Eršov 65 proved that the theory of p-adic fields is decidable, P. Cohen found a decision procedure for this theory which uses the quantifier elimination method.

9) See Tarski-Mostowski-Robinson 53, p. 22. Theorem 7.

The scope of the undecidability proofs was greatly increased when some finitely axiomatizable and essentially undecidable fragments of Peano's number theory were discovered [1]). This enabled several logicians to prove, by various means, the undecidability of many theories, such as the elementary theories of groups, rings, fields, lattices, commutative rings, ordered rings, non-densely-ordered rings, modular lattices, complemented modular lattices, and distributive lattices, as well as the first-order theories of a single binary relation, a single symmetric relation, a single symmetric and reflexive relation, two equivalence relations, and two unary operations [2]).

Of special interest for us is the fact that by this method the essential undecidability of a small fragment of set theory could be shown [3]). This fragment contains '$\in$' as the only extra-logical constant and just three extra-logical axioms, viz. the axiom of extensionality (I, on p.27), an axiom stating the existence of a null-set, and an axiom guaranteeing, for any given sets x and y, the existence of a set z whose members are all the members of x as well as y itself [4]). This is a stronger result than the one mentioned two paragraphs before.

For theories whose undecidability has been proved or for which decidability has so far been neither proved nor refuted, *restricted decision problems* arise, i.e. problems as to whether certain proper subsets of the set of all their valid sentences are general recursive. For Skolem's arithmetic, e.g., which is a complete, non-axiomatizable and essentially undecidable formalized theory, the problem as to the general recursiveness of the set of all valid sentences of the form

$$(1) \qquad \exists\xi_1\exists\xi_2\ldots\exists\xi_n\,(\alpha=\beta),$$

where α and β are terms (polynoms) all whose free variables are among the $\xi_1, \xi_2, \ldots, \xi_n$, is nothing but the famous tenth problem of Hilbert [5]). This problem has been solved only recently by Matijasevič, using results of Julia Robinson, Davis and Putnam [6]). The answer is negative, i.e., the set of all

1) Tarski-Mostowski-Robinson 53. The theory given there has the axioms (a), (b), (d), (e) of p. 298 together with the axiom $x \neq 0 \rightarrow \exists y(x=y')$. See also the theory N of Shoenfield 67, p. 22 and Ch. 6.

2) For these and other results see Tarski-Mostowski-Robinson 53, Rabin 65, Eršov 66, Shoenfield 67 (§ 6.9 and the problems at the end of Ch. 6) and their bibliographies.

3) Tarski-Mostowski-Robinson 53, p. 34.

4) See footnote 3 on p. 32.

5) Raised in his talk to the International Congress of Mathematicians in Paris, 1900 (Hilbert 00).

6) Matijasevič 70, Julia Robinson 52, 69, Davis-Putnam-Robinson 61.

valid sentences of the form (1) is not recursive. This is proved by showing that for every recursively enumerable set A of natural numbers there are terms $\alpha(\xi,\xi_1,\dots,\xi_n)$ and $\beta(\xi,\xi_1,\dots,\xi_n)$ such that for every natural number k, $k \in A$ if and only if

$$(2) \qquad \exists\xi_1 \exists\xi_2 \dots \exists\xi_n [\alpha(\mathbf{k},\xi \quad \dots,\xi_n) = \beta(\mathbf{k},\xi_1,\dots,\xi_n)]$$

is valid (where **k** is the standard term denoting k). If the set of all valid sentences of the form (1) were recursive then the set of all valid sentences (2) would also be recursive, since the sentence (2) is of the form (1), for every value of k. This would imply, in turn, that the recursively enumerable set A is necessarily recursive, but we know that not every recursively enumerable set is recursive (p. 309 above).

For elementary group theory, the problem of whether the set of all valid sentences of the form

$$\forall\xi_1 \forall\xi_2 \dots \forall\xi_n \phi,$$

where ϕ is a formula without quantifiers, is general recursive has only recently been proved to be unsolvable. This problem, better known in a different but equivalent formulation [1]), is the famous *word-problem* which was proposed by Thue in 1914 and had defied mathematicians for almost 40 years. Already in 1947, Post and Markov had independently shown its unsolvability for semi-groups [2]). In 1950, Turing [3]) showed that this problem is unsolvable for semi-groups with cancellation. But only later did Novikov and Boone, independently, succeed in proving the recursive unsolvability of the word-problem for groups [4]).

As mentioned on p.315, the question of the logical validity of the sentences of the first-order predicate calculus is undecidable. It is therefore appropriate to consider various natural subsets B of the set of all sentences of the first-order predicate calculus, and ask whether the question of the logical validity of the sentences of B is decidable or not. A simple example of such a set B is the set of all existential sentences, i.e., the sentences which consist of

1) Cf. McKinsey 44, p. 68.

2) Post 47, Markov 47; Kleene 52, pp. 382 ff. For stronger results see Shepherdson 65.

3) Turing 50.

4) Novikov 55, Boone 54–57. At present the simplest proof is that of Britton 63, while the stronger result is that of Higman 61 (its proof is also given in the appendix of Shoenfield 67). For other related results see Chapham 64.

a string of existential quantifiers followed by a quantifierless part. Positive and negative answers have been obtained for various sets B[1]).

The reader might have already asked himself whether there might not exist a decision method by which it could be effectively determined for every given formalized theory, whether it is decidable or not. However, a rather simple argument shows that, at any rate for finitely axiomatizable first-order theories, no such general method could possibly exist, hence that this second-degree decision problem, so to speak, is unsolvable [2]). Other higher-degree decision problems deal, for instance, with the existence of an effective procedure of deciding, for every given *presentation* of a certain algebraic structure, such as of a semi-group, of a semi-group with cancellation, or of a group, whether the structures defined by these presentations have certain algebraic properties such as cyclicity, finiteness, simplicity, decomposability into a finite product, etc. By combining Markov's methods with Novikov's mentioned result, Rabin has been able to prove the recursive unsolvability of a large number of group-theoretic problems [3]). The recursive unsolvability of the topological homeomorphism of four-dimensional manifolds, given explicitly as polyhedra, was established by Markov [4]).

Aspects of relative decidability problems were treated by Post who introduced [5]) the term *'degree of recursive unsolvability'* which served as the basis for a classification of functions, properties, relations, and sets into equivalence classes created by the reflexive, symmetric, and transitive relation 'A is recursive in B and B is recursive in A' [6]). The least degree is that of the general recursive functions etc., the decision problem of which is recursively solvable, of course [7]). This issue is too intricate to be treated here in any detail [8]).

1) Ackermann 54, Klaua 55, Church 56 (pp. 246 ff.) and Kahr-Moore-Wang 62.

2) Tarski-Mostowski-Robinson 53, p. 35.

3) Rabin 58. For a survey of more recent results see Boone 68.

4) Markov 60; for additional results, see Boone-Haken-Poénaru 68.

5) Post 44.

6) For the notion of (general) recursiveness of a function in other functions, see Kleene 52, p. 275; for the other notions, see *ibid.*, pp. 276 and 307. Another excellent book dealing with this subject is Rogers 67.

7) The word 'unsolvability' is therefore slightly misleading in this context, as is, incidentally, the word 'degree', the "degrees" being *classes* and not, as one might have thought, numerical measures of some sort.

8) It is treated in Rogers 67, Sacks 63, Yates 70. In 1956, a question raised in Post 44, which has become known as *Post's problem* and which has been considered as one of the most important open foundational problems, was almost simultaneously, but independently, answered in the negative by a 19-year old American student,

§ 8. THE METAMATHEMATICS AND SEMANTICS OF SET THEORY

We are now ready to present in a systematic fashion the major results of the metamathematical and semantical investigations into the foundations of the various set theories treated in the preceding chapters. First, however, we shall summarize the generic relationships of some metatheoretical terms introduced in this chapter with the help of a few tree-diagrams. This should be helpful, in view of the fact that the terminological situation in this field is in general still rather confused, that the technical terms preferred in this book are not always the most customary ones, and that the meaning of most terms is by no means self-explanatory.

Among others, we made the following distinctions.

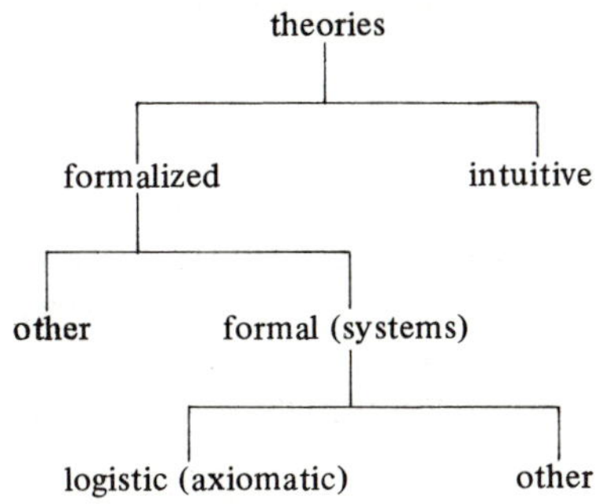

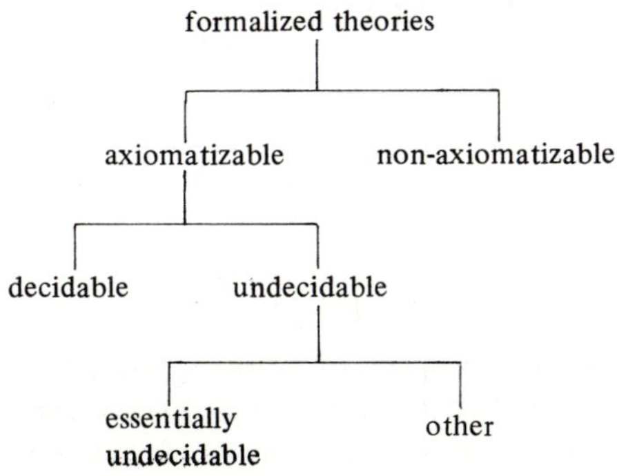

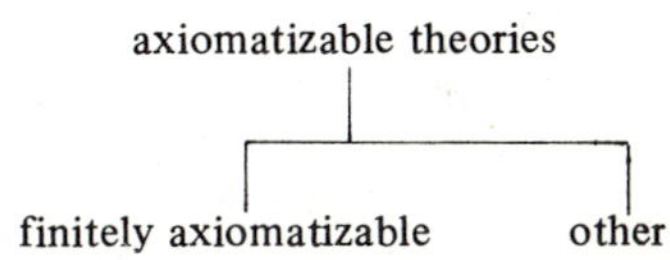

R. Friedberg and a 17-year old Russian student, A.A. Mučnik (see Rogers 67, § 10.2). Post's problem was whether all degrees of unsolvability of recursively enumerable sets could be linearly ordered by the relation is-recursive-in. Friedberg and Mucnik were able to construct two recursively enumerable sets whose degrees of unsolvability are incomparable with respect to this relation.

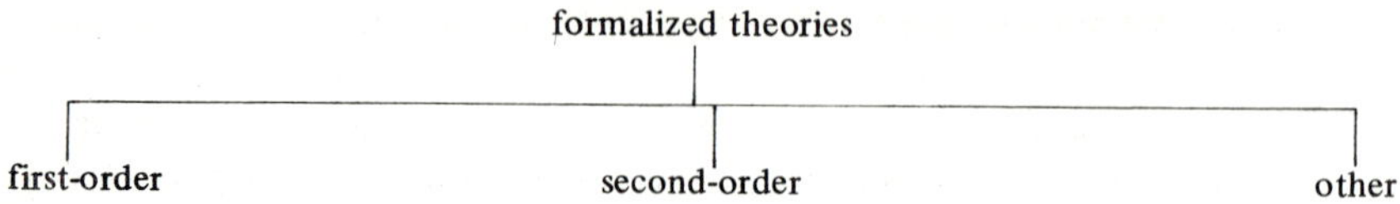

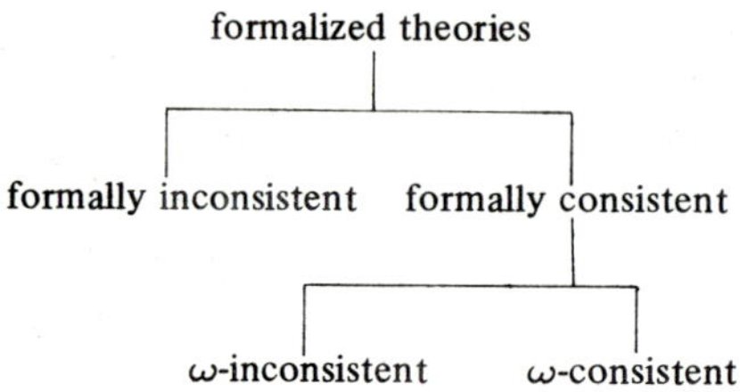

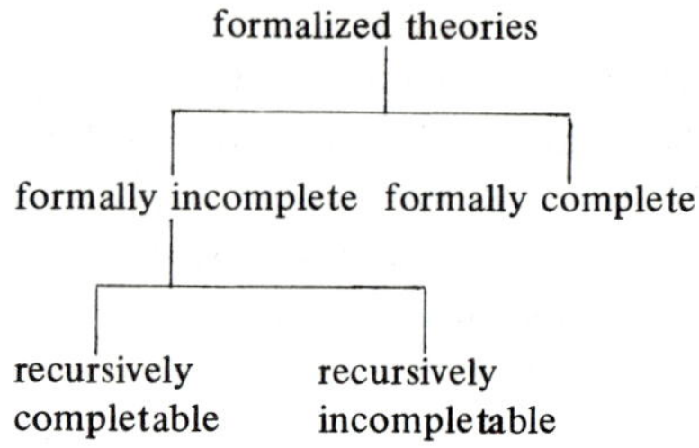

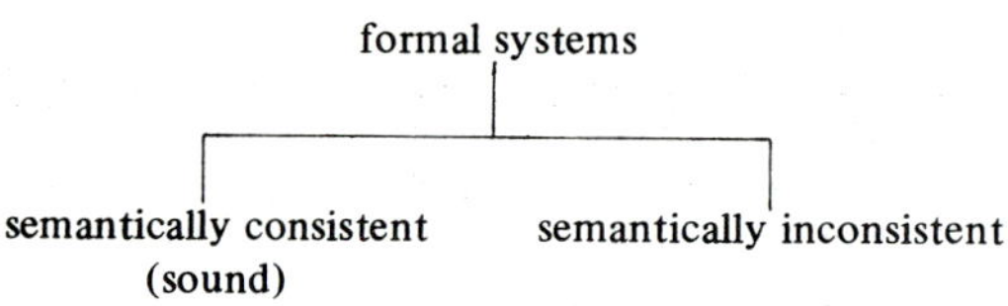

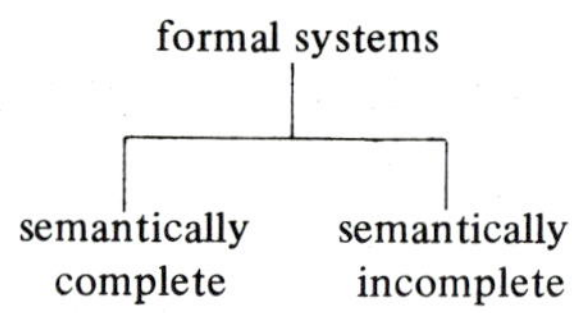

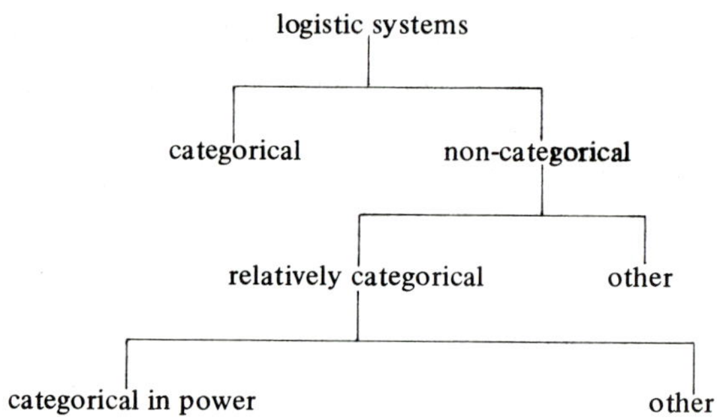

The first question that confronts us with respect to intuitive theories is whether these theories can and should be formalized. It is well known that one of the basic tenets of the intuitionistic approach is that no formalized theory can do full justice to intuitive (which is for them intuitionistic) mathematics or any of its subtheories. Heyting, who formalized intuitionistic propositional and predicate logic (cf. Chapter IV, § 4), has always been very insistent that his own formalization should by no means been treated as an adequate representation of intuitionistic modes of argumentation, and that in general:

> no formal system can be proved to represent adequately an intuitionistic theory. There always remains a residue of ambiguity in the interpretation of the signs, and it can never be proved with mathematical rigour that the system of axioms really embraces every valid method of proof [1]).

It is therefore the more interesting that Beth has succeeded in proving [2]), in spite of this explicit disavowal of its originator, that Heyting's propositional and first-order predicate calculi are complete with respect to intuitionistic arguments, in a very pregnant sense, hence that this part of intuitionistic logic at least can be exhaustively formalized and even axiomatized. A notion of (intuitionistic) truth can be satisfactorily defined for intuitionistic elementary logic under which the resulting formalized theory is complete and Heyting's logistic system semantically complete.

There are very many mathematicians, and even more so other scientists, who doubt it very much whether mathematical (and other) theories should be formalized even if they can be so in principle, suspecting that the fruits of formalization are not worth the effort. It is very difficult to discuss this issue *in abstracto*, but the present book will not have attained its aim unless the reader has become convinced that only through formalization can many important problems be given a formulation which makes it worthwhile to attempt their solution. The two outstanding examples discussed in this book are the formalization of intuitionistic logic just mentioned which enabled many mathematicians and logicians, who had no intuitionistic inclinations, to react to intuitionistic mathematics otherwise than by just shrugging their shoulders, and the formalization of Zermelo's concept of "definiteness" (cf. Chapter II, pp.36 ff) [3]).

Unless a formalized theory is presented *ab initio* as an axiomatic system, so that its set of valid sentences is from the start identified with the set of sentences provable in this system, or – to put it positively – whenever the set of valid sentences of a theory is determined semantically rather than syntactically, the question arises whether this set is recursively enumerable, i.e. whether the theory is axiomatizable.

With regard to Skolem's arithmetic, the answer is of course negative: the (recursive) non-axiomatizability of this theory is an immediate consequence of Gödel's incompleteness theorem. Skolem's arithmetic possesses a well-

1) Heyting 56, p. 102.

2) Beth 56a, cf. also Kripke 65.

3) For an interesting balanced discussion of the merits and demerits of formalization, see Wang 55a; cf. also Dubarle 55.

determined semantic notion of validity and is complete under this notion, its completeness being a simple consequence of a precise truth-definition; but no sufficiently rich axiomatic subtheory is complete.

With regard to set theory, the situation seems to be different. There exists at present no formalized theory having the same status relative to (intuitive) Set Theory as Skolem's arithmetic has relative to (intuitive) Number Theory. There exists no generally accepted truth definition for a formalized set theory from which its completeness could be deduced, since there exists no universe of "standard" sets that could in any serious way be compared with the universe of standard natural numbers. This point, seemingly of the greatest importance for the understanding of what set theory is about, has hardly been discussed in the literature, and the few existing discussions are incompatible and inconclusive. We shall later return to it.

Among the axiomatizable theories, those that are from the beginning based upon a finite number of axioms form an interesting subclass, though the exact significance of this feature is not quite clear, in all its generality. We know, for instance, that the classical propositional calculus is finitely axiomatizable, if a rule of substitution is admitted among its rules of inference, but is no longer so if *modus ponens* is its sole rule of inference. But here the explanation is relatively trivial. It is only with regard to first-order theories that the issue becomes both important and complex. Finite axiomatizability, for such calculi, means that the number of their *specific* axioms is finite, disregarding the way in which their basic logic is formalized. The schema of mathematical induction – standing for an infinity of axioms – in elementary arithmetic cannot be replaced by a finite number of axioms (formulated in the original vocabulary) [1]). Among the set theories treated in previous chapters some, such as VNB and G, were finitely axiomatized from the start by their originators, others that had an infinite number of axioms in their original presentation, such as NF, were later shown to be finitely axiomatizable, of others it could be proved that they were not finitely axiomatizable – this is the case for Zermelo's original theory, for ZF, QM and ML [2]). The latter results are based, of course, on the assumption that the systems in question are consistent.

We already saw above, p.315, the import of finite axiomatizability in connection with undecidability proofs. This feature is equally important for other metamathematical investigations. Finitizability of a given infinite axiom

1) Ryll-Nardzewski 53, Montague 61, Kreisel-Lévy 68.

2) Wang 52, Montague 61, Kreisel-Levy 68.

system is therefore always an interesting problem. The question whether finitely axiomatizable systems have any real *epistemological* advantage over other axiomatic systems is more complex. So long as an axiomatic system is axiomatizable with the help of a finite number of axiom-schemata (in addition, perhaps, to a finite number of axioms proper), no serious advantage, from an epistemological point of view, can apparently be claimed. All systems of set theory which we may come across are indeed axiomatizable by a finite number of axiom achemata of one kind or another [1]).

The proof of the existence of non-axiomatizable formalized theories is doubtless one of the most important achievements of recent foundational research. The philosophical and epistemological implications of this result have not yet been exhaustively evaluated.

Let us conclude this section with a study of the role of results concerning the existence of consistency proofs for various set theories. There are finitary proofs of the consistency of some very weak systems of set theory, such as the simple theory of types without an axiom of infinity, but this means, by Gödel's theorem on consistency proofs (p. 313), that these systems of set theory are not adequate even for Peano's number theory. Because of Gödel's theorem we know that the consistency of a system of set theory sufficient for any reasonable part of mathematics cannot be proved by finitary means, since it cannot be proved even in the theory itself. There is still an extremely remote possibility that one may prove the consistency of some strong set theory by using in the proofs non-finitary means which are not available in that theory but which have some kind of independent justification; nothing of this kind has been discovered so far for systems of set theory such as ZF [2]).

What about non-finitary consistency proofs for systems of set theory? Suppose we want to prove the consistency of type theory with an axiom of infinity – which is the theory we denoted in Chapter III (p.159) by T*. Assuming, on the intuitive level, the existence of an infinite set U, the infinite sequence U,**P**U,**PP**U,... is a model for T* once we interpret the variables of level 1 as varying over U, the variables of level 2 as varying over **P**U, and so on. Thus T* has a model and it is therefore consistent. Such a consistency proof will be taken seriously only by a determined Platonist who considers the

1) See Vaught 67. Kleene 52a proved that first-order theories can always be finitely axiomatized through enlarging their original vocabulary; see also Craig-Vaught 58. Hermes 51 proved that finite axiomatizability can also be achieved through adding (artificial and opaque) rules of inference to the underlying logic.

2) Another unsuccessful attempt to obtain such a consistency proof for ZF has been made by Esenin-Volpin 61. Cf. also the consistency proofs of Peano's number theory mentioned on p. 314 and footnote 6 on that page.

existence of an infinite set U and its iterative power sets as absolute mathematical facts independent of any particular axiom system for set theory. According to this point of view, an axiom system like T * is given just in order to describe what happens in the "real world" and it is not the case that the "real world" is the way it is just because there is an axiom system which asserts this. A non-Platonist cannot accept this argument; to him the purpose of T * is to "set up a world of sets" and one is not justified in using this very world to prove the consistency of T *. To an extreme Platonist the question of the consistency of a system of set theory is not really a central foundational problem. He will discard a theory at the point when it turns out not to fit the "true facts" about the universe, even if it is consistent. To put it more strongly, an extreme Platonist, like a physicist, will prefer an inconsistent theory some parts of which give a faithful description of the real situation to a demonstrably consistent theory which gives "wrong information" about the universe. Notwithstanding these fundamentally different philosophical attitudes, there is little disagreement among mathematicians as to the mathematical importance of the finitary consistency proofs on the one hand and of the consistency proofs by means of models on the other hand. The reason for this unanimity is that proofs of the consistency of various systems of set theory and other theories turned out to yield many additional results which are interesting to Platonists and formalists alike [1]).

For example, the proof that if ZF is consistent then so is VNB immediately yielded the stronger and much more interesting result that the statements of the language of ZF which are provable in VNB are exactly the theorems of ZF. Another example is the proof in QM that ZF is consistent. This proof turned out to yield the result that there are infinitely many different statements of number theory which are provable in QM but not in ZF, and even stronger results. [2]) We shall not discuss here the stronger results obtained by the methods of the consistency proofs. What we shall study is the relationship between the existence of certain natural models for systems of set theory and the existence of consistency proofs for these systems.

As a consequence of Gödel's theorem on consistency proofs and according to the accumulated experience in this field one can say that the available consistency proofs for systems C of set theory turn out to be of the following two kinds. First, there are the proofs of the consistency of C in some stronger system D of set theory; these are based, essentially, on the construction in D of a model of C; An example of such a proof is the proof in QM that ZF is

1) For a similar situation in number theory, see Shoenfield 67, pp. 214, 223.

2) Kreisel-Levy 68, Theorems 10 and 11.

consistent. Second, there are proofs, which use only finitary or otherwise relatively weak means, that if some system B of set theory is consistent then so is C. It was by such a method that the consistency of the axiom of choice, i.e., of ZFC was proved, viz. by proving (see p.60) that if ZF is consistent then so is ZFC.

Let us observe the following hierarchy of systems of set theory, all of which are formulated either in the language of ZF or in that of VNB.

A_1. ZF without the axiom of infinity VI

A_2. QM without the axiom of infinity VI

A_3. ZF without the axiom schema of replacement VII and with VIc as its axiom of infinity (this is close to Zermelo's original system.)

A_4. QM without the axiom of replacement and with VIc as its axiom of infinity

A_5. ZF

A_6. QM

A_7. ZF strengthened by an axiom IN asserting the existence of an inaccessible number

A_8. QM strengthened by IN

A_9. ZF strengthened by an axiom IN_2 asserting that there are at least two different inaccessible numbers.

Let us recall from Chapter II, § 5.3, that $R(\alpha)$ is the α-th iteration of the power-set operation applied to the null-set. We shall now be interested in models of the systems A_i, $1 \leqslant i \leqslant 9$, where the only extra-logical symbol '$\in$' is interpreted as the membership relation and where the universe of the sets is $R(\alpha)$ and the universe of the classes, for the systems formulated in the language of VNB, is $\mathbf{PR}(\alpha) = R(\alpha+1)$. We shall refer to such models as *natural models.* We shall now list, next to the symbol of each system, that $R(\alpha)$ (and $R(\alpha+1)$ where applicable) with the smallest α which constitutes a model of the theory [1]). In the following list, ϑ_1, ϑ_2, and ϑ_3 will denote the first three inaccessible numbers in that order.

A_1. $R(\omega)$

A_2. $R(\omega)$ and $R(\omega+1)$

A_3. $R(\omega+\omega)$

A_4. $R(\omega+\omega)$ and $R(\omega+\omega+1)$

A_5. $R(\xi_1)$, where ξ_1 is a certain ordinal $<\vartheta_1$

A_6. $R(\vartheta_1)$ and $R(\vartheta_1+1)$

A_7. $R(\xi_2)$, where ξ_2 is a certain ordinal between ϑ_1 and ϑ_2

1) Tarski 56a, Montague-Vaught 59.

A_8. $R(\vartheta_2+1)$

A_9. $R(\xi_3)$, where ξ_3 is a certain ordinal between ϑ_2 and ϑ_3.

We denote by $\mathrm{Con}(A_i)$ the set-theoretical statement, formulated by means of set variables only, which asserts that A_i is consistent. Let us notice that no arithmetization of the syntax is needed for the formulation of $\mathrm{Con}(A_i)$. Since the languages of ZF and VNB use only denumerably many symbols, these symbols can be identified with some distinct sets, even in A_1. All the A_i's admit the existence of finite sequences of sets and have mathematical induction; thus we can define in all the A_i's the notions of a formula and a proof.

Throughout the following discussion we shall use i and j for two indices such that $1 \leqslant i < j \leqslant 9$. The existence of the universe (or universes) of the smallest natural model of A_i is provable in A_j; this universe will be a proper class in A_j when $j = 2,4$ and $i = j-1$). One can therefore prove in A_j the formal statement $\mathrm{Con}(A_i)$ which asserts the consistency of A_i [1]).

It is, of course, easy to give a finitary proof of $\mathrm{Con}(A_j) \to \mathrm{Con}(A_i)$ since in some of the cases all the axioms of A_i are also axioms of A_j, and in the other cases one can prove in A_j the existence of a model of A_i and this yields directly an interpretation of A_i in A_j. However, the interesting question is whether one can prove $\mathrm{Con}(A_i) \to \mathrm{Con}(A_j)$. Such a proof would reduce the question of the consistency of the stronger theory A_j to that of the weaker theory A_i. The answer here is negative (if A_j is consistent). Suppose there were a finitary proof of $\mathrm{Con}(A_i) \to \mathrm{Con}(A_j)$, or even a proof of this statement in A_j, then since $\mathrm{Con}(A_i)$ is provable in A_j we could also prove $\mathrm{Con}(A_j)$ in A_j, and, by Gödel's theorem on consistency proofs, A_j would be inconsistent. What does this mean in terms of our approach to the question of the consistency of strong axioms of set theory? Given a system B of set theory, such as ZF, we shall call a *strengthening axiom* an axiom such as IN, which when added to the system B yields a system C of set theory whose relation to B is like that of A_j to A_i in the hierarchy above. Suppose we add to a system B of set theory a strengthening axiom ψ to obtain a system which we denote by B+ψ. We would like to have a proof that if B is consistent so is B+ψ; but as we have seen just now, no such proof is possible even in B itself (if B is consistent). We do have a proof of Con(ZF) → Con(ZF+IN) in ZF+IN_2 but this is of no avail since if we doubt the consistency of ZF+IN we cannot believe in the consistency of ZF+IN_2, let alone in the truth of its theorems.

1) In most cases, the model of A_i is a set and we can apply the formal version of the easy general theorem that if a theory has a model then it is consistent. For the other cases, we use the methods of Mostowski 51 and Montague-Vaught 59.

Therefore, if we add a strengthening axiom to a system of set theory we cannot establish consistency of the new system relative to the old one by means of a relative consistency proof. It seems that the only thing to do is to develop the new system, and as more and more of its theorems are discovered while no contradiction is derived we can become more and more convinced that the new system is practically consistent. By the system being "practically consistent" we mean that in it there is no derivation of feasible length for a contradiction; if there "exists" in it a derivation of a contradiction whose "length" (say, its total number of symbol occurrences) is greater than the number of atoms in the universe, there is no chance of finding it while proving theorems of the system.

We can also make a general statement about the independence of strengthening axioms. We shall prove that if B is a consistent system of set theory and ψ is a strengthening axiom with respect to B then ψ is independent of B, i.e., ψ is not a theorem of B. This can be shown in two different ways. The first and straightforward way is to get an interpretation of $\mathsf{B}+\neg\psi$ in B. In fact, in each one of the interpretations of A_i in A_j above which are used to show that A_j is "stronger" that A_i one can easily see that $\neg\psi$ "holds" in that interpretation. The second way is as follows. Since ψ is a strengthening axiom, Con(B) is a theorem $\mathsf{B}+\psi$. If ψ were provable in B then Con(B) would already be a theorem of B and, by Gödel's theorem on consistency proofs, it would follow that B is inconsistent.

Let us remark that the list A_1–A_9 above is by no means complete. Our choice was mostly arbitrary and our sole intention was to point out the relationship between stronger and weaker systems of set theory.

In the list $\mathsf{A}_1,\ldots,\mathsf{A}_9$ of systems of set theory the special relation of A_j to A_i, for $i<j$, is consistent with the fact that the smallest natural model of A_j consists of a larger $\mathrm{R}(\alpha)$ than that of A_i. Even though this is a very distinct phenomenon it is difficult to formulate a natural general principle to this effect. To see the difficulty let us observe the following example. If we replace A_2, which is QM without the axiom of infinity, by A_2', which is VNB without the axiom of infinity, we still have that the smallest natural model of A_2', like the smallest natural model of A_2, consists of the universes $\mathrm{R}(\omega)$ and $\mathrm{R}(\omega+1)$ but, assuming that A_1 is consistent, Con(A_1) is not provable in A_2'. The latter statement is proved as follows. In Chapter II, §7.2, we showed that every statement of the language of ZF which is provable in VNB is already provable in ZF. In the same way we get here that every statement of the language of ZF which is provable in A_2' is already provable in A_1. Since Con(A_1) is a statement of the language of ZF, if it were provable in A_2' it would also be provable in A_1 and, by Gödel's theorem on consistency proofs, A_1 would be inconsistent.

An even more extreme example is the following. Let A_1^* be the system A_1 strengthened by the additional axiom Con(ZF). The universes of the smallest natural models of A_1^* and ZF are $R(\omega)$ and $R(\xi_1)$, respectively, with ξ_1 much larger than ω (it is provable in every A_i with $i \geqslant 6$ that $R(\omega)$ is the universe of a model of A_1^*), and yet Con(ZF) is obviously a theorem of A_1^* while Con(A_1^*) is not a theorem of ZF (assuming that ZF is consistent – this follows from Gödel's theorem on consistency proofs since there is a simple finitary proof of Con(A_1^*) $\rightarrow$ Con(ZF)).

Let us finally consider one more example that will throw much light on the question of the existence of a consistency proof for one theory in another and on the role of impredicativity in these matters.

For $n \geqslant 1$, let each of K_n, I_n, and T_n be number-theory formulated in n-th order logic, i.e., the language of each of those systems is the language of the type theory T (of Chapter III, §2) with variables of the first n levels only. The variables of level 1 are taken to be number variables. The axioms are the usual axioms of Peano's number theory (p. 298) – the number variables being variables of level 1. For $n = 1$, we take as an axiom of induction the axiom-schema of induction (c) of p. 298; for $n > 1$, we use the axiom

$$\forall y^2 [0 \in y^2 \wedge \forall x^1 (x^1 \in y^2 \rightarrow x^1 + 1 \in x^2) \rightarrow \forall x^1 (x^1 \in y^2)] .$$

The axiom of comprehension is

$\exists y^{i+1} \forall x^i (x^i \in y^{i+1} \leftrightarrow \varphi(x^i))$, where $i < n$ and y^{i+1} does not occur free in $\varphi(x^i)$.

K_n, I_n, and T_n differ in the restriction on the formula $\varphi(x^i)$ in the axiom-schema of comprehension. In T_n, $\varphi(x^i)$ can be any formula of the language we use here; in I_n, $\varphi(x^i)$ is required to be a formula such that the levels of all variables in it are at most i+1; in K_n, $\varphi(x_i)$ is required to be a formula such that the levels of all the free variables in it are at most i+1 and the levels of the bound variables in it are at most i. T_n can be called *n-th order impredicative number theory*, since in the axiom-schema of comprehension the object y^{i+1} is determined by the formula $\varphi(x^i)$ which refers to the totalities of all objects of arbitrary levels greater than i. K_n can be called *n-th order predicative number theory* since the object y^{i+1} is determined in the axiom of comprehension by means of the totalities of all objects of levels 1 to i and also by finitely many objects of level $i+1$, but the totality of all objects of level j, for $j > i$, is not referred to in $\varphi(x^i)$. I_n is already impredicative, since y^{i+1} is determined by $\varphi(x^i)$ which may refer to the totality of all objects of level $i+1$, but it is "less impredicative" than T_n, since $\varphi(x^i)$ does not refer to the totalities of all objects of level j, for $j > i+1$.

We denote by K_ω, I_ω, and T_ω number theories formulated in ω-th order logic, i.e., in simple type theory. They have the same respective axioms as K_n, I_n, and T_n above, without the restriction $i<n$ in the axiom-schema of comprehension. The relations between the various systems are as follows. K_1, I_1, and T_1 are just Peano's number theory. I_2 and T_2 are both just (impredicative) second-order number theory. In T_2, which is identical with I_2, one can prove $\mathrm{Con}(\mathsf{K}_\omega)$ (but, by Gödel's theorem, if T_1 is consistent then one cannot even prove $\mathrm{Con}(\mathsf{K}_1)$ in T_1); in T_3 one can prove $\mathrm{Con}(\mathsf{I}_\omega)$ (but, by Gödel's theorem, if T_2 is consistent then one cannot even prove $\mathrm{Con}(\mathsf{I}_2)$ in T_2) [1]). $\mathsf{T}_2 \vdash \mathrm{Con}(\mathsf{K}_1)$ is obtained in the same way as $\mathsf{QM} \vdash \mathrm{Con}(\mathsf{ZF})$ [2]), there is a finitary proof of $\mathrm{Con}(\mathsf{K}_n) \to \mathrm{Con}(\mathsf{K}_{n+1})$ which is like the finitary proof of $\mathrm{Con}(\mathsf{ZF}) \to \mathrm{Con}(\mathsf{VNB})$ (see p. 132 and footnote 2 on that page), and $\forall n\, \mathrm{Con}(\mathsf{K}_n) \leftrightarrow \mathrm{Con}(\mathsf{K}_\omega)$ is obviously finitarily provable; hence we have $\mathsf{T}_2 \vdash \mathrm{Con}(\mathsf{K}_\omega)$. $\mathsf{T}_3 \vdash \mathrm{Con}(\mathsf{I}_n) \to \mathrm{Con}(\mathsf{I}_{n+1})$ is obtained like the proof that $\mathrm{Con}(\mathsf{NF}) \to \mathrm{Con}(\mathsf{ML})$ (p. 168), and since $\forall n\, \mathrm{Con}(\mathsf{I}_n) \leftrightarrow \mathrm{Con}(\mathsf{I}_\omega)$ is finitarily provable we have $\mathsf{T}_3 \vdash \mathrm{Con}(\mathsf{I}_\omega)$. The fact that I_2 is already "stronger" than K_ω and that T_3 is "stronger" than I_ω is explained by the differences in the impredicativity of the respective systems [3]).

§9. PHILOSOPHICAL REMARKS

On many occasions, when our discussions reached a certain ticklish, "philosophical" stage, they were disrupted by the remark that the issue will be taken up "later on". It is now high time that we pay our accumulated debts. Not that the reader is likely to rise, after the reading of this last section, with the feeling that all his problems have now found their final solution. Very few judgments will be passed here, and the only progress that might possibly be made will consist in formulating some of these problems and the various views on them in a more systematic fashion which could contribute to a better understanding.

Our first problem regards the ontological status of sets – not of this or the

1) McNaughton 53; see also A. Levy 60c.

2) Mostowski 51.

3) Using the methods mentioned in the last footnote it is easy to verify that the statements of first-order number theory provable in any of the K_n's and in K_ω are exactly the theorems of T_1, and that the statements of first-order number theory provable in any of the I_n's and in I_ω are just those which are provable in T_2. On the other hand, for every $n \geqslant 1$, there are infinitely many different statements of first-order number theory which are provable in T_{n+1} but not in T_n – see Kreisel-Lévy 68, Th. 10.

other set, but of sets in general. Since sets, as ordinarily understood, are what philosophers call *universals*, our present problem is part of the well-known and amply discussed classical problem of the *ontological status of the universals.* The three main traditional answers to the general problem of universals, stemming from medieval discussions, are known as *realism, nominalism,* and *conceptualism.* We shall not deal here with these lines of thought in their traditional version [1]) but only with their modern counterparts, known as *Platonism* [2]), *neo-nominalism*, and *neo-conceptualism* (though we shall mostly omit the prefix 'neo-' since we shall have no opportunity to deal with the older versions). In addition, we shall deal with a fourth attitude which regards the whole problem of the ontological status of universals in general and of sets in particular as a metaphysical pseudo-problem.

A *Platonist* is convinced that corresponding to each well-defined (monadic) condition there exists, in general, a set, or class, which comprises all and only those entities that fulfil this condition and which is an entity on its own right of an ontological status similar to that of its members. Were it not for the antinomies, the calculus that would best represent his intuitions would be the ideal calculus K (p. 155) or something of this kind, whose main feature is an unrestricted axiom-schema of comprehension. Things being as they are, he reluctantly admits that his vision of what constitutes a well-defined condition might be slightly blurred and declares himself ready to accept certain restrictions in the use of the axiom-schema of comprehension, temporarily working with a type theory or a set theory of a Zermelian brand, but hoping that sooner or later someone will be able to show that much less radical interventions will do the trick. Of course, some Platonists may convince themselves, or become convinced by others, that the objects of the world they live in are *really* stratified into types and orders and, as a consequence, accept type theory not as an *ad hoc* advice but as an expression of hard fact.

A *neo-nominalist* declares himself unable to understand what other people mean when they are talking about sets unless he is able to interpret their talk as a *façon de parler*. The only language he professes to understand is a calculus of individuals, constructed as first-order theory. With regard to many locutions used in scientific or ordinary discourse, which *prima facie* involve

1) For an able, modernized description of these classical views, see Stegmüller 56–57; these papers present equally well some of the contemporary views.

2) This term, in the present sense, seems to have been used first in Bernays 35. Whether Plato was, or even would have been, a Platonist is a moot question. Cf., e.g., Henle 52.

sets, he has little trouble in translating them adequately into his restricted language. This is the case, for instance, for such a common statement as 'the set of the *a*'s is a subset of the set of the *b*'s', which he renders as 'for all x, if x is a, x is b'. With regard to other locutions and devices he has greater trouble. The quite common kind of concept formation by which the ancestral of a given asymmetric and intransitive relation is formed – the resulting relation then being transitive – is easily formulable in set theory. Assuming, e.g., that the relation is-greater-by-one-than in the domain of integers is available (but not yet is-greater-than), one defines: x is-greater-than y if and only if x is-different-from y and x belongs to *all sets*, which contain y and all integers greater-by-one-than any of its members. The corresponding concept formation within a calculus of individuals calls, in certain cases, for a considerable amount of ingenuity and seems to be hardly feasible in other cases [1]). It is well known that expressions of the kind "the cardinal number of the set a is 17" (or "... at most 17", or " ... at least 17", or " ... between 12 and 21" etc.) can be readily rendered in first-order predicate calculus with equality. But a sentence like "There are more cats than dogs" causes again grave difficulties, and though these can be overcome in this and any other particular case, no general method is available for a nominalistic rendering of "There are more *a*'s than *b*'s" [2]).

The difficulties in rephrasing *all* of classical mathematics in nominalistic terms seem, and probably are, insurmountable. Inasmuch as Cantorian set theory, the theory of transfinite cardinals, and similar theories are concerned, nominalists are only too happy to get rid of them and will regard the "loss" incurred with equanimity. But they have a healthy respect for those parts of mathematics which are used in the sciences and many would rather renounce their philosophic intuitions than curtail the useful mathematics. The only serious ways out of their predicament are either to go on using all the useful parts of mathematics in the hope – admittedly not too well founded [3]) – that one day someone will produce an adequate rephrasing in nominalistic terms, or else to declare that all higher mathematics is an uninterpreted calculus which remains manageable despite its lack of interpretation through the fact that its syntax is formulated, or formulable, in a well-understood nominalistic metalanguage [4]). How exactly an uninterpreted (and directly

1) Cf. Goodman-Quine 47, N. Goodman 51, 56, Quine 53.

2) See N. Goodman 51, pp. 37 ff.

3) For reasons why it is hopeless to find an interpretation for an axiom of infinity which would be palatable for a finitistic nominalist, see Henkin 53a, p. 27.

4) Cf. Goodman-Quine 47.

uninterpretable) calculus is able to perform its useful function of mediating between interpreted empirical statements is an issue that is still far from being definitely clarified, in spite of the great efforts put into this task by many philosophers of science [1]). We recognize here a relationship with the formalistic (Hilbertian) approach which regards a certain part of mathematics – essentially recursive number theory – as being interpretable and the remainder as an uninterpreted calculus useful as a means of transformation of meaningful statements into other meaningful statements and compares this status of the "ideal" parts of mathematics to the status of the "ideal" points in affine geometry.

It is only one step from here to the adoption of an "as-if" philosophy, and Henkin [2]) intimates that a finitistic nominalist, i.e. one who believes that the universe which for him is always just one homogeneous domain of individuals – whatever these individuals may be – comprises only finitely many elements, could very well assume the existence of infinitely many objects as a useful pretense (the older word was 'fiction'). He sees, of course, that as soon as one is ready to pretend one might as well pretend that there are universals and use a full-fledged Platonistic language – while still denying that one thereby accepts the ontological commitments usually connected with such languages – but feels that there is some difference between these two pretenses, a difference which makes it easier for a conscientious nominalist to accept the first than the second pretense; Henkin admits that he knows of no objective criterion for this distinction. He is certainly right that this kind of behavior, using linguistic forms without accepting the conjugate ontological commitments, does look somewhat frivolous and is therefore in need of further clarification [3]).

There are authors who are attracted neither by the luscious jungle flora of Platonism nor by the ascetic desert landscape of neo-nominalism. They prefer to live in the well-designed and perspicuous orchards of *neo-conceptualism.* They claim to understand what sets are, though the metaphor they prefer is that of *constructing* (or *inventing*) rather than of *singling out* (or *discovering*), which is the one cherished by the Platonists, these metaphors replacing the older antithesis of *existence in the mind* versus *existence in some outside* (*real or ideal*) *world.* They are ready to admit that any well-determined and perspicuous condition indeed determines a corresponding set – since they are

1) For a thorough, recent discussion of a closely related topic, namely the status of theoretical terms in empirical science, see Carnap 56 and Hempel 58.

2) Henkin 53a, p. 28.

3) See Carnap 50a, 56, Alston 58, Issman 58.

able in this case to "construct" this set out of a stock of sets whose existence is either intuitively obvious or which have been constructed previously – but they are not ready to accept axioms or theorems that would force them to admit the existence of sets which are not constructively characterizable [1]). Therefore, they do not accept sets that correspond to impredicative conditions (unless, of course, these conditions are demonstrably equivalent to predicative ones) and deny the validity of Cantor's theorem in its naive, absolute interpretation as endowing the power-set of a given set with a higher cardinal than that of the given set itself. Absolute non-denumerability is declared to be void of sense, though an infinite set may not be enumerable with certain given means.

A nominalistic interpreted set theory, with '$\in$' interpreted as 'is-a-member-of', is, of course, a *contradictio in adiecto.* But we already mentioned that some nominalists are ready to use set theory as an uninterpreted calculus fulfilling transformational functions. Both Platonists and conceptualists insist that set theory (and mathematics in general) must be interpretable and understood as such and have no use for uninterpretable calculi. They differ in their conception of intelligibility.

It goes without saying that each of these broad philosophical views splits into many narrower ones, that their borders are blurred, and that it will often be very difficult to pin some author down to one of them. Logicism is usually regarded as one brand of Platonism, but Russell himself, during his 70 years of philosophic activity, expressed many ideas which were conceptualistic and even nominalistic. Ramified type theory has a definite conceptualistic flavor, but the axiom of reducibility is obviously Platonistic. When he professed a *no-class theory,* this was understood by many as a strictly nominalistic continuation of the use of Occam's razor. (This was, however, definitely a misunderstanding, partly created by the ambiguity in Russell's use of the term 'propositional function' for 'open formula' and 'attribute' simultaneously. Russell indeed showed how to eliminate classes in favor of "propositional functions", but these functions were just attributes (properties or relations), hence at least as "universal" as classes; Russell, due to his ambiguous usage, deceived himself in thinking that they were linguistic forms [2]).) Gödel is now usually regarded to be a Platonist, but his first publications were strongly influenced by the Hilbert school and even by Skolem's still more radically conceptualistic thinking. His postulate of constructibility (p. 60) is clearly

1) For a discussion of this point, as well as of the whole issue treated in this subsection, see Beth 56, pp. 41 ff.

2) Cf. Quine 53, pp. 122–123.

conceptualistic and has been hailed and accepted as such by conceptualists, but Gödel himself refuses to regard it as a true set-theoretical statement. Hilbert is the father of modern formalism, but his metamathematics is strongly conceptualistic and the talk about the "ideal" nature of most higher mathematical notions is far from being unambiguously classifiable into any of the standard views. Lorenzen's operationism must be dubbed a blend between conceptualism and nominalism of the "as-if" brand, but this characterization is of only little help in revealing the idiosyncrasies of his approach. Quine, starting as a logicist, has for many years tried to uphold a nominalistic position but he now feels that conceptualism is a position into which he can lapse when tired of his quixotic attempts at nominalistic reconstruction, while allaying "his puritanic conscience with the reflection that he has not quite taken to eating lotus with the platonists" [1]). Tarski's first publications exhibited an attitude, derived from Leśniewski, which he characterized as *intuitionistic formalism,* but this is no longer his present attitude [2]). Whereas he formerly had troubles in justifying operating with infinite sets of sentences, he now operates, apparently with few pangs of conscience, with languages whose set of individual constants is of any cardinality.

It would be easy, far too easy, to continue in this vein. There are very few contemporary logicians and mathematicians who have consistently and unflinchingly adhered throughout all their lifetime to one philosophic view. Among the exceptions we may count Brouwer who has been a whole-hearted and uncompromising conceptualist all his life (though this attitude is occasionally bracketed in some of his "classical" contributions to topology), Church who has always professed a straightforward, though never dogmatic, Platonism, and Goodman who so far has not yielded to the conceptualist temptation and continues to adhere to a stead-fast extreme nominalism, which if anything is growing more radical in time. It must be noted, however, that his nominalism is of a special brand which has very little in common with classical nominalism. It is what we might call purely *syntactical nominalism,* insisting that the only legitimate language form is a first-order predicate calculus but putting no restrictions, at least no official ones, on the ontological status of the individuals themselves which, for all he cares, might even be intimations of immortality, numbers, or sets, which would, however, be rather "sets" since such sets could not be said to contain members. To put it in slogan form: Goodman has no objections against sets, he is only unable to understand sets-of [3]).

1) *Ibid.,* p. 129.

2) Cf., e.g., Tarski 56, p. 62.

3) For the clearest description of this brand of nominalism and for a very able defense of its many unusual contentions against various objections, see N. Goodman 56.

Most authors who occupied themselves with the foundations of mathematics have exhibited a curious unsteadiness in matters philosophical. It was only natural for them to ascribe these changes of mind to their increasing maturity of thinking and to regard their later positions as better justified than their earlier ones, in whatever direction the shift might have gone.

It is understandable, on the other hand, that some thinkers should have seen in these vagaries a confirmation of their view that the three major ontological conceptions treated above should all of them be objectively irrelevant to the foundations problem, whatever those who upheld these conceptions thought about the matter and however strong their feelings were in this respect. Set theories, so these authors came to think, should not be judged by their ontologies (in Quine's sense) but by their fruits. Whether there are impredicative sets or not is not a matter to be decided by theoretical arguments nor a matter of (irrational?) belief based upon intuition or conscience. The prevalent opinions to the contrary are caused by a fusion of, and confusion between, two different questions: the one whether certain existential sentences can be proved, or disproved, or shown to be undecidable, *within a given theory*, the other whether this theory as a whole should be accepted. Whether the existence of a set which is the union of three given sets is provable in ZF is a serious question though easily answerable in the affirmative, as we know. Whether the non-existence of a non-trivial inaccessible number is provable in ZF is an even more serious question which is so difficult that we don't know the answer. The same question with respect to A_7 (of p. 327) is trivially answerable in the negative. For still other theories the answer is affirmative, trivially or deeply so. Whether Z or B or A_7 or T* or ML or Σ or what have you should be accepted is another very serious question but of an entirely different kind. It is a matter of practical decision, based upon such (theoretical) considerations as likelihood of being consistent, ease of maneuverability, effectiveness in deriving classical analysis, teachability, perhaps possession of standard models, etc. It is by fusing these two questions that such pseudo-problems as whether there exist non-denumerable sets (as such, absolutely, not within a given theory) are posed, leading either to futile pseudo-theoretical discussions or to the feeling that such a question is answerable only through an appeal to intuition and philosophical conscience, on the basis of which the Platonist would answer this question with a clear 'yes', the conceptualist and nominalist with an equally clear 'no', though out of entirely different intuitions.

The foremost exponent of this fourth, anti-ontological view was Carnap. In one of his later formulations [1]), he coined the terms *internal* and *external*

1) See Carnap 50a.

existence questions for the two kinds of questions mentioned above, though he did not directly apply this distinction to the foundations of set theory. This application does look to us, however, rather straightforward and we are convinced not to have misrepresented Carnap's view on this point.

This view is not without its own difficulties. We shall not discuss them here. Let us only stress that the presentation given here probably exaggerates the degree of disagreement between such authors as Quine and Carnap. Though Quine is used to say that by accepting a certain theory you have taken upon yourself certain absolute ontological commitments and Carnap denies just this, it is not yet clear to what degree this clash is not wholly or mostly verbal [1]).

Among the neo-conceptualists, we already had opportunity (Chapter III, p. 196) to mention those who reject not only impredicative concept-formations but the more extensive class of indefinite (in Carnap's sense) concept-formations, who reject – to formulate it metalinguistically – languages with unlimited quantification. These authors, among whom may be reckoned Poincaré, Brouwer, Wittgenstein, Skolem and Goodstein, arrive at their rejection of these transfinite operations from the observation that there exists no decision procedure for the truth of quantified statements. Identifying meaningfulness with effective verifiability [2]), they immediately arrive at the conclusion that sentences containing unlimited quantifiers are in general meaningless.

Though this position, *qua* philosophical attitude, is highly questionable – the major counter-argument being that it would cripple mathematics just as the parallel view concerning empirical statements would cripple empirical science – theories complying with it have, of course, their attractions. An arithmetic, for instance, that starts with relations (or operations) which are effectively decidable in each specific instance and proscribes the use of unlimited quantifiers in further concept-formations, remains intuitive all the way and is one of the safest and least doubt-ridden theories dealing with an infinite universe. It is understandable that Hilbert should have wanted to prove in this highly intuitive *recursive number theory* that mathematics is formally consistent. Skolem was able to develop a great part of classical

1) Cf. the last sections of Carnap 50a and Quine 53, Essay II (p. 46), respectively.

2) The corresponding identification in respect of empirical statements stems from Peirce and played, as the *verifiability criterion of meaning*, a central role in the early stages of logical empiricism. For the story, see, e.g., Carnap 36–37.

arithmetic in this theory and Gödel succeeded in showing that it suffices for the arithmetization of the elementary syntax of any formal system [1]).

In definite languages, in spite of the fact that classical propositional logic holds in them, those uses of the principle of the excluded middle to which intuitionists object and which are partly responsible for the antinomies are obviated in that they just cannot be formulated. Unrestricted generality is expressible by means of free variables, but unrestricted existentiality is not expressible at all: asserting '$F(x)$' is to assert that all x are F, but asserting '$\sim F(x)$' is not to assert that not all x are F but rather that all x are not-F or that no x are F.

That the ban on ordinary (unlimited) quantifiers does not have those grave restrictive effects which one might have expected can be illustrated by the following rather trivial example. Assuming that the binary predicate 'D' ("divides") has already been defined in some theory of natural numbers, one would normally define the unary predicate 'P' ("is-prime") in something like the following fashion:

$$P(x) =_{\mathrm{Df}} x > 1 \;\&\; \forall y\; [D(y,x) \leftrightarrow (y = 1 \vee y = x)]\,.$$

Simply replacing '$\forall y$' with '$\forall y \leqslant x$' – read: 'for all y from 0 up to and including x' – will now turn the trick. In general, whenever a decidable attribute is introduced, one only needs to find some upper bound of the numbers involved in order to be able to replace the unlimited quantifiers by limited ones.

It has therefore been proposed [2]) to see in a definite language, such as Language I of Carnap, the realization "in a certain sense" of the more radical among the conceptualistic tendencies, sometimes called 'finitary' or 'constructivist'. While such authors as Skolem or Goodstein would probably agree to such a formulation of their views, the intuitionists would not, though perhaps for no other reason than that no formalization adequately expresses their intuitions.

Lorenzen, on the other hand, while he strongly rejects impredicative concept-formations, most definitely accepts unlimited quantification [3]), refusing to be hampered by the verifiability criterion. His philosophy is not conceptualistic – sets for him are nothing but propositional forms [4]), conditions with free variables, and not, as with a normal conceptualist, the extra-linguistic entities corresponding to these forms. Nor is Lorenzen a syntactical nominalist and least of all a Platonist. But he is not a Carnapian

1) In Goodstein 57, this theory has found its authoritative textbook. Nothing is presupposed, not even propositional calculus. For the philosophy behind it, see Goodstein 52.

2) See Carnap 37, p. 46.

3) Lorenzen 55, p. 6.

4) Or rather, they are obtained by abstraction from equivalent propositional forms, the same set corresponding to all equivalent forms.

either. Mathematics for him is not an uninterpreted linguistic framework to be judged by its properties of fruitfulness etc., but is an interpreted theory of schematic operations with uninterpreted calculi.

In spite of many divergencies, Curry's philosophy of mathematics [1]) is most closely related to that of Carnap. Like him, he rejects any ontological commitments [2]) and stresses *acceptability* [3]) as the criterion by which mathematical theories should be judged. He calls his view *empirical formalism* to distinguish it from Hilbert's version of formalism, from which it indeed diverges considerably; *pragmatical formalism* would probably be a better label. Curry's insistence that the formalist definition of mathematics (as he gives it) requires no philosophical presuppositions and that philosophical differences should be transferred rather to the level of acceptability squares well with Carnap's latest views, and his distinction between discussions around the truth of some given mathematical statement within a given system and that of the acceptability of the system as a whole is probably equivalent to that between internal and external existence questions of Carnap.

Curry goes on to deflate the importance of provable consistency for acceptability, in contradistinction to Hilbertian formalism. This difference in attitude is admittedly only a matter of degree, since Hilbert himself did not see in consistency a sufficient condition for acceptability [4]). And intuitive evidence is, in Curry's view, a luxury which mathematics can easily afford to forego. "So far as acceptability for physics is concerned, analysis has no more need for a consistency proof than it has of intuitive evidence" [5]).

Curry's final plea for tolerance in matters of acceptability [6]) mirrors, probably on purpose, Carnap's famous tolerance principle [7]). Any author who, for reasons of intuitive convictions, insists that only mathematical systems of a certain kind have a *raison d'être* would do well to ponder once more whether his intolerance does not hamper the progress of science rather

1) Curry's views underwent many changes with the years. In addition, many of his publications appeared – due to World War II – years after they were written, sometimes after the appearance of later compositions. This, and frequent changes of terminology, tends to blur the evaluation of Curry's contributions to the foundations of mathematics. The present passage is based mostly on Curry 51 (originally written in 1939), which was later condensed in Curry 54.

2) Curry 51, p. 31.

3) *Ibid.*, pp. 59 ff.

4) Cf. Hilbert 25, p. 163.

5) Curry 51, p. 62.

6) *Ibid.*, p. 64.

7) Carnap 37, p. 51.

than channel it into the only promising road. Whereas constructibility could well be a necessary condition for the acceptability of a mathematical theory for certain purposes, say for metamathematics or for electronic computation, so that *theories of the constructible* – to use a confrontation made by Heyting [1]) – deserve to be studied by mathematicians of any philosophical conviction and has indeed been studied by authors with strongly divergent convictions as well as by authors with no philosophical convictions, the claim that the only legitimate mathematics is *constructible mathematics* has little chance of convincing anybody who does not share the specific convictions of the intuitionists.

We have had no intention to present here a summary of all current philosophies of mathematics. A few more remarks on this topic are however appropriate.

We had not mentioned at all that conception of mathematics which sees in it an empirical science, distinguished at most in degree from other empirical sciences. We had not done it so far, because we cannot imagine what justification there might be for the belief that "the source and ultimate *raison d'etre* of the notion of number, both natural and real, is experience and practical applicability" [2]), though this is the belief of Mostowski, and similar formulations are to be found in many other publications, starting with John Stuart Mill. Unless, of course, by this formulation nothing beyond the trivial view is meant that experience has led humanity to develop mathematics. This trivialization is however very unlikely, since it is hard to see how from this interpretation one can "draw the conclusion that there exists only one arithmetic of natural numbers, one arithmetic of real numbers and one theory of sets" [3]). But in what other sense can infinite sets be said to have their source in experience? (We have little quarrel with the view that the ultimate *raison d'être* of the notions of number and set are practical applicability, but we fail again to see how from this view the uniqueness of number theory and set theory can be derived.)

This attempt to abolish the qualitative distinctness of the formal sciences (logic and mathematics) from the real (empirical) sciences, which does not seem to us to have been substantiated [4]), should not be confused with

1) In the Symposium on Constructivity in Mathematics, Amsterdam 1957; see Heyting 59a.

2) Mostowski 55, p. 16.

3) *Ibid.*

4) For the latest attempt in this direction, see Kalmár 67 and the ensuing discussion.

another recent attempt, undertaken by Quine [1]) and others, to abolish the borderline. It differs from the first attempt, to put it in slogan form, in claiming that the empirical sciences are less "empirical" than one usually thinks rather than in claiming that the formal sciences are less "formal". The arguments for this view are rather convincing, but the conclusion is not necessarily forthcoming. One could as well, perhaps even better, draw the conclusion that in an empirical theory a theoretical sub-theory should be distinguished from an observational sub-theory, so that mathematics, or appropriate sections of it, would form together with the specific theoretical sub-theory a calculus which is not directly interpreted at all but receives a partial and indirect interpretation through rules of correspondence which connect the theoretical terms with the observational terms of the observational sub-theory [2]).

Many attempts have been made to interpret some metamathematical theorems such as the Löwenheim-Skolem theorem or Gödel's incompleteness theorem as discrediting certain ontological views and bolstering others. We do not believe that these attempts were succesful. We had already opportunity to express our doubts in this respect with regard to Löwenheim-Skolem (§5). With regard to Gödel's theorem, we would like to endorse Myhill's penetrating critique [3]) of the argument from the divergence of 'provable' and 'true' and insist, like him, that this argument does not disprove nominalism (though we would by no means concur with Myhill's psychological interpretation of the limitative theorems of Gödel, Church, etc.). On the contrary, we believe it to be unlikely that any new mathematical or metamathematical results will ever definitely refute any ontological standpoint, though they might conceivably have some influence on the readiness to adopt such a standpoint, for reasons which are extra-rational. Should one want to go on from here and conclude that all ontological views on mathematics, since irrefutable, are thereby also irrelevant for mathematics, though not necessarily for mathematicians, we see no good reasons against such a conclusion.

We are now in a position better to evaluate, though perhaps not to solve, a problem which was raised above (p. 323). There we saw that the incompleteness of certain logistic systems can sometimes be interpreted as the non-axiomatizability of certain formalized theories. But whereas this interpretation was rather natural with regard to arithmetical theories, it was quite

1) The *locus classicus* is Quine 53, I.

2) This is the view of Carnap 56.

3) Myhill 52a, cf. Turquette 50.

dubious with regard to set theories. While there exists at least one formalized arithmetic which is complete under a perspicuous and natural notion of validity, viz. Skolem's arithmetic, nothing of this kind seems to hold for set theory.

In what sense then, if in any, does there exist a unique notion of set (natural number) governed by a unique Theory of Sets (Theory of Natural Numbers), of which existing axiomatic set theories (arithmetics) are incomplete approximations? We already saw that some empirical realists, such as Mostowski, would answer this question by claiming that there exist sets and natural numbers in (approximately) the same sense in which there exist animals and stones and that Set Theory and Arithmetic are unique in the same sense in which Zoology [1]) and Mineralogy are unique. It is conceivable that other empirical realists would want to make here a distinction and assert reality and uniqueness only of numbers and their theory but not of sets. We already declared ourselves unable to understand either stand.

All Platonistic realists believe in the uniqueness of numbers, not as empirical entities but as Platonic ideas, and of their theory. (It is unimportant, for our purposes, what terms are used to denote the specific "mode of being" of these entities to distinguish it from the mode of being of animals and stones. Some use qualifying adjectives, others distinguish between 'being', 'existence', 'subsistence', 'reality', etc.) Gödel, for instance, believes "that the assumption of such objects (classes and concepts) is quite as legitimate as the assumption of physical bodies and there is quite as much reason to believe in their existence" [2]). But it is not clear whether this view entails the uniqueness of classes and concepts, or whether various, perhaps even mutually incompatible, systems of such entities could fulfil the task of allowing to "obtain a satisfactory system of mathematics". We are not convinced that the distance between the *pragmatic Platonism* of Gödel and the *pragmatic formalism* of Carnap and Curry is as great as the customary formulations would make one think. Believing in the existence of sets because they are necessary for obtaining some satisfactory system and accepting some set theory because it is helpful for obtaining some satisfactory system is the abyss between these views really so deep?

1) Notice that even Russell, for some time, expressed himself in a similar vein. In Russell 19, p. 169, we read: "Logic is concerned with the real world just as truly as zoology, though with its more abstract and general features". The last clause, of course, raises some doubts about the seriousness of this mode of expression, which he abandoned very soon in any case.

2) Gödel 44, p. 137; cf. also 47, from which we already quoted on p. 106 of Chapter II.

Conceptualists and nominalists have little reason to believe in the uniqueness of the notion of set, though most conceptualists would believe in the uniqueness of the natural numbers series which serves them as the major basis for their constructions. But the constructions themselves need not proceed in a unique fashion.

For the anti-ontologists, the whole problem does not arise.

It is easy to understand the urge for the belief in the uniqueness of Set Theory. Set-theoretical notions enter everywhere in non-elementary theories, and if set theory itself is treated as an elementary axiomatic theory, every non-elementary theory can as well be regarded as the union of two elementary theories, an elementary set theory and some elementary theory which is specific for the discipline treated. The decisive notion of an "absolute model" of a non-elementary theory, i.e. a model in which all set-theoretical notions receive their standard interpretation, is unique only to the degree that there exists one unique standard interpretation of these notions. It is true, therefore, that the notion of an absolute model "will gain essentially in value only when the difficult problems of the foundations of the theory of sets are solved; this will enable mathematicians to agree to one method of establishing that theory" [1]). But so far we don't see any reason compelling us to believe that there will be a unique solution to the foundational problems of set theory which will induce all mathematicians to accept one such theory as *the* Set Theory. It is doubtful whether such a belief is pragmatically necessary in the sense that otherwise a chaotic situation would arise in which every mathematician would work with his own set theory. The pragmatic criterion of acceptability should suffice to keep the situation under control. The existence of many competing set theories, at least so long as they induce little changes in the day-by-day work of the mathematician and physicist, is hardly harmful enough to justify the imposition of some *credo* or other in this respect. So long as the belief in the objective reality (whatever this may mean) and the resulting uniqueness of the notion of set and its theory is a kind of tranquilizer and does not lead to the dogmatic rejection of proposed set theories – and notice that even Mostowski states in no uncertain terms that "there are no criteria indicating the proper choice among all these numerous [set theories]" [2]) – it remains a harmless, and in a certain sense even helpful, metaphysical act of faith. But there is often only one step from the belief in the existence of an objective criterion that would uniquely determine the issue between competing

1) Mostowski 55, p. 12.
2) *Ibid.*, p. 19.

theories and the belief that one has found this criterion and is therefore entitled to disqualify all these theories, except possibly one, in the name of some earthly or heavenly reality. There are many authors who prefer perturbation out of freedom to tranquility out of external coercion.

The attitudes on how set theory might be given a satisfactory foundation are as yet widely divergent, and a host of problems connected herewith are far from being solved. Nevertheless, the great majority of mathematicians refuse to accept the thesis that Cantor's ideas were but a pathological fancy. Though the foundations of set theory are still somewhat shaky, these mathematicians continue to apply successfully its concepts, methods, and results in most branches of analysis and geometry as well as in some parts of arithmetic and algebra, confident that future foundational research will converge towards a vindication of set theory to an extent that will be identical with, or at least close to, its classical one. This attitude is compatible with a readiness to interpret set theory in a way which might diverge considerably from the customary ones, in line with the apparently existing need for a reinterpretation of logic and mathematics in general.

BIBLIOGRAPHY

Each item contained in the bibliography is quoted in this book at one or several places, for its connection with a certain subject.

For many publications, especially those with whose languages the authors were not familiar, they relied on the (almost always highly authoritative) reviews in the *Journal of Symbolic Logic,* without indicating this in every case.

Regarding the data of the publication, differences of one year may occur, since in general the year of the appearance of the *Volume* is stated, yet in particular cases the year indicated on the fascicle or the reprint.

In those cases where we refer in the listing of one publication of an author to another publication of the same author the name of the author is omitted in the reference, and the other publication is referred to by the last two digits of the year number only. For example, when in the listing of Hilbert 22 it says "also in 35" this means "also in Hilbert 35".

ABBREVIATIONS USED IN THE BIBLIOGRAPHY

(Self-evident abbreviations such as Acad., Akad., Biblioth., Enzykl., Intern., Mat., Math., Philos., Ph(ys)., Psychol., Scientif., Soc., Univ., etc. are not mentioned. "The" is mostly omitted.)

A.M.S. = American Mathematical Society

Abh. = Abhandlungen

Abh. Hamburg = Abhandlungen aus dem mathematischen Seminar der Hamburgischen Universität.

Acad. U.S.A. = Proceedings of the National Academy of Sciences (U.S.A.)

Act. Sc. Ind. = Actualités Scientifiques et Industrielles

Acta Szeged = Acta litterarum ac scientiarum Regiae Universitatis Hungaricae Francisco-Josephinae, Sectio scientiarum mathematicarum

Afd. = Afdeling

Alg. = Algemeen

Am. = American

Ann. = Annales

Anzeiger Akad. Wien = Akademie der Wissenschaften in Wien, Mathematisch-Naturwissenschatliche Klasse, Anzeiger

appl. = applied

Archiv f. math. Logik = Archiv für mathematische Logik und Grundlagenforschung

Ber. = Bericht

Ber. Leipzig = Berichte über die Verhandlungen der Sächsischen Akademie der Wissenschaften zu Leipzig, Math. Ph. Klasse.
Bull. = Bulletin
Bull. Acad. Polon. Sc. = Bulletin de l'Académie Polonaise des Sciences, Série des Sciences Mathématiques, Astronomiques et Physiques.
C.N.R.S. = Centre National de la Recherche Scientifique
C.R. = Comptes Rendus
C.R. Paris = Comptes Rendus Hebdomadaires des Séances de l'Académie des Sciences (Paris)
C.R. Varsovie = Comptes Rendus des Séances de la Société des Sciences et des Lettres de Varsovie, Classe III
Časopis = Časopis pro Pěstování Matematiky a Fysiky
Cl. = Classe
Colloq. = Colloquium, Colloque, etc.
Comm. = Commentarii, Commentationes
Congr. = Congress(o), Congrès, etc.
Congr. Amsterdam 1954 = Proceedings of the International Congress of Mathematicians, Amsterdam 1954
Congr. Cambridge Mass. 1950 = Proceedings of the International Congress of Mathematicians, Cambridge (Massachusetts) 1950
D.M.V. = Deutsche Mathematiker-Vereinigung
Ens. = Enseignement
f. = for, für, etc.
Fac. = Faculty, etc.
Fund. = Fundamenta
I.M. = Indagationes Mathematicae
Int. Enc. Un. Sc. = International Encyclopedy of Unified Science
J. = Journal
J.f.Math. = Journal für die reine und angewandte Mathematik (Crelle)
J.S.L. = Journal of Symbolic Logic
Jahrb. = Jahrbuch
Jahresb. = Jahresbericht
Koll. = Ergebnisse eines mathematischen Kolloquiums (herausgegeben von K. Menger)
Kon. = Koninklijke
Kongr. Heidelberg 1904 = Verhandlungen des Dritten Intern. Mathematiker-Kongresses in Heidelberg, 1904.
Kongr. Zürich 1932 = Verhandlungen des Internationalen Mathematiker-Kongresses, Zürich 1932
Les Entretiens de Zurich = Les Entretiens de Zurich sur les Fondements et la Méthode des Sciences Mathématiques 1938
Monatsh. = Monatshefte
N. = Nieuw
N.S. = New Series, Neue Folge, Nieuwe Reeks, etc.
Nachr. Göttingen = Nachrichten der Gesellschaft der Wissenschaften zu Göttingen, Mathematisch-Physikalische Klasse
Nat. = National, etc.; Natural, etc.
Ned. = Nederlandse
phänomen. = phänomenologisch

Phenomen. = Phenomenological
Polon. = Polonaise
Proc. = Proceedings
Proc. Amsterdam = Kon. Nederlandse Akademie van Wetenschappen te Amsterdam, Proceedings of the section of sciences
Publ. = Publications
R. = Royal(e)
R.M.M. = Revue de Métaphysique et de Morale
Rendic. = Rendiconti
Rendic. Palermo = Rendiconti del Circolo Matematico di Palermo
Rev. = Revue, Review, etc.
Sc. = Science(s), etc.
Scand. = Scandinavica
Sem. = Seminar, etc.
Semesterberichte Münster = Semesterberichte zur Pflege des Zusammenhangs von Universität und Schule (Math. Seminar, Münster i.W.)
Sitz. Berlin = Sitzungsberichte der Preusischen Akademie der Wissenschaften, Phys.-Mathemat. Klasse
Tr. = Transactions
u. = und
v. = van, voor
Verh. = Verhandlungen, Verhandelingen
Vid. = Videnskab
Wetensch. = Wetenschappen
Wiss. = Wissenschaft(en)
Ztschr. = Zeitschrift
Ztschr. f. Math. Logik = Zeitschrift für mathematische Logik und Grundlagen der Mathematik
(2) = second series (and similarly for other numerals and letters in parentheses)

BIBLIOGRAPHY

ACKERMANN, W.

1937. Die Widerspruchsfreiheit der allgemeinen Mengenlehre. *Math. Annalen* **114**, 305–315.

1940. Zur Widerspruchsfreiheit der Zahlentheorie. *Ibid.* **117**, 162–194.

1950. Widerspruchsfreier Aufbau der Logik I. (Typenfreies System ohne tertium non datur.) *J.S.L.* **15**, 33–57.

1952. Review of Fitch 52. *J.S.L.* **17**, 266–268.

1952–53. Widerspruchsfreier Aufbau einer typenfreien Logik. (Erweitertes System.) *Math. Ztschr.* **55**, 364–384, 1952; **57**, 155–166, 1953.

1954. Solvable cases of the decision problem. Amsterdam. 114 pp.

1956. Zur Axiomatik der Mengenlehre. *Math. Annalen* **131**, 336–345.

ADDISON, J. W.

1959. Some consequences of the axiom of constructibility. *Fund. Math.* **46**, 337–357.

1962. The theory of hierarchies. Nagel et al. 62, 26–37.

ADDISON, J. W., HENKIN, L. and TARSKI, A.

1965 (editors). The theory of models, Proc. of the 1963 intern. symposium at Berkeley. Amsterdam. 494 pp.

ADDISON, J. W. and MOSCHOVAKIS, Y. N.

1968. Some consequences of the axiom of definable determinateness. *Acad. U.S.A.* **59**, 708–712.

AJDUKIEWICZ, K.

1935. Die syntaktische Konnexität. *Studia Philos.* **1**, 1–27.

ALSTON, W. P.

1958. Ontological commitments. *Philos. Studies* **9**, 8–17.

AMBROSE, ALICE.

1935. Finitism in mathematics. *Mind,* N.S. **44**, 186–203, 317–340.

1936. The nature of the question "are there three consecutive 7's in the expansion of π?" *Papers of the Michigan Acad. of Sc., Arts and Letters* **22**, 505–513. (Cf. *Mind,* N.S. **46**, 379–385, 1937.)

ANDREWS, P. B.

1965. A transfinite type theory with type variables. Amsterdam. 143 pp.

ASHWINIKUMAR.

1966. Hilbert spaces in intuitionism. Ph.D. thesis. Amsterdam.

AVEY, A. E.

1929. The law of contradiction: its logical status. *J. Philos.* **26**, 519–526.

AX, J.

1968. The elementary theory of finite fields. *Ann. of Math.* (2) **88**, 239–271.

AX, J. and KOCHEN, S.

1965. Diophantine problems over local fields. *Am. J. of Math.* **87**, 605–648.

BACHMANN, H.

1955. Transfinite Zahlen. (*Ergebnisse der Math.* etc., N.S., Heft 1.) Berlin. 204 pp. (2nd ed. 1967, 228 pp.)

BAER, R.

1929. Zur Axiomatik der Kardinalzahlarithmetik. *Math. Ztschr.* **29**, 381–396.

BANACH, S. and TARSKI, A.

1924. Sur la décomposition des ensembles de points en parties respectivement congruentes. *Fund. Math.* **6**, 244–277.

BAR-HILLEL, Y.

1950. On syntactical categories. *J.S.L.* **15**, 1–16; also in 64, 19–37.

1953. A quasi-arithmetical notation for syntactic description. *Language* **29**, 47–58; also in 64, 61–74. French translation: *Langages* **9**, 9–22, 1968.

1957. Husserl's conception of a purely logical grammar. *Philos. and Phenomen. Research* **17**, 362–369; also in 70, 89–97.

1957a. New light on the Liar. *Analysis* **18**, 1–6; also in 70, 253–257.

1964. Language and information: selected essays on their theory and application. Jerusalem and Reading, Mass. 388 pp.

1965 (editor). Logic, methodology and philosophy of science, Proc. of the 1964 intern. congr. Jerusalem. Amsterdam. 440 pp.

1966. Do natural languages contain paradoxes? *Studium Generale* **19**, 391–397; also in 70, 273–285.

1967. Syntactical and semantical categories. Encycl. of philos., vol. VIII. New York. 57–61.

1970. Aspects of language. Jerusalem and Amsterdam. 381 pp.

1970a (editor). Mathematical logic and foundations of set theory, Proc. of an intern. colloq., 11–14 November 1968, Jerusalem. Amsterdam. 145 pp.

BAR-HILLEL, Y., POZNANSKI, E. I. J., RABIN, M. O. and ROBINSON, A.

1961 (editors). Essays on the foundations of mathematics, dedicated to A. A. Fraenkel. Jerusalem. 351 pp.

BARWISE, J.

1968 (editor). The syntax and semantics of infinitary languages. Lecture notes in math. 72. Berlin. 268 pp.

1969. Infinitary logic and admissible sets. *J.S.L.* **34**, 226–252.

BARZIN, M. and ERRERA, A.

1927. Sur la logique de M. Brouwer. *Acad. R. de Belgique, Bull. de la Cl. des Sc.* (5) **13**, 56–71.

BECKER, A.

1936. Bestreitet Aristoteles die Gültigkeit des "tertium non datur" für Zukunftsaussagen? *Act. Sc. Ind.* 393, 69–74.

BECKER, O.

1927. Mathematische Existenz. Untersuchungen zur Logik und Ontologie mathematischer Phänomene. Jahrb. f. Philos. und phänomen. Forschung 8, 439–809. Reprinted as a book; Halle a.S., 1927.

1929. Über den sogenannten "Anthropologismus" in der Philosophie der Mathematik. *Philos. Anzeiger* **3**, 369–387.

1930. Zur Logik der Modalitäten. *Jahrb. f. Philos. u. phänomen. Forschung* **11**, 497–548. Reprinted as a book; Halle a.S., 1930. (Cf. *Blätter f. deutsche Philos.* **16**, 387–422, 1943, and **18**, 82–93, 1944.)

1933. Eudoxos-Studien. II. Warum haben die Griechen die Existenz der vierten Proportionale angenommen? *Quellen und Studien zur Geschichte der Math.* **B2**, 369–387. (Cf. *ibid.* **B3**, 236–244, 1936.)

1936. Eudoxos-Studien. IV. Das Prinzip des ausgeschlossenen Dritten in der griechischen Mathematik. *Ibid.* **B3**, 370–388.

BEHMANN, H.

1931. Zu den Widersprüchen der Logik und Mengenlehre. *Jahresb. D.M.V.* **40**, 37–48.

BELINFANTE, H. J.

1929. Zur intuitionistischen Theorie der unendlichen Reihen. *Sitz. Berlin* 1929, No. XXIX. 24 pp.

1930. Über eine besondere Klasse von non-oszillierenden Reihen. *Proc. Amsterdam* **33**, 1170–1179.

1931. Über die Elemente der Funktionentheorie und die Picardschen Sätze in der intuitionistischen Mathematik. *Proc. Amsterdam* **34**, 1395–1397. (Cf. *Kongr. Zürich* 1932, II, 345–346, 1933; *Proc. Amsterdam* **44**, 173–185, 276–285, 420–425, 563–567, 711–717, 1941.)

1938. Das Riemannsche Umordnungsprinzip in der intuitionistischen Theorie der unendlichen Reihen. *Compositio Math.* **6**, 118–123.

1938a. Der Lévysche Umordnungssatz und seine intuitionistische Übertragung. *Ibid.*, 124–135.

BENACERRAF, P. and PUTNAM, H.

1964 (editors). Philosophy of mathematics. Englewood Cliffs, N. J. 536 pp.

BENEŠ, V. E.

1952. Review of Quine 51a. *J.S.L.* **17**, 212.

BERNAYS, P.

1935. Sur le platonisme dans les mathématiques. *L'Ens. Math.* **34**, 52–69. English translation in Benacerraf-Putnam 64, 274–286.

1935a. Quelques points essentiels de la métamathématique. *L'Ens. Math.* **34**, 70–95.

1937–54. A system of axiomatic set theory. I–VII. *J.S.L.* I: **2**, 65–77. 1937; II: **6**, 1–17, 1941; III: **7**, 65–89, 1942; IV: **7**, 133–145, 1942; V: **8**, 89–106, 1943; VI: **13**, 65–79, 1948; VII: **19**, 81–96, 1954.

1946. Review of Gödel 44. *J.S.L.* **11**, 75–79.

1950. Mathematische Existenz und Widerspruchsfreiheit. *Études de philos. des sc., en hommage a F. Gonseth a l'occasion de son 60me anniversaire,* Neuchatel, 11–25.

1958. Axiomatic set theory (with a historical introduction by A. A. Fraenkel). Amsterdam. 226 pp.

1961. Die hohen Unendlichkeiten und die Axiomatik der Mengenlehre. Infinitistic Methods, Warsaw. 11–20.

1961a. Zur Frage der Unendlichkeitsschemata in der axiomatischen Mengenlehre. Bar-Hillel et al. 61, 3–49.

1967. What do some recent results in set theory suggest? Lakatos 67, 109–117.

BERNSTEIN, F.

1919. Die Mengenlehre Georg Cantors und der Finitismus. *Jahresb. D.M.V.* **28**, 63–78.

BETH, E. W.

1953. On Padoa's method in the theory of definition. *Proc. Amsterdam* **A56** (= *I.M.* **15**), 330–339. (Cf. A. Robinson: *J.S.L.* **21**, 220–221, 1956.)

1956. L'existence en mathématiques. Paris. 60 pp.

1956a. Semantic construction of intuitionistic logic. *Mededelingen der Kon. Ned. Akad. v. Wetensch., Afd. Letterkunde,* N.S. **19**, no. **13**. (Originally in French: *Colloque de Logique Math. du C.N.R.S.* Paris 1955.)

1959. The foundations of mathematics. Amsterdam. 741 pp.

BILLING, J.

1949. The principle of excluded third and the Bolzano-Weierstrass lemma. *Arkiv f. Mat. etc.* (2) **1**, p. 59. (Cf. *ibid.* **34B**, No. 11, 1947. 2 pp.)

BISHOP, E.

1967. Foundations of constructive analysis. New York. 370 pp.

BLEICHER, M. N.

1965. Multiple choice axioms and axioms of choice for finite sets. *Fund. Math.* **57**, 247–252.

BOCHEŃSKI, I. M.

1949. On the syntactical categories. *The New Scholasticism* **23**, 257–280.

1956. Formale Logik. Freiburg. 639 pp.

BOCKSTAELE, P.

1949. Het intuïtionisme bij de Franse wiskundigen. (*Verh. Kon. Vlaamse Acad. v. Wetensch.* etc.) Brussels. 123 pp.

BOČVAR, D. A.

1939. On a three-valued logical calculus and its application to the analysis of the paradoxes of the extended functional calculus (Russian). *Mat. Sbornik,* N.S. **4**, 287–308.

1943. To the question of the consistency of a three-valued logical calculus (Russian). *Ibid.,* **12**, 353–369.

BOFFA, M.

1968. Graphes extensionnelles et axiome d'universalité. *Ztschr. f. math. Logik* **14**, 329–334.

1969. Axiome et schéma de fondement dans le système de Zermelo. *Bull. Acad. Polon. Sc.* **17**, 113–115.

BOLZANO, B.

1851. Paradoxien des Unendlichen. (Originally ed. posthumously by F. Přihonsky, 1851.) New edition by A. Höfler, with notes by H. Hahn. Philos. Bibliothek, vol. 99. Leipzig, 1920. English ed. 1059. 189 pp.

BOONE, W. W.

1954–57. Certain simple, unsolvable problems of group theory, I–IV. *Proc. Amsterdam* **A57** (= *I.M.* **16**), 1954, 231–237, 492–497; **A58** (= *I.M.* **17**), 1955, 252–256, 571–577; **A60** (= *I.M.* **19**), 1957, 22–27, 227–232.

1968. Decision problems about algebraic and logical systems as a whole and recursively enumerable degrees of unsolvability. Schmidt et al. 68, 13–33.

BOONE, W. W., HAKEN, W. and POÉNARU, V.

1968. On recursively unsolvable problems in topology and their classification. Schmidt et al. 68, 37–74.

BOREL, É.

1914. Leçons sur la théorie des fonctions. Paris 1898. 2nd ed. 1914. 260 pp; 3rd (4th) ed. 1928 (1950). (Cf. *Math. Annalen* **60**, 194–195, 1905, and *Rev. du Mois* **1**, 248–250, 1906.)

1946. Les paradoxes de l'infini. Paris. 237 pp.

1947. Les paradoxes de l'axiome du choix. *C.R. Paris* **224**, 1537–1538. (Cf. *ibid.,* 1597–1599.)

BORGERS, A.

1949. Development of the notion of set and of the axioms of sets. *Synthese* **7**, 374–390.

BOULIGAND, G.

1947. Une épistémologie conforme à l'esprit de l'analyse classique. *C.R. Paris* **224**, 1747–1749.

BOURBAKI, N.

1949. Foundations of mathematics for the working mathematician. *J.S.L.* **14**, 1–8.

1954. Théorie des ensembles, Ch. 1–2. Paris. 136 pp. (2nd ed. 1960.)

1956. Théorie des ensembles, Ch. 3. Paris. 118 pp. (2nd ed. 1963. 148 pp.)

BOUTROUX, P.

1920. L'idéal scientifique des mathématiciens dans l'antiquité et dans les temps modernes, Paris. (German ed. 1927. 253 pp.)

BRIDGMAN, P. W.

1927. The logic of modern physics. New York. 242 pp. (German ed. 1932.)

BRITTON, J. L.

1963. The word problem. *Ann. of Math.* (2) **77**, 16–32.

BROUWER, L. E. J.

1907. Over de grondslagen der wiskunde. Thesis, Amsterdam & Leipzig. 183 pp. (Cf. *N. Archief v. Wiskunde* (2)**8**, 326–328, 1908.)

1908. De onbetrouwbaarheid der logische principes. *Tijdschrift v. Wijsbegeerte* **2**, 152–158; also in **19**.

1912. Intuïtionisme en formalisme. Groningen. 32 pp. Also in **19**. English translation in *Bull. A.M.S.* **20**, 81–96, 1914, and in Benacerraf-Putnam **64**, 66–77.

1918–19. Begründung der Mengenlehre unabhängig vom logischen Satz vom ausgeschlossenen Dritten. I and II. *Verh. der Kon. Akad. v. Wetensch. te Amsterdam* (eerste sectie) **12**, Nos. 5 and 7. 43 + 33 pp.

1919. Wiskunde, waarheid, werkelijkheid. Groningen.

1923. Begründung der Funktionenlehre unabhängig vom logischen Satz vom ausgeschlossenen Dritten. I. Teil. *Verh. der Kon. Akad. v. Wetensch. te Amsterdam* (eerste sectie), **13**, No. 2. 24 pp. (Cf. *Proc. Amsterdam* **27**, 189–193, 644–646, 1924.)

1924. Intuitionistische Zerlegung mathematischer Grundbegriffe. *Jahresb. D.M.V.* **33**, 251–256. (Cf. *ibid.* **36**, 127–129, 1927.)

1924a. Beweis dass jede Funktion gleichmässig stetig ist. *Proc. Amsterdam* **27**, 189–194.

1925. Über die Bedeutung des Satzes vom ausgeschlossenen Dritten in der Mathematik, insbesondere in der Funktionentheorie. *J. f. Math.* **154**, 1–7. (Cf. *Proc. Amsterdam* **29**, 866–867, 1926.)

1925a. Intuitionistischer Beweis des Jordanschen Kurvensatzes. *Proc. Amsterdam* **28**, 503–508.

1925–27. Zur Begründung der intuitionistischen Mathematik. I–III. *Math. Annalen* **93**, 244–257, 1925; **95**, 453–472, 1926; **96**, 451–488, 1927. (Cf. *J. f. Math.* **157**, 255–257, 1927.)

1926. Intuitionistische Einführung des Dimensionsbegriffs. *Proc. Amsterdam* **29**, 855–863. (Cf. *ibid.* **27**, 636–638, 1924.)

1927. Über Definitionsbereiche von Funktionen. *Math. Annalen* **97**, 60–75. English translation in van Heijenoort **67**, 446–463.

1928. Intuitionistische Betrachtungen über den Formalismus. *Sitz. Berlin* 1928, 48–52 (also *Proc. Amsterdam* **31**, 374–379, 1928). English translation in van Heijenoort **67**, 490–492.

1929. Mathematik, Wissenschaft und Sprache. *Monatsh. Math. Ph.* **36**, 153–164. (Cf. *Proc. Amsterdam* **50**, 339, 1947.)

1948. Essentieel-negatieve eigenschappen. *Proc. Amsterdam* **51**, 963–964 (= *I.M.* **10**, 322–323). (Cf. *ibid.* **52**, 122–124, 315–316 (= *I.M.* **11**, 37–39, 89–90), 1949.)

1949. Consciousness, philosophy, and mathematics. *Library of the Xth Intern. Congr. of Philos., Amsterdam* 1948, I, 1235–1249. (Cf. *Proc. Amsterdam* **51**, 1239–1243 (= *I.M.* **10**, 383–387), 1948.)

1950. Sur la possibilité d'ordonner le continu. *C.R. Paris* **230**, 349–350. (Cf. *ibid.*, 263–265; *Proc. Amsterdam* **A54** (= *I.M.* **13**), 357–358, 1951, and **A55** (= *I.M.* **14**), p. 79, 1952; F. Loonstra: Math. Centrum Amsterdam, Rapport Z.W. 1949,014. 4 pp. 1949.)

1952. An intuitionistic correction of the fixed-point theorem on the sphere. *Proc. R. Soc. London* (A) **213**, 1–2.

1952a. Over accumulatiekernen van oneindige kernsoorten *I.M.* **14**, 439–442.

1952–53. Historical background, principles and methods of intuitionism. *South African J. of Sc.* **49**, 139–146. (Cf. *Proc. of the R. Irish Acad.* A57, 113–116, 1955.)

1954. Points and spaces. *Canadian J. of Math.* **6**, 1–17.

BROUWER, L. E. J. and DE LOOR, B.

1924. Intuitionistischer Beweis des Fundamentalsatzes der Algebra. *Proc. Amsterdam* **27**, 186–188. (Cf. *ibid.*, 631–634.)

BÜCHI, J. R.

1953. Investigation of the equivalence of the axiom of choice and Zorn's lemma from the viewpoint of the hierarchy of types. *J.S.L.* **18**, 125–135.

BUKOVSKÝ, L.

1965. The continuum problem and the powers of alephs. *Comment. Math. Univ. Carolinae* **6**, 181–197.

1965a. An elementary proof of normality of the class of accessible cardinals. *Ibid.*, 409–412.

BULOFF, J. J., HOLYOKE, T. C. and HAHN, S. W.

1969 (editors). Foundations of mathematics. Symposium papers commemorating the sixtieth birthday of Kurt Gödel. Berlin. 195 pp.

BURALI-FORTI, C.

1897. Una questione sui numeri transfiniti. *Rendic. Palermo* **11**, 154–164, 260. English translation in van Heijenoort 67, 105–111. (Cf. P.E.B. Jourdain: *Philos. Magazine* (6) 7, 70, 1904.)

CANTOR, G.

1878. Ein Beitrag zur Mannigfaltigkeitslehre. *J.f. Math.* **84**, 242–258; also in 32, 119–138.

1879–84. Über unendliche, lineare Punktmannigfaltigkeiten. *Math. Annalen:* I. **15**, 1–7, 1879; II. **17**, 355–358, 1880; III. **20**, 113–121, 1882; IV. **21**, 51–58, 1883; V. *Ibid.*, 545–591 (appeared also as a book: Grundlagen einer allgemeinen Mannigfaltigkeitslehre. Ein mathematisch-philosophischer Versuch in der Lehre des Unendlichen. Leipzig, 1883. 47 pp.); VI. **23**, 453–488, 1884, also in 32, 139–246 (I–V appeared in French, partly in form of extracts, in *Acta Math.* **2**, 1883).

1895–97. Beiträge zur Begründung der transfiniten Mengenlehre. *Math. Annalen:* I. **46**, 481–512; II. **49**, 207–246, also in 32, 282–356. (English ed. by P.E.B. Jourdain, Chicago, 1915, reprinted 1941. French ed. by F. Marotte in *Mémoires de la Soc. des Sc. Phys. et Nat. de Bordeaux* (5) **3** (1899), 343–437; also separately, Paris, 1899.)

1932. Gesammelte Abhandlungen mathematischen und philosophischen Inhalts. Ed. by E. Zermelo. Berlin. 486 pp. Reprinted, Hildesheim, 1962.

CARNAP, R.

1936–37. Testability and meaning. *Philos. of Sc.* **3**, 419–471 and **4**, 1–40.

1937. The logical syntax of language. New York & London. 352 pp.

1939. Foundations of logic and mathematics. *Int. Enc. Un. Sc.*, vol. I, No. 3. 71 pp.

1947. Meaning and necessity. Chicago. 210 pp. – 2nd enlarged ed., 1956. 258 pp.

1950. Logical foundations of probability. Chicago. 607 pp.

1950a. Empiricism, semantics and ontology. *Rev. Intern. de Philos.* 4, 20–40. Reprinted in L. Linsky, ed.: Semantics and the philosophy of language, Urbana 1952; (with slight revisions) in P. Wiener, ed.: Readings in philosophy of science, New York, 1953; in J. L. Jarrett and S. L. McMurrin, eds.: Contemporary philosophy, New York, 1954; in Benacerraf-Putnam 64, 233–248; and in the second edition of 47.

1954. Einführung in die symbolische Logik. Wien. 209 pp. English translation: Introduction to symbolic logic, New York, 1958. 241 pp.

1956. The methodological character of theoretical concepts. Minnesota Studies in the Philos. of Sc., I, 38–75.

CARNAP, R. and BACHMANN, F.

1936. Über Extremalaxiome. *Erkenntnis* 6, 166–188.

CASSINA, U.

1936. Sul principio della scelta ed alcuni problemi dell' infinito. *Rendic. Sem. Mat. Fis. Milano* 10, 53–81.

CASSIRER, E.

1929. Philosphie der symbolischen Formen. III. Teil: Phänomenologie der Erkenntnis. Berlin.

CAVAILLÈS, J.

1947. Transfini et continu. *Act. Sc. Ind.* 1020. Paris. 24 pp.

CHANG, C. C.

1965. A simple proof of the Rabin-Keisler theorem. *Bull. A.M.S.* 71, 642–643.

1965a. Infinite valued logic as a basis for set theory. Bar-Hillel 65, 93–100.

CHEVALIER, J.

1929. Continu et discontinu (by Chevalier and others). *Cahiers de la Nouvelle Journée*, No. 15. Paris.

CHOMSKY, N.

1957. Syntactic structures. 's-Gravenhage. 116 pp.

CHURCH, A.

1927. Alternatives to Zermelo's assumption. *Tr. A.M.S.* 29, 178–208.

1934. The Richard paradox. *Am. Math. Monthly* 41, 356–361.

1936. An unsolvable problem of elementary number theory. *Am. J. of Math.* 58, 345–363.

1936a. A note on the *Entscheidungsproblem. J.S.L.* 1, 40–41.

1939. Schröder's anticipation of the simple theory of types. Preprinted from *J. of Unified Sc.* 9, 4 pp.

1949. A formulation of the simple theory of types. *J.S.L.* 5, 56–68.

1951. Review of Copi 50. *J.S.L.* 16, 154–155.

1956. Introduction to mathematical logic. I. Princeton. 376 pp.

CHWISTEK, L.

1948. The limits of science. With an introduction and appendix by Helen C. Brodie. London & New York. 347 pp.

CLAPHAM, C. R. J.

1964. Finitely presented groups with word problems of arbitrary degrees of insolubility. *Proc. London Math. Soc.* 56, 633–676.

COGAN, E. J.

1955. A formalization of the theory of sets from the point of view of combinatory logic. *Ztschr. f. math. Logik* 1, 198–240.

COHEN, P. J.

1963. A minimal model for set theory. *Bull. A.M.S.* **69**, 537–540.

1963–4. The independence of the continuum hypothesis. *Acad. U.S.A.*: I. **50**, 1143–1148, 1963; II. **51**, 105–110, 1964.

1965. Independence results in set theory. Addison et al. 65, 39–54.

1966. Set theory and the continuum hypothesis. New York. 160 pp.

COLLINGWOOD, R. G.

1933. An essay on philosophical method. Oxford. 226 pp.

COPI, I. M.

1950. The inconsistency or redundancy of Principia Mathematica. *Philos. and Phenomen. Research* **11**, 190–199.

1954. Symbolic logic. New York. 355 pp.

VAN DER CORPUT, J. G.

1946. On the fundamental theorem of algebra. III. *Proc. Amsterdam* **49**, 985–994. (Cf. *ibid.,* 722–732, 878–886.) Also in *I.M.* **8**, 605–614, 1946. (Cf. *ibid.,* 430–440, 549–557.)

CRAIG, W.

1953. On axiomatizability within a system. *J.S.L.* **18**, 30–32.

1957. The uses of the Herbrand-Gentzen theorem in relating model theory and proof theory. *Ibid.,* 269–285.

CRAIG, W. and VAUGHT, R. L.

1958. Finite axiomatizability using additional predicates. *J.S.L.* **23**, 289–308.

CROSSLEY, J. N.

1967 (editor). Sets, models and recursion theory. Proc. of the summer school in math. logic and tenth logic colloq., Leicester, August-September 1965. Amsterdam. 331 pp.

CROSSLEY, J. N. and DUMMETT, M. A. E.

1965 (editors). Formal systems and recursive functions. Proc. of the eighth logic colloq., Oxford, July 1963. Amsterdam. 320 pp.

CURRY, H. B.

1930 Grundlagen der kombinatorischen Logik. I and II. *Am. J. of Math.* **52**, 509–536, 789–834. (Cf. *ibid.* **54**, 551–558, 1932.)

1942. The inconsistency of certain formal logics. *J.S.L.* **7**, 115–117.

1950. A theory of formal deducibility. (*Notre Dame Math. lectures,* No. 6.) 126 pp. (Cf. *J.S.L.* **17**, 245–248, 249–265, 1952.)

1951. Outlines of a formalist philosophy of mathematics. Amsterdam. 75 pp. (Cf. *Mind* **62**, 172–183, 1953; *Philos. Rev.* **59**, 346–353, 1950.)

1954. Remarks on the definition and nature of mathematics. *Dialectica* **8**, 228–233; also in Benacerraf-Putnam 64, 152–156.

1954a. Review of Rosser 53. *Bull. A.M.S.* **60**, 266–272.

1963. Foundations of mathematical logic. New York. 408 pp.

CURRY, H. B. and FEYS, R.

1958. Combinatory logic. Amsterdam. 390 pp.

VAN DALEN, D.

1963. Extension problems in intuitionistic plane projective geometry. *I.M.* **25**, 349–383.

1967. A note on spread cardinals. *Compositio Math.* **20**, 21–28.

1968. Reducibilities in intuitionistic topology. *J.S.L.* **33**, 412–417.

VAN DANTZIG, D.

1932. Over de elementen van het wiskundig denken. Groningen, 15 pp. (Also in *Euclides* 9, 102–116, 1933.)

1946. Wiskunde, logica en ervaringswetenschappen. *Studium Generale: Syllabus van Technische Hoogeschool te Delft*, pp. 71–104. (Cf. *Synthese* **5**, 441–455, 1947.)

1947. On the principles of intuitionistic and affirmative mathematics. *Proc. Amsterdam* **50**, 918–929, 1092–1103 (= *I.M.* 9, 429–440, 506–517). (Cf. *Actes Sc. Ind.* **1134**, 23–135, 1951; also Gilmore 56.)

1949. Comments on Brouwer's theorem on essentially-negative predicates. *Ibid.* **52**, 949–957 (= *I.M.* **11**, 347–355).

1949a. Significs and its relation to semiotics. *Library of the Xth Intern. Congr. of Philos., Amsterdam* 1948, II. 176–189.

1951. Mathématique stable et mathématique affirmative. Congrès intern. de philos. des sc., 1949, II Logique. Paris. 123–135.

1956. Is $10^{10^{10}}$ a finite number? *Dialectica* **9**, 273–277.

1957. Gerrit Mannoury's significance for mathematics and its foundations. *N. Archief v. Wiskunde* (3) **5**, 1–18.

DAVIS, M.

1958. Computability and unsolvability. New York. 210 pp.

1965 (editor). The undecidable. Hewlett, N.Y. 440 pp.

DAVIS, M., PUTNAM, H. and ROBINSON, JULIA

1961. The decision problem for exponential diophantine equations. *Ann. of Math.* (2) 74, 425–436.

DEDEKIND, R.

1888. Was sind und was sollen die Zahlen? 6th ed.: Braunschweig 1930, 58 pp.; also in 30–32 III. English ed. by W. W. Beman in: Essays on the theory of numbers by R. Dedekind, Chicago and London 1901.

1930–32. Gesammelte mathematische Werke. Ed. by R. Fricke, E. Noether, O. Ore. 3 vols. Braunschweig.

DENJOY, A.

1946–54. L'énumeration transfinie. Livres I–IV. Paris. 971 pp.

DERRICK, J. and DRAKE, F. R.

1967. Independence of the axiom of choice from variants of the generalized continuum hypothesis. Crossley 67, 75–84.

DESTOUCHES-FÉVRIER, PAULETTE

1951. La structure des théories physiques. Paris. 423 pp.

DE SUA, F.

1956. Consistency and completeness – a résumé. *Amer. Math. Monthly* **63**, 295–305.

DIENES, Z. P.

1952. Sulla definizione dei gradi di rigore. *Univ. e Politecnico Torino, Rend. Sem. Mat.* **11**, 223–253.

DIEUDONNÉ, J.

1951. L'axiomatique dans les mathématiques modernes. *Act. Sc. Ind.* **1137**, 47–53.

DIJKMAN, J. G.

1946. Einige Sätze über mehrfach negativ-konvergente Reihen in der intuitionistischen Mathematik. *Proc. Amsterdam* **49**, 829–833. (Cf. **51**, 681–692, 1948.) Also in *I.M.* **8**, 532–536 (**10**, 232–243).

1952. Convergentie en divergentie in de intuitionistische wiskunde. Thesis, Univ. of Amsterdam, 98 pp.

1962. A note on intuitionistic divergence theory. *N. Archief v. Wiskunde* (3) **10**, 17–19.

DINGLER, H.

1913. Über die logischen Paradoxien der Mengenlehre und eine paradoxienfreie Mengendefinition. *Jahresb. D.M.V.* **22**, 307–315. (Also in *Z. f. positivistische Philos.* **1**, 143–150, 1913.) Cf. *ibid.* **28**, 138–158, 1919.

1931. Der Zusammenbruch der Wissenschaft und der Primat der Philosophie. 2nd ed. München.

1931a. Philosophie der Logik und Arithmetik. München. 198 pp.

DOSS, R.

1963. On Gödel's proof that V=L implies the generalized continuum hypothesis. *Notre Dame J. of Formal Logic* **4**, 283–287.

DUBARLE, D.

1955. Remarque sur la philosophie de la formalisation logico-mathematique. *R.M.M.* **60**, 352–390.

EASTON, W. B.

1964. Powers of regular cardinals. Ph.D. thesis, Princeton Univ. 66 pp.

1970. Powers of regular cardinals. *Ann. of Math. Logic* **1**, 139–178.

ELLENTUCK, E.

1965. The universal properties of Dedekind finite cardinals. *Ann. of Math.* (2) **82**, 223–248.

1966. Generalized idempotence in cardinal arithmetic. *Fund. Math.* **58**, 241–258.

ERDÖS, P.

1964. An interpolation problem associated with the continuum hypothesis. *Michigan Math. J.* **11**, 9–10.

ERDÖS, P. and HAJNAL, A.

1966. On a problem of Jonsson. *Bull. Acad. Polon. Sc.* **14**, 19–23.

ERDÖS, P., HAJNAL, A. and RADO, R.

1965. Partition relations for cardinal numbers. *Acta Math. Acad. Sc. Hungaricae* **16**, 93–196.

ERŠOV, YU. L.

1965. On the elementary theory of maximal normed fields. *Doklady Akad. Nauk SSSR* **165**, 21–23. English translation, with an additional note, in *Soviet Mathematics Doklady* **6**, 1390–1393, 1965.

1966. New examples of undecidable theories (Russian). *Algebra i Logika Seminar* **5**, 37–47.

1968. Numbered fields. van Rootselaar-Staal 68, 31–34.

ESENIN-VOLPIN, A.S.

1961. Le programme ultra-intuitionniste des fondements des mathématiques. Infinitistic Methods, Warsaw. 201–223. (Cf. the review by Kreisel and Ehrenfeucht, *J.S.L.* **32**, 517, 1967.)

1970. The ultra-intuitionistic criticism and the antitraditional program for foundations of mathematics. Kino-Myhill-Vesley 70, 3–45.

FEFERMAN, S.

1960. Arithmetization of metamathematics in a general setting. *Fund. Math.* **49**, 35–92.

1964. Independence of the axiom of choice from the axiom of dependent choices (abstract). *J.S.L.* **29**, 226.

1965. Some applications of the notions of forcing and generic sets. *Fund. Math.* **56**, 325–345. (Summary in Addison et al. 65, 89–95.)

FELGNER, U.

1967. Untersuchungen über das Zornsche Lemma. *Compositio Math.* **18**, 170–180.

1969. Die Existenz wohlgeordneter, konfinaler Teilmengen in Ketten und das Auswahlaxiom. *Math. Ztschr.* **111**, 221-232. Correction, *ibid.* **115**, 392, 1970.

1971. Comparison of the axioms of local and universal choice. *Fund. Math.* **71**, 43–62.

1971a. Über das Ordnungstheorem. *Ztschr. math. Logik,* **17**, 257–272.

FINDLAY, J.

1942. Goedelian sentences: a non-numerical approach. *Mind,* N.S. 51, 259–265.

FINE, H. B.

1914. An unpublished theorem of Kronecker respecting numerical equations. *Bull. A.M.S.* **20**, 339–358.

FINSLER, P.

1926. Über die Grundlegung der Mengenlehre. I. Teil. Die Mengen und ihre Axiome. *Math. Ztschr.* **25**, 683–713.

FIRESTONE, C. D. and ROSSER, J. B.

1949. The consistency of the hypothesis of accessibility (abstract). *J.S.L.* **14**, p. 79.

FITCH, F. B.

1938. The consistency of the ramified Principia. *J.S.L.* **3**, 140–149.

1942. A basic logic. *Ibid.* **7**, 105–114. (Cf. *ibid.* **13**, 95–106; **14**, 9–15, 209–218; **15**, 17–24.)

1944. Representations of calculi. *Ibid.* **9**, 57–62.

1949. Intuitionistic modal logic with quantifiers. *Portugaliae Math.* **7**, 113–118.

1952. Symbolic logic. New York. 238 pp.

1956. Recursive functions in basic logic. *J.S.L.* **21**, 337–346.

FRAENKEL, A. A.

1922. Zu den Grundlagen der Cantor-Zermeloschen Mengenlehre. *Math. Annalen* **86**, 230–237. (Cf. already *Jahresber. D.M.V.* **30**, p. 97 ital., 1921.)

1922a. Über den Begriff "definit" und die Unabhängigkeit des Auswahlaxioms. *Sitz. Berlin,* 253–257. English translation in van Heijenoort 67, 284–289.

1925. Untersuchungen über die Grundlagen der Mengenlehre. *Math. Ztschr.* **22**, 250–273.

1926. Axiomatische Begründung der geordneten Mengen. *J. f. Math.* **155**, 129–158. (Cf. *Fund. Math.* 7, 308–310, 1925.)

1927. Zehn Vorlesungen über die Grundlegung der Mengenlehre. Leipzig & Berlin. 182 pp.

1927a. Über die Gleichheitsbeziehung in der Mengenlehre. *J. f. Math.* **157**, 79–81.

1928. Einleitung in die Mengenlehre. 3rd ed. Berlin. 424 pp. (Reprinted New York 1946.)

1932. Axiomatische Theorie der Wohlordnung. *J. f. Math.* **167**, 1–11.

1935. Sur l'axiome du choix. *L'Ens. Math.* **34**, 32–51. (Cf. Kongr. Zürich 1932, II, 341–342, 1933; *Rev. Philos. de Louvain* **50**, 429–459, 1952.)

1937. Über eine abgeschwächte Fassung des Auswahlaxioms. *J.S.L.* **2**, 1–25.

FRAENKEL, A. A. and BAR-HILLEL, Y.

1958. Foundations of set theory. (The first edition of the present book.) Amsterdam, 415 pp.

FRASCELLA, W. J.

1965. A generalization of Sierpiński's theorem on Steiner triples and the axiom of choice. *Notre Dame J. of Formal Logic* **6**, 163–179.

FREGE, G.

1884. Die Grundlagen der Arithmetik. Eine logisch-mathematische Untersuchung über den Begriff der Zahl. Breslau. 119 pp. (Reprinted: Breslau 1934; English-German ed. by J. L. Austin, 1950.)

1893–1903. Grundgesetze der Arithmetik, begriffsschriftlich abgeleitet. Jena, vol. I, 1893, 254 pp.; vol. II, 1903, 265 pp. Reprinted: Hildesheim, 1962. Partial English translations: with an introduction by M. Furth, Berkeley, 1964; Geach-Black 52, 137–244.

FREUDENTHAL, H.

1932. Qualität in der Mathematik. *Euclides* **8**. 89–98.

1935. Review of Heyting 30 and 30a, *Jahrb. über die Fortschritte der Mathematik* **56 II**, 823–824.

1936. Zum intuitionistischen Raumbegriff. *Compositio Math.* **4**, 82–111.

1936a. Zur intuitionistischen Deutung logischer Formeln. *Ibid.*, 112–116, 118. (Cf. A. Heyting: *ibid.*, 117–118.)

FRIEDMAN, J.

1971. The generalized continuum hypothesis is equivalent to the generalized maximization principle. *J.S.L.* **36**, 39–54.

FRÖHLICH, A. and SHEPHERDSON, J. C.

1956. Effective procedures in field theory. *Philos. Tr. of the R. Soc. London* A **248**, 407–432.

FUCHS, L.

1958. Abelian groups. Budapest. 267 pp.

GAIFMAN, H.

1967. A generalization of Mahlo's method for obtaining large cardinal numbers. *Israel J. of Math.* **5**, 188–201.

1973. Self-extending models and measurable cardinals. *Ann. of Math. Logic*, to appear.

GANDY, R. O. and YATES, C. E. M.

1971 (editors). Logic colloquium '69; proceedings of the summer school and colloquium in mathematical logic, Manchester, August 1969. Amsterdam. 451 pp.

GAUNTT, R. J.

1970. Axiom of choice for finite sets – a solution to a problem of Mostowski (abstract). *Notices A.M.S.* **17**, 454.

GEACH, P. T.

1970. A program for syntax. *Synthese* **22**, 3–17.

GEACH, P. and BLACK, M.

1952. Translations from the philosophical writings of Frege. Oxford. 244 pp.

GENTZEN, G.

1934. Untersuchungen über das logische Schliessen. *Math. Ztschr.* **39**, 176–210, 405–431. (French translation with commentary by R. Feys and J. Ladrière, Paris 1955.)

1936. Die Widerspruchsfreiheit der reinen Zahlentheorie. *Math. Annalen* **112**, 493–565. (Cf. *Semesterberichte Münster* 9. Sem., 65–80, 1937.)

1938. Neue Fassung des Widerspruchsfreiheitsbeweises für die reine Zahlentheorie. *Forschungen zur Logik* etc., N.S. No. 4, 19–44.

GIBSON, C. G.

1967. The Radon integral in intuitionism. Ph. D. thesis. Amsterdam. 49 pp.

GILMORE, P. C.

1953. The effect of Griss' criticism of the intuitionistic logic on deductive theories formalized within the intuitionistic logic. Thesis. Univ. of Amsterdam. 24 pp. Also in *Proc. Amsterdam* A**56** (= *I.M.* **15**), 162–174, 175–186, 1953. (Cf. *Actes du XIme Congr. Intern. de Philos., Bruxelles* 1953. V, 98–104.)

1956. Review of van Dantzig 51. *J.S.L.* **21**, 323–324.

GLIVENKO, V.

1929. Sur quelques points de la logique de M. Brouwer. *Acad. R. de Belgique, Bull. de la Cl. des Sc.* **15**, 183–188.

GODDARD, L.

1958. "True" and "Provable". *Mind* **67**, 13–31.

GÖDEL, K.

1930. Die Vollständigkeit der Axiome des logischen Funktionenkalküls. *Monatsh. Math. Ph.* 37, 349–360. English translation in van Heijenoort 67, 582–591.

1931. Über formal unentscheidbare Sätze der Principia Mathematica und verwandter Systeme I. *Ibid.* **38**, 173–198. (See already *Anzeiger Akad. Wien* **67**, 214–215, 1930.) English translation, with an introduction by R. B. Braithwaite: On formally undecidable propositions of Principia Mathematica and related systems I; Edinburgh, 1962, 72 pp. Other English translations in Davis 65, 5–38, and in van Heijenoort 67, 596–616.

1932. Zum intuitionistischen Aussagenkalkül. *Anzeiger Akad. Wien* **69**, 65–66. Reprinted in *Koll.*, Heft **4**, p. 40, 1933.

1933. Zur intuitionistischen Arithmetik und Zahlentheorie. *Koll.*, Heft **4**, 34–38.

1933a. Eine Interpretation des intuitionistischen Aussagenkalküls. *Koll.*, Heft **4**, 39–40.

1934. On undecidable propositions of formal mathematical systems (mimeographed). Princeton. 30 pp.

1936. Über die Länge von Beweisen. *Koll.*, Heft 7, 23–34.

1938. The consistency of the axiom of choice and of the generalized continuum-hypothesis. *Acad. U.S.A.* **24**, 556–557.

1939. Consistency-proof for the generalized continuum-hypothesis. *Ibid.* **25**, 220–224.

1940. The consistency of the axiom of choice and of the generalized continuum hypothesis. Annals of Math. Studies 3. Princeton. 66 pp. (2nd printing, 1951, 74 pp.)

1944. Russell's mathematical logic. Schilpp 44, 125–153; also in Benacerraf-Putnam 64, 211–232.

1947. What is Cantor's continuum problem? *Am. Math. Monthly* **54**, 515–525. An extended version in Benacerraf-Putnam 64, 258–273.

1958. Über eine bisher noch nicht benutzte Erweiterung des finiten Standpunktes. *Dialectica* **12**, 280–287.

1965. Remarks before the Princeton bicentennial conference on problems in mathematics. Davis 65, 84–88.

GONSETH, F.

1926. Les fondements des mathématiques. Avec préface de J. Hadamard. Paris. 243 pp.

GOODMAN, NELSON

1951. The structure of appearance. Cambridge, Mass. 315 pp.

1956. A world of individuals. The problem of universals. Notre Dame, Indiana. 13–31; also in Benacerraf-Putnam 64, 197–210.

GOODMAN, NELSON and QUINE W. V. O.

1947. Steps towards a constructive nominalism. *J.S.L.* **12**, 105–122.

GOODMAN, NICHOLAS D.

1968. Intuitionistic arithmetic as a theory construction. Ph. D. thesis. Stanford Univ.

1970. A theory of constructions equivalent to arithmetic. Kino-Myhill-Vesley 70, 101–120.

GOODSTEIN, R. L.

1948. Proof by reduction ad absurdum. *Math. Gazette* **32**, 198–204.

1951. Constructive formalism. Essays on the foundations of mathematics. Leicester. 91 pp.

1952. The foundations of mathematics (inaugural lecture). Leicester. 27 pp.

1957. Recursive number theory. Amsterdam. 190 pp.

GRÄTZER, G.

1967. A new equivalent of the axiom of choice (abstract). *Notices A.M.S.* **14**, 133.

GREENWOOD, T.

1930. Invention and description in mathematics. *Proc. Arist. Soc.*, N.S. **30**, 79–90.

GRELLING, K.

1937. Identitas indiscernibilium. *Erkenntnis* **6**, 252–259.

GRELLING, K and NELSON, L.

1908. Bemerkungen zu den Paradoxien von Russell und Burali-Forti. *Abh. der Fries-schen Schule*, N.S. **2**, 301–324. (Cf. *ibid.*, 331–334, and H. Goesch: *ibid.*, 324–328.)

GRIŠIN, V. N.

1969. Consistency of a fragment of Quine's NF system (Russian). *Doklady Akad. Nauk SSSR* **189**, No. 2. English translation in *Soviet Math. Doklady* **10**, 1387–1389, 1969.

GRISS, G. F. C.

1944–51. Negatieloze intuitionistische wiskunde. *Ned. Akad. v. Wetensch., Verslagen, Afd. Natuurkunde* **53**, 261–268. (Cf. the English version and continuation [Negationless intuitionistic mathematics, I–IV]. *Proc. Amsterdam* **49**, 1127–1133 (= *I.M.* 8, 675–681), 1946; **53**, 456–463 (= *I.M.* **12**, 108–115), 1950; **A54** (= *I.M.* **13**), 41–49, 193–199, 452–471, 1951.)

GRIZE, J. B.

1955. L'implication et la négation vues au travers des méthodes de Gentzen et de Fitch. *Dialectica* **9**, 363–381.

GRÜNBAUM, A.

1967. Modern science and Zeno's paradoxes. Middletown, Conn. 148 pp. Revised ed.: London, 1968.

GRZEGORCZYK, A.

1954. Elementarily definable analysis. *Fund. Math.* **41**, 311–338. (Cf. M. Kondô: *C.R. Paris* **242**, 1841–1843, 1956.)

1955. The systems of Leśniewski in relation to contemporary logical research. *Studia Logica* **3**, 77–95.

1959. Some approaches to constructive analysis. Heyting 59a, 43–61.

1964. A philosophically plausible formal interpretation of intuitionistic logic. *I.M.* **26**, 596–601.

HADAMARD. J.

1905. Cinq lettres sur la théorie des ensembles, de MM. Hadamard, Borel, Baire. *Bull. de la Soc. Math. de France* **33**, 261–273. Reprinted in Borel 14.

1945. An essay on the psychology of invention in the mathematical field. Princeton, 1945. 143 pp. Revised ed. 1949.

HAHN, H., NEURATH, O. and CARNAP, R.

1929. Wissenschaftliche Weltauffassung. Der Wiener Kreis. Wien. 59 pp.

HAILPERIN, T.

1944. A set of axioms for logic. *J.S.L.* **9**, 1–19.

1954. Remarks on identity and description in first-order axiom systems. *Ibid.* **19**, 14–20.

1957. A theory of restricted quantification. *Ibid.* **22**, 19–35, 113–129.

HÁJEK, P.

1965. Modelle der Mengenlehre in denen Mengen gegebener Gestalt existieren. *Ztschr. f. math. Logik* **11**, 103–115.

1966. The consistency of the Church's alternatives. *Bull. Acad. Polon. Sc.* **14**, 423–430.

HAJNAL, A.

1961. On a consistency theorem connected with the generalized continuum hypothesis. *Acta Math. Acad. Sc. Hungaricae* **12**, 321–376.

HALMOS, P. R.

1950. Measure theory. New York. 304 pp.

1960. Naive set theory. Princeton. 104 pp.

HALPERN, J. D. and LÄUCHLI, H.

1966. A partition theorem. *Tr. A.M.S.* **124**, 360–367.

HALPERN, J. D. and LEVY, A.

1971. The Boolean prime ideal theorem does not imply the axiom of choice. Scott 71, 83–134.

HARDY, G. H.

1904. A theorem concerning the infinite cardinal numbers. *Quarterly J. Pure Appl. Math.* **35**, 87–94.

HARROP, R.

1956. On disjunctions and existential statements in intuitionistic systems of logic. *Math. Annalen* **132**, 347–361.

HARTOGS, F.

1915. Über das Problem der Wohlordnung. *Math. Annalen* **76**, 438–443.

HAUSDORFF, F.

1908. Grundzüge einer Theorie der geordneten Mengen. *Math. Annalen* **65**, 435–505.

1914. Grundzüge der Mengenlehre. Leipzig. 476 pp. Reprinted New York 1949.

1927. Mengenlehre (2nd revised ed. of 14). Berlin & Leipzig. 185 pp. 3rd ed. 1935; reprinted New York 1944. English ed., New York 1957, 352 pp.

VAN HEIJENOORT, J.

1967. From Frege to Gödel – a source book in mathematical logic 1879–1931. Cambridge, Mass. 660 pp.

HEMPEL, C. G.

1958. The theoretician's dilemma. Concepts, Theories and the Mind-Body Problem. Minnesota Studies in the Philos. of Sc. II, 37–98.

HENKIN, L.

1950. Completeness in the theory of types. *J.S.L.* **15**, 81–91.

1953. On the primitive symbols of Quine's "Mathematical Logic". *Rev. Philos. de Louvain* **51**, 591–593.

1953a. Some notes on nominalism. *J.S.L.* **18**, 19–29.

1954. Mathematical theorems equivalent to the prime ideals theorem for Boolean algebras (abstract). *Bull. A.M.S.* **60**, 387–388.

1955. On a theorem of Vaught. *J.S.L.* **20**, 92–93. (Cf. *Proc. Amsterdam* A **58** (= *I.M.* 17), 326–328, 1955.)

HENKIN, L., SUPPES, P. and TARSKI, A.

1959 (editors). The axiomatic method, with special reference to geometry and physics. Proc. of an intern. symposium held at the Univ. of California, Berkeley, December 26, 1957 – January 4, 1958. Amsterdam. 488 pp.

HENLE, P.

1952. Review of N. Goodman 51. *J.S.L.* **17**, 130–133.

HERBRAND, J.

1931. Sur le problème fondamental de la logique mathématique. *C.R. Varsovie* **24**, 12–56.

1932. Sur la non-contradiction de l'arithmétique. *J.f.Math.* **166**, 1–8. English translation in van Heijenoort 67, 618–628.

HERMES, H.

1937. Definite Begriffe und berechenbare Zahlen. *Semesterber. Münster* 10. Sem., 1937, 110–123.

1951. Zum Begriff der Axiomatisierbarkeit. *Math. Nachrichten* **4**, 343–347.

1971. A simplified proof for the unsolvability of the decision problem in the case $\wedge\vee\wedge$, Gandy-Yates 71, 307–310.

HERMES, H. and SCHOLZ, H.

1952. Mathematische Logik. *Enzykl. der math. Wiss.,* Band I, 1. Teil, Heft 1, Teil I. Leipzig. 82 pp.

HESSENBERG, G.

1906. Grundbegriffe der Mengenlehre. (*Abh. der Friesschen Schule,* N.S. [1], Heft 4.) Göttingen. 220 pp.

HEYTING, A.

1928. Die Theorie der linearen Gleichungen in einer Zahlenspezies mit nichtkommutativer Multiplikation. *Math. Annalen* **98**, 465–490.

1928a. Zur intuitionistischen Axiomatik der projektiven Geometrie. *Math. Annalen* **98**, 491–538.

1930. Die formalen Regeln der intuitionistischen Logik. *Sitz. Berlin* 1930, 42–56.

1930a. Die formalen Regeln der intuitionistischen Mathematik. *Ibid.,* 57–71, 158–169. (To 1930 and 1930a cf. H. Scholz: *Deutsche Literaturzeitung* 1930, 2202–2208.)

1930b. Sur la logique intuitionniste. *Acad. R. de Belgique, Bull. de la Cl. des Sc.* (5) **16**, 957–963.

1933. Discussion avec M. M. Barzin et Errera, *L'Ens. Math.* **31**, 21–122, 271–272, 274–275.

1934. Mathematische Grundlagenforschung. Intuitionismus. Beweistheorie. (*Ergebnisse der Math.* etc. III, 4.) Berlin. 73 pp. Revised French ed.: 55.

1936. Bemerkungen zu dem Aufsatz von Herrn Freudenthal: "Zur intuitionistischen Deutung logischer Formeln". *Compositio Math.* **4**, 117–118.

1941. Untersuchungen über intuitionistische Algebra. *Verh. der Ned. Akad. van Wetensch., Afd. Natuurkunde,* 1. Sectie, 18, No. 2. 36 pp.

1946. On weakened quantification. *J.S.L.* **11**, 119–121. (Cf. F. B. Fitch: *Portugaliae Math.* 7, 113–118, 1949.)

1948. Formal logic and mathematics. *Synthese* **6**, 275–282. (Cf. *Alg. Ned. Tijdschrift v. Wijsbegeerte en Psychol.* **40**, 121–131, 1948. Dutch, with résumé in French.)

1951. Note on the Riesz-Fischer theorem. *Proc. Amsterdam A***54** (= *I.M.* **13**), 35–40.

1953. Espace de Hilbert et intuitionnisme. *Les Methodes Formelles en Axiomatique, Colloq. Intern. du C.N.R.S.,* No. **26** (1950), 59–64. Paris.

1954. Logique et intuitionnisme. *Actes du 2me Colloq. Intern. de Logique Math., Paris* 1952, 75–82. See also the discussion, *ibid.,* 82–83.

1955. Les fondements des mathématiques. Intuitionnisme. Théorie de la démonstration. (Revised French ed. of 34.) Paris. 91 pp.

1956. Intuitionism. An introduction. Amsterdam. 133 pp.

1958. Blick von der intuitionistischen Warte. *Dialectica* **12**, 332–345.

1959. Axioms for intuitionistic plane affine geometry. Henkin et al. 59, 237–247.

1959a (editor). Constructivity in mathematics, Proc. of the colloq. held at Amsterdam 1957. Amsterdam. 297 pp.

1967. In memoriam L.E.J. Brouwer (Dutch). *Alg. Tijdschrift voor Wijsbegeerte en Psychologie* **59**, 1–13.

HIGMAN, G.

1961. Subgroups of finitely presented groups. *Proc. R. Soc. London A* **262**, 455–475.

HILBERT, D.

1899. Grundlagen der Geometrie. 9th ed., revised by P. Bernays, 1962. English translation: The foundations of geometry. La Salle, III., 1962. 143 pp.

1900. Mathematische Probleme. Nachr. Göttingen, 1900, 253–297. Reprinted with some additions in *Archiv der Math. und Phys.* (3) **1**, 44–63, 213–237, 1901, and in 35, 290–329. In English: *Bull. A.M.S.* **8**, 437–479, 1902. In French: C.R. du 2me Congr. Intern. des Math. Paris 1900, pp. 58–114, 1902.

1905. Über die Grundlagen der Logik und Arithmetik. Kongr. Heidelberg 1904, 174–185. In English: *The Monist* **15**, 338–352, 1905; and in van Heijenoort 67, 129–138. In French: *L'Ens. Math.* 7, 89–103, 1905.

1918. Axiomatisches Denken. *Math. Annalen* 78, 405–419; also in 35, 146–156. In French: *L'Ens. Math.* **20**, 122–136, 1918.

1922. Neubegründung der Mathematik. Erste Mitteilung. Abh. Hamburg I, 157–177; also in 35, 157–177.

1923. Die logischen Grundlagen der Mathematik. *Math. Annalen* **88**, 151–165; also in 35, 178–195.

1925. Über das Unendliche. *Ibid.* **95**, 161–190. In English: Benacerraf-Putnam 64, 134–151 and van Heijenoort 67, 367–392. In French: *Acta Math.* **48**, 91–122, 1926.

1928. Die Grundlagen der Mathematik. Mit Diskussionsbemerkungen von H. Weyl und einem Zusatz von P. Bernays. *Abh. Hamburg* **6**, 65–92. Appeared also as

No. 5 of Hamburger Math. Einzelschriften; Leipzig 1928; 28 pp. English translation in van Heijenoort 67, 464–479, 485–489.

1931. Beweis des *Tertium non datur. Nachr. Göttingen* 1931, 120–125.

1935. Gesammelte Abhandlungen. Vol. 3. Berlin. 435 pp.

HILBERT, D. and ACKERMANN, W.

1928–49. Grundzüge der theoretischen Logik. (Appeared first in 1928.) 3rd revised ed., Berlin 1949. 155 pp.; English ed., New York 1950.

HILBERT, D. and BERNAYS, P.

1934–39. Grundlagen der Mathematik. Vol. I, 1934; Vol. II, 1939, Berlin. 471 + 498 pp.

HINTIKKA, K. J. J.

1955. Form and content in quantification theory. *Acta Philos. Fennica* 8, 1–56.

1956. Identity, variables, and impredicative definitions. *J.S.L.* **21**, 225–245.

1957. Vicious circle principle and the paradoxes. *Ibid.* **22**, 245–249.

HÖLDER, O.

1924. Die mathematische Methode. Berlin. 563 pp.

1926. Der angebliche circulus vitiosus und die sogenannte Grundlagenkrise in der Analysis. *Ber. Leipzig* 78, 243–250.

HOWARD, W. A.

1968. Functional interpretation of bar induction by bar recursion. *Compositio Math.* **20**, 117–124.

HOWARD, W. A. and KREISEL, G.

1966. Transfinite induction and bar induction of types zero and one, and the role of continuity in intuitionistic analysis. *J.S.L.* **31**, 325–358.

HULL, R.

1969. Counterexamples in intuitionistic analysis using Kripke's schema. *Ztschr. f. math. Logik* **15**, 241–246.

HUSSERL. E.

1921–28. Logische Untersuchungen. Vol. I (4th ed.), II 1 (4th ed.), II 2 (3rd ed.). Halle a.S.

DE IONGH, J. J.

1949. Restricted forms of intuitionistic mathematics. *Library of the Xth Intern. Congr. of Philos., Amsterdam* 1948, I, 744–748.

ISBELL, J. R.

1963. Two set theoretical theorems in categories. *Fund. Math.* **53**, 43–49.

ISBELL, J. R. and WRIGHT, F. B.

1966. Another equivalent form of the axiom of choice. *Proc. A.M.S.* **17**, 174.

ISSMAN, S.

1958. Le nominalisme contemporain et la notion d'existence. *Logique et Analyse* N.S. **1**, 61–73.

JAŚKOWSKI, S.

1936. Recherches sur le système de la logique intuitionniste. *Act. Sc. Ind.* **393**, 58–61. (Cf. H. Hiż: *C.R. Paris* **223**, 973–974, 1946.)

JECH, T.

1966. On ordering of cardinalities. *Bull. Acad. Polon. Sc.* **14**, 293–296.

1966a. On cardinals and their successors. *Bull. Acad. Polon. Sc.* **14**, 533–537.

1967. Non-provability of Souslin's hypothesis. *Comm. Math. Univ. Carolinae* 8, 291–305.

1971. Lectures in set theory with particular emphasis on the method of forcing. Lecture notes in Math. 217. Berlin. 137 pp.

1971a. Trees. *J. S. L.* **36**, 1–14.

∞ The axiom of choice. Amsterdam.

JECH, T. and SOCHOR, A.

1966. Applications of the θ-model. *Bull. Acad. Polon. Sc.* **14**, 351–355.

JENSEN, R. B.

1966. Independence of the axiom of dependent choices from the countable axiom of choice (abstract). *J. S. L.* **31**, 294.

1967. Modelle der Mengenlehre. Lecture Notes in Math. 37. Berlin. 176 pp.

1967a. Ramsey cardinals and the generalized continuum hypothesis (abstract). *Notices A.M.S.* **14**, 253.

1967b. Concrete models of set theory. Crossley 67, 44–74.

1969. A slight (?) modification of "New Foundations". Words and objections – On the work of W.V. Quine; Davidson and Hintikka, eds.; Dordrecht – Holland, 278–291. (Abstract in *Notices A.M.S.* **14**, 407, 1967.) Cf. Quine's reply; ibid., 349–352.

1970. Definable sets of minimal degree. Bar-Hillel 70a, 122–128.

∞ The axiom of constructibility refutes Souslin's hypothesis. To appear.

∞ a. Some combinatorial properties of L. To appear.

∞ b. Souslin's hypothesis is compatible with the continuum hypothesis. To appear.

∞ c. Souslin's hypothesis equals weak compactness in L. To appear.

JENSEN, R. B. and SCHRÖDER, M. E.

1969. Mengeninduktion und Fundierungsaxiom. *Archiv f. math. Logik* **12**, 119–133.

JENSEN, R. B. and SOLOVAY, R. M.

1970. Some applications of almost disjoint sets. Bar-Hillel 70a, 84–104.

JOHANSSON, I.

1936. Der Minimalkalkül, ein reduzierter intuitionistischer Formalismus. *Compositio Math.* **4**, 119–136. (Cf. Paulette Destouches–Février: *C.R. Paris* **224**, 545–547, 1947; **227**, 1192–1193, 1948.)

JØRGENSEN, J.

1932. Remarques sur les principales implications métaphysiques des théories et des idées récentes de la physique. *R.M.M.* **39**, 323–351.

JOURDAIN, P. E. B.

1918. The philosophy of Mr. B*rtr*nd R*ss*ll, with an appendix of leading passages from certain other works. London. 96 pp.

KAHR, A. S., MOORE, E. F. and WANG, H.

1962. Entscheidungsproblem reduced to the AEA case. *Acad. U.S.A.* **48**, 365–377.

KALMÁR, L.

1967. Foundations of mathematics – whither now? Lakatos 67, 187–207.

KAMKE, E.

1939. Allgemeine Mengenlehre. Enzykl. der math. Wiss., Band I, 1. Teil, Heft 2, Art. A5. Leipzig. 56 pp.

KARP, CAROL R.

1964. Languages with expressions of infinite length. Amsterdam. 183 pp.

1967. A proof of the relative consistency of the continuum hypothesis. Crossley 67, 1–32.

KEISLER, H. J.

1961. Ultraproducts and elementary classes. *Proc. Amsterdam* **A64** (= *I.M.* **23**), 477–495.

1964. Complete theories of algebraically closed fields with distinguished subfields. *Michigan Math. J.* **11**, 71–81.

1968. Review of Erdös-Hajnal 66. *Math. Reviews* **35**, # 64.

KEISLER, H. J. and TARSKI, A.

1964. From accessible to inaccessible cardinals. *Fund. Math.* **53**, 225–308. Corrections, *ibid.* **57**, 119, 1965.

KELLEY, J. L.

1955. General topology. New York. 298 pp.

KEMENY, J. G.

1956. A new approach to semantics. *J.S.L.* **21**, 1–27, 149–161.

KEYSER, C. J.

1941. Charles Sanders Peirce as a pioneer. *Galois Lectures* (*Ser. Math. Library* **5**), 87–112. (Cf. *Scripta Math.* **3**, 11–37, 1935; also the review by A. Church: *J.S.L.* **6**, 161–162, 1941.)

KINNA, W. and WAGNER, K.

1955. Über eine Abschwächung des Auswahlaxioms. *Fund. Math.* **42**, 75–82.

KINO, A., MYHILL, J. and VESLEY, R.

1970 (editors). Intuitionism and proof theory, Proc. of the summer conference at Buffalo, N.Y., 1968. Amsterdam. 516 pp.

KLAUA, D.

1955. Systematische Behandlung der lösbaren Fälle des Entscheidungsproblems für den Prädikatenkalkül der ersten Stufe. *Ztschr. f. math. Logik* **1**, 264–269.

1961. Konstruktive Analysis. Math. Forschungsberichte XI. Berlin.

1964. Allgemeine Mengenlehre. Berlin. 581 pp.

KLEENE, S. C.

1936. λ-definability and recursiveness. *Duke Math. J.* **2**, 340–353.

1943. Recursive predicates and quantifiers. *Tr. A.M.S.* **53**, 41–73. (Cf. Mostowski 1947; *Ann. de la Soc. Polon. de Math.* **21**, 114–119, 1948.)

1945. On the interpretation of intuitionistic number theory. *J.S.L.* **10**, 109–124. (Cf. L.E.J. Brouwer: *Proc. Amsterdam* **50**, 307, 1947.)

1952. Introduction to metamathematics. Amsterdam, 550 pp. (2nd printing, 1957.)

1952a. Finite axiomatizability of theories in the predicate calculus using additional predicate symbols. Two papers on the predicate calculus. *Memoirs A.M.S.* **10**, 1–26. Providence R.I.

1962. Disjunction and existence under implication in elementary intuitionistic formalisms. *J.S.L.* **27**, 11–18.

1967. Mathematical logic. New York. 398 pp.

KLEENE, S. C. and VESLEY, R. E.

1965. The foundations of intuitionistic mathematics. Amsterdam. 206 pp.

KOLMOGOROFF, A.

1932. Zur Deutung der intuitionistischen Logik. *Math. Ztschr.* **35**, 58–65. (Cf. N. A. Šanin: *Doklady Akad. Nauk SSSR* **93**, 779–782, 1953; **94**, 193–196, 1954.)

KONDÔ, M.

1937. Sur l'hypothèse de M.B. Knaster dans la théorie des ensembles de points. *J. For. of Sc. Hokkaido Univ.*, Ser. I **6**, 1–20.

KÖNIG, J.

1914. Neue Grundlagen der Logik, Arithmetik und Mengenlehre (edited by D. König). Leipzig. 259 pp.

KÖRNER, S.

1967. On the relevance of post-Gödelian mathematics to philosophy. Lakatos 67, 118–137.

KOTARBIŃSKI, T.

1929. Elements of epistemology, formal logic, and methodology of the sciences (Polish). Lwów. 483 pp.

KÖTHE, G.

1960. Topologische lineare Räume I. Berlin. 456 pp.

KOŹNIEWSKI, A. and LINDENBAUM, A.

1930. Sur les opérations d'addition et de multiplication dans les classes d'ensembles. *Fund. Math.* **15**, 342–355.

KREISEL, G.

1950. Note on arithmetic models for consistent formulae of the predicate calculus. *Fund. Math.* **37**, 265–285.

1959. Interpretation of analysis by means of constructive functionals of finite type. Heyting 59a, 101–128.

1962. Foundations of intuitionistic logic. Nagel et al. 62, 198–210.

1962a. On weak completeness of intuitionistic predicate logic. *J.S.L.* **27**, 139–158.

1964. Hilbert's programme. Benacerraf-Putnam 64, 157–180. An earlier version in *Dialectica* **12**, 346–372, 1958.

1965. Mathematical logic. Saaty 65, 95–195.

1967. Informal rigour and compleness proofs. Lakatos **67**, 138–186.

1968. Lawless sequences of natural numbers. *Compositio Math.* **20**, 222–248.

1970. Church's Thesis: a kind of reducibility axiom for constructive mathematics. Kino-Myhill-Vesley 70, 121–150.

KREISEL, G. and KRIVINE, J. L.

1967. Eléments de logique mathématique (Théorie des modèles). Paris. 213 pp. English ed.: Elements of mathematical logic (Model Theory). Amsterdam. 222 pp. 1967.

KREISEL, G. and LEVY, A.

1968. Reflection principles and their use for establishing the complexity of axiomatic systems. *Ztschr. f. math. Logik* **14**, 97–191.

KREISEL, G. and NEWMAN, M. H. A.

1969. Luitzen Egbert Jan Brouwer. Biographical memoirs of fellows of the R. Soc. 15, 39–68.

KREISEL, G. and PUTNAM, H.

1957. Eine Unableitbarkeitsbeweismethode für den intuitionistischen Aussagenkalkül. *Arch. f. math. Logik* **3**, 74–78.

KREISEL, G. and TROELSTRA, A. S.

1970. Formal systems for some branches of intuitionistic analysis. *Ann. of Math. Logic* **1**, 229–387.

KRIPKE, S. A.

1965. Semantical analysis of intuitionistic logic I. Crossley-Dummett 65, 92–130.

KRONECKER, L.

1882. Grundzüge einer arithmetischen Theorie der algebraischen Grössen (1882). *Kroneckers Werke,* II, 238–387. Leipzig 1899.

1883. Die Zerlegung der ganzen Grössen eines natürlichen Rationalitätsbereiches in ihre irreductiblen Factoren (1883). *Ibid.*, 411–416. (Cf. G. Hermann: *Math. Annalen* **95**, 736–788, 1925.)

1887. Über den Zahlbegriff. *J. f. Math.* **101**, 337–355. (Also in *Kroneckers Werke*, III 1, 249–274.)

KRUSE, A. H.

1963. A problem on the axiom of choice. *Ztschr. f. math. Logik* **9**, 207–218.

1963a. A method of modelling the formalism of set theory in axiomatic set theory. *J.S.L.* **28**, 20–34.

1966. Grothendieck universes and the super-complete models of Shepherdson. *Compositio Math.* **17**, 96–101.

KÜHNRICH, M.

1966. Über den Begriff des Universums. *Ztschr. f math. Logik* **12**, 37–50.

KUNEN, K.

1970. Some applications of iterated ultrapowers in set theory. *Ann. of Math. Logic* **1**, 179–227.

∞ Saturated ideals. To appear.

KURATA, R.

1964. On the existence of a proper complete model of set theory. *Comm. Math. Univ. Sancti Pauli* **13**, 35–43.

KURATOWSKI, K.

1921. Sur la notion d'ordre dans la théorie des ensembles. *Fund. Math.* **2**, 161–171.

1925. Sur l'état actuel de l'axiomatique de la théorie des ensembles. *Ann. de la Soc. Polon. de Math.* **3**, 146–147.

1958. Topologie I. 4th ed. Warsaw. 494 pp. English translation: Topology, vol. 1, New York, 1966.

KURATOWSKI, K. and MOSTOWSKI, A.

1968. Set theory. Warsaw. 417 pp.

KUREPA, G.

1953. Über das Auswahlaxiom. *Math. Annalen* **126**, 381–384.

KURODA, S.

1956. Review of Heyting 56. *J.S.L.* **21**, 367–371.

L'ABBÉ, M.

1953. Systems of transfinite types involving λ-conversion. *J.S.L.* **18**, 209–224.

LADRIÈRE, J.

1957. Les limitations internes des formalismes. Paris. 715 pp.

LAKATOS, I.

1967 (editor). Problems in the philosophy of mathematics, Proc. of the intern. colloq. in the philos. of sc., London, 1965, vol. I. Amsterdam. 241 pp.

LAMBERT, W. M.

1968. A notion of effectiveness in arbitrary structures. *J.S.L.* **33**, 577–602.

LÄUCHLI, H.

1961. Ein Beitrag zur Kardinalzahlarithmetik ohne Auswahlaxiom. *Ztschr. f. math. Logik* **7**, 141–145.

1962. Auswahlaxiom in der Algebra. *Comm. Math. Helvetici* **37**, 1–19.

1964. The independence of the ordering principle from a restricted axiom of choice. *Fund. Math.* **54**, 31–43.

1968. A decision procedure for the weak second order theory of linear order. Schmidt et al. 68, 189–197.

LÄUCHLI, H. and LEONARD, J.

1966. On the elementary theory of linear order. *Fund. Math.* **59**, 109–116.

LAWVERE, F. W.

1964. An elementary theory of the category of sets. *Acad. U.S.A.* **52**, 1506–1511.

1966. The category of categories as a foundation for mathematics. Proc. conf. categorical algebra, La Jolla, Calif., 1965. New York. 1–20. (See the review by J.R. Isbell in *Math. Reviews* **34** (1967), # 7332.)

LEBESGUE, H.

1905. Sur les fonctions representables analytiquement. *J. Math. Pures Appliquées* **60**, 139–216.

1941. Les controverses sur la théorie des ensembles et la question des fondements, *Les Entretiens de Zurich*, 109–122.

LEŚNIEWSKI, S.

1929. Grundzüge eines neuen Systems der Grundlagen der Mathematik. *Fund. Math.* **14**, 1–81.

LEVI, B.

1902. Intorno alla teoria degli aggregati. *R. Istituto Lombardo di Sc. e Lettere, Rendic.* (2) **35**, 863–868. (Cf. F. Bernstein: Nachr. Göttingen, 1904, 557–560.)

LEVY, A.

1958. The independence of various definitions of finiteness. *Fund. Math.* **46**, 1–13.

1959. On Ackermann's set theory. *J.S.L.* **24**, 154–166.

1960. Axiom schemata of strong infinity in axiomatic set theory. *Pacific J. Math.* **10**, 223–238.

1960a. A generalization of Gödel's notion of constructibility. *J.S.L.* **25**, 147–155.

1960b. Principles of reflection in axiomatic set theory. *Fund. Math.* **49**, 1–10.

1960c. Review of McNaughton 53. *J.S.L.* **25**, 353–354.

1964. The interdependence of certain consequences of the axiom of choice. *Fund. Math.* **54**, 135–157.

1965. The Fraenkel-Mostowski method for independence proofs in set theory. Addison et al. 65, 221–228.

1965a. Definability in axiomatic set theory I. Bar-Hillel 65, 127–151.

1965b. A hierarchy of formulas in set theory. *Memoirs A.M.S.* **57**, Providence, R.I. 76 pp.

1968. On von Neumann's axiom system for set theory. *Am. Math. Monthly* **75**, 762–763.

1969. The effectivity of existential statements in axiomatic set theory. *Information Sc.* **1**, 119–130.

1969a. The definability of cardinal numbers. Bulloff et al. 69, 15–38.

1970. Definability in axiomatic set theory II. Bar-Hillel 70a, 129–145.

1971. The sizes of the indescribable cardinals. Scott 71, 205–218.

LEVY, A. and SOLOVAY, R. M.

1967. Measurable cardinals and the continuum hypothesis. *Israel J. of Math.* **5**, 234–248.

LEVY, A. and VAUGHT, R. L.

1961. Principles of partial reflection in the set theories of Zermelo and Ackermann. *Pacific J. of Math.* **11**, 1045–1062.

LÉVY, P.

1926. Sur le principe du tiers exclu et sur les théorèmes non-susceptibles de démonstration. *R.M.M.* **33**, 253–258.

1926a. Critique de la logique empirique. Réponse à M. R. Wavre. *Ibid.*, 545–551.

1927. Logique classique, logique brouwerienne, et logique mixte. *Acad. R. de Belgique, Bull. de la Cl. des Sc.* (5)**13**, 256–266.

1930. Le raisonnement et l'expérience dans les fondements des mathématiques. *Scientia* **47**, 325–334.

1950. Axiome de Zermelo et nombres transfinis. *Ann. Scient. de l'Ecole Norm. Sup.* (3)**67**, 15–49.

LINDENBAUM, A. and MOSTOWSKI, A.

1938. Über die Unabhängigkeit des Auswahlaxioms und einiger seiner Folgerungen. *C. R. Varsovie* **31**, 27–32.

LITTLEWOOD, J. E.

1954. The elements of the theory of real functions. 3rd ed. New York. 71 pp.

LORENZEN, P.

1950. Konstruktive Begründung der Mathematik. *Math. Ztschr.* **53**, 162–202. (Cf. *Math. Annalen* **123**, 331–338, 1951.)

1951. Die Widerspruchsfreiheit der klassischen Analysis. *Ibid.* **54**, 1–24.

1955. Einführung in die operative Logik und Mathematik. Berlin. 298 pp. (2nd ed., Berlin, 1969.)

1962. Metamathematik. Mannheim. 173 pp.

1965. Differential und Integral, eine konstruktive Einführung in die klassische Analysis. Frankfurt a.M. 293 pp.

ŁOŚ, J. and RYLL-NARDZEWSKI, C.

1951. On the application of Tychonoff's theorem in mathematical proofs. *Fund. Math.* **38**, 233–237.

1954. Effectiveness of the representation theory for Boolean algebras. *Fund. Math.* **41**, 49–56.

ŁUKASIEWICZ, J.

1930. Philosophische Bemerkungen zu mehrwertigen Systemen des Aussagenkalküls. *C. R. Varsovie* **23**, 51–77.

LUSCHEI, E. C.

1962. The logical systems of Leśniewski. Amsterdam. 361 pp.

LUSIN, N.

1927. Sur les ensembles analytiques. *Fund. Math.* **10**, 1–95. (Cf. *Math. Sbornik* **33**, 237–289, 1926; *Ann. de la Soc. Polon. de Math.* **5**, 104–106, 1927.)

1930. Leçons sur les ensembles et leurs applications. (Collection Borel). Paris, 1930. 328 pp.

1934. Sur les suites stationnaires. *Act. Sc. Ind.* **149**. Paris. 19 pp.

LUXEMBURG, W. A. J.

1964. A remark on Sikorski's extension theorem for homomorphisms in the theory of Boolean algebras. *Fund. Math.* **55**, 239–247.

LYAPUNOW, A. A., STSCHEGOLKOW, E. A. and ARSENIN, W. J.

1955. Arbeiten zur deskriptiven Mengenlehre. Berlin. 108 pp.

McALOON, K.

1966. Doctoral dissertation. Univ. of California, Berkeley.

McGILL, V. J.

1939. Concerning the laws of contradiction and excluded middle. *Philos. of Sc.* **6**, 196–211.

MACHOVER, M.

1961, The theory of transfinite recursion. *Bull. A. M. S.* **67**, 575–578.

McKINSEY, J. C. C.

1939. Proof of the independence of the primitive symbols of Heyting's calculus of propositions. *J. S. L.* **4**, 155–158.

1944. On the number of complete extensions of the Lewis systems of sentential calculus. *J. S. L.* **9**, 42–45.

McKINSEY, J. C. C. and SUPPES, P.

1954. Review of Destouches-Février 51. *J.S.L.* **19**, 52–55.

McKINSEY, J. C. C. and TARSKI, A.

1946. On closed elements in closure algebras. *Annals of Math.* (2)**47**, 122–162. (Cf. *ibid.* **45**, 141–191, 1944; A. Tarski: *ibid.* **47**, 163–165; S. Halldén: *Norsk Math. Tidsskrift* **31**, 89–94, 1949.)

1948. Some theorems about the sentential calculi of Lewis and Heyting. *J.S.L.* **13**, 1–15. (Cf. L. Simons: *ibid.* **18**, 309–316, 1953; H. Rasiowa and R. Sikorski, *Fund. Math.* **41**, 21–28, 1954; also Ruth B. Marcus: *J.S.L.* **18**, 234–236, 1953.)

MacLANE, S.

1961. Locally small categories and the foundations of set theory. Infinitistic methods. Warsaw. 25–43.

McNAUGHTON, R.

1953. Some formal consistency proofs. *J.S.L.* **18**, 136–144.

MAGIDOR, M.

1971. There are many normal ultrafilters corresponding to a supercompact cardinal. *Israel J. Math.* **9**, 186–192.

1971a. On the role of supercompact and extendible cardinals in logic. *Israel J. of Math.* **10**, 147–157.

MAHLO, P.

1911. Über lineare transfinite Mengen. *Ber. Leipzig* **63**, 187–225. (Cf. *ibid.*, 319–347.)

1912–13. Zur Theorie und Anwendung der ρ_0-Zahlen. *Ber. Leipzig.* I: **64**, 108–112. II: **65**, 268–282.

MALCEV, A.

1936. Untersuchungen aus dem Gebiet der mathematischen Logik. *Mat. Sbornik* (2)**1**, 323–336.

1961. Constructive Algebras I. *Uspechi Mat. Nauk* 16, No. 3, 3–60. English translation: *Russian Math. Surveys* 16, No. 3, 77–129, 1961.

MANNOURY, G.

1925. Mathesis en mystiek. Een signifiese studie van kommunisties standpunt. Amsterdam, s.a. 116 pp. French ed.: Les deux pôles de l'esprit. Étude de psychologie linguistique du point de vue communiste. Paris 1933. (Cf. *Tijdschrift v. Wijsbegeerte* 17, 32–45, 1923.)

1931. Woord en gedachte. Een inleiding tot de signifika, inzonderheid met het oog op het onderwijs in de wiskunde. Groningen. Also in *Euclides* 7, 1–61, 1931.

1934. Die signifischen Grundlagen der Mathematik. *Erkenntnis* **4**, 288–309, 317–345. (Cf. *Synthese* **2**, 195–201, 406–418, 1937; French ed. *Biblioth. Scientif.*, No. 74, Neuchatel 1946.)

MANSFIELD, R.

1971. A Souslin operation for Π^1_2. *Israel J. Math.* **9**, 367–379.

MAREK, W.

1966. Well orderings in powers of cardinals. *Bull. Acad. Polon. Sc.* **14**, 477–478.

1966a. A remark on independence proofs. *Ibid.* 543–545.

MAREK, W. and ONYSZKIEWICZ, J.

1966. Representation of partial ordering in cardinals of the model of Zermelo-Fraenkel set theory. I: *Ibid.* 357–358. II: *Ibid.* 479–481.

MARKOV, A. A.

1947. On the impossibility of certain algorithms in the theory of associative systems. *Doklady Akad. Nauk SSSR* **55**, 587–590. English translation: *ibid.* 583–586.

1954. Téoriá algorifmov. Moskva. 375 pp. English translation: The theory of algorithms. Jerusalem 1961. 444 pp.

1958. On constructive functions. *Trudy Mat. Inst. Steklov* **52**, 66–74. English translation in *A.M.S. Translations* (2) **29**, 163–197.

1960. Insolubility of the problem of homeomorphy (Russian). Proc. of the intern. congr. of math. 1958. Cambridge. 300–306.

MARTIN, D. A.

1968. The axiom of determinateness and reduction principles in the analytical hierarchy. *Bull. A.M.S.* 74, 687–689.

1970. Measurable cardinals and analytic games. *Fund. Math.* **66**, 287–291.

MARTIN, D. A. and SOLOVAY, R.

1969. A basic theorem for Σ^1_3 sets of reals. *Ann. of Math.* (2) **89**, 138–159.

1970. Internal Cohen extensions. *Ann. of Math. Logic* **2**, 143–178.

MASLOV, S. Yu., MINTS, G. E. and OREVKOV, V. P.

1965. Undecidability of certain classes of formulas containing only single-place predicate variables, in the constructive predicate calculus (Russian). *Doklady Akad. Nauk SSSR* **163**, No. 2, 295–297. English translation in *Soviet Math. Doklady* **6**, 918–920, 1965.

MATHIAS, A. R. D.

1967. On the order extension principle (abstract). *Notices A.M.S.* **14**, 410.

MATIJASEVIČ, Yu. V.

1970. Enumerable sets are diophantine. *Doklady Akad. Nauk SSSR* **191**, 279–282. Revised English translation in *Soviet Math. Doklady* **11**, 354–358, 1970.

MAZUR, S.

1963. Computable analysis. Rozprawy Mat. 33. 111 pp.

MENDELSON, E.

1956. Some proofs of independence in axiomatic set theory. *J.S.L.* **21**, 291–303.

1956a. The independence of a weak axiom of choice. *Ibid.*, 350–366.

1958. The axiom of Fundierung and the axiom of choice. *Arch. f. math. Logik* **4**, 67–70.

1964. Introduction to mathematical logic. Princeton. 300 pp.

MENGER, K.

1928. Bemerkungen zu Grundlagenfragen. *Jahresb. D.M.V.* **37**, 213–226, 298–325.

1930. Der Intuitionismus. *Blätter f. deutsche Philos.* **4**, 311–325.

1931. Über den Konstruktivitätsbegriff. Zweite Mitteilung. *Anzeiger Akad. Wien* **68**, 7–9. (Cf. already *ibid.* **67**, 257–258, 1930; *Annals of Math.* **(2)32**, 757–760, 1931.)

MESERVE, B. E.

1955. Decision methods for elementary algebra. *Am. Math. Monthly* **62**, 1–8.

MIRIMANOFF, D.

1917. Les antinomies de Russell et de Burali-Forti et le problème fondamental de la théorie des ensembles. *L'Ens. Math.* **19**, 37–52.

MONTAGUE, R.

1955. Well-founded relations; generalizations of principles of induction and recursion (abstract). *Bull. A.M.S.* **61**, 442.

1957. Non-finite axiomatizability (abstract). Summaries of talks at the Summer Institute for Symbolic Logic (mimeographed). Ithaca, N. Y. 256–259. (2nd ed., Princeton, 1960.)

1961. Semantical closure and non-finite axiomatizability I. Infinitistic methods. Warsaw. 45-69.

1961a. Fraenkel's addition to the axioms of Zermelo. Bar-Hillel et al. 61, 91–114.

1962. Two contributions to set theory. Nagel et al. 62, 94–110.

MONTAGUE, R. and VAUGHT, R. L.

1959. Natural models of set theories. *Fund. Math.* **47**, 219–242.

1959a. A note on theories with selectors. *Ibid.* 243–247.

MOOY, J. J. A.

1966. La philosophie des mathématiques de Henri Poincaré. Collection de logique mathematique A 20. Paris.

MORLEY, M.

1965. Categoricity in power. *Tr. A.M.S.* **114**, 514–538.

MORRIS, C. W.

1938. Foundations of the theory of signs. Int. Enc. Un. Sc. I, No. 2. Chicago. 59 pp.

MORSE, A. P.

1965. A theory of sets. New York. 130 pp.

MOSCHOVAKIS, JOAN R.

1967. Disjunction and existence in formalized intuitionistic analysis. Crossley 67, 309–331.

MOSCHOVAKIS, Y. N.

1964. Recursive metric spaces. *Fund. Math.* **55**, 215–238.

1970. Determinacy and prewellorderings of the continuum. Bar-Hillel 70a, 24–62.

MOSTOWSKI, A.

1938. On the independence of the definitions of finiteness in a system of logic. Polish Thesis. *Dodatek do Roczn. Polsk. Tow. Mat.* **11**, 1–54. (Cf. A. Tarski: *J.S.L.* 3, 115–116, 1938.)

1939. Über die Unabhängigkeit des Wohlordnungssatzes vom Ordnungsprinzip. *Fund. Math.* **32**, 201–252.

1945. Axiom of choice for finite sets. *Ibid.* **33**, 137–168.

1949. An undecidable arithmetical statement. *Fund. Math.* **36**, 143–164.

1951. Some impredicative definitions in the axiomatic set theory. *Fund. Math.* **37**, 111–124. (Correction, *ibid.* **38**, p. 238, 1952.)

1952. Sentences undecidable in formalized arithmetic. An exposition of the theory of Kurt Gödel. Amsterdam. 117 pp. (2nd printing, 1957.)

1955. (In collaboration with other Polish scholars). The present state of investigations on the foundations of mathematics. *Rozprawy Mat.* IX. Warszawa. 48 pp. (Appeared also in German: Deutscher Verlag der Wiss., Berlin 1954.)

1958. On a problem of W. Kinna and K. Wagner. *Colloq. Math.* **6**, 207–208.

1959. On various degrees of constructivism. Heyting 59a, 178–195.

1966. Thirty years of foundational studies; lectures on the development of mathematical logic and the study of the foundations of mathematics in 1930–1964. Oxford. 180 pp. Also in *Acta Philos. Fennica* **17**, 1967.

1967. Recent results in set theory. Lakatos 67, 82–108.

1969. Constructible sets with applications. Amsterdam. 269 pp.

MUNROE, M. E.

1953. Introduction to measure and integration. Reading, Mass. 310 pp.

MYCIELSKI, J.

1964–66. On the axiom of determinateness. *Fund. Math.* **54**, 205–224, and **59**, 203–212.

MYCIELSKI, J. and STEINHAUS, H.

1962. A mathematical axiom contradicting the axiom of choice. *Bull. Acad. Polon. Sc.* **10**, 1–3.

MYCIELSKI, J. and ŚWIERCZKOWSKI, S.

1964. On the Lebesgue measurability and the axiom of determinateness. *Fund. Math.* **54**, 67–71.

MYHILL, J. R.

1949. Review of Chwistek 48. *J.S.L.* **14**, 119–125.

1950. A system which can define its own truth. *Fund. Math.* **37**, 190–192.

1950a. A complete theory of natural, rational and real numbers. *J.S.L.* **15**, 185–196.

1951. Report on investigations concerning the consistency of the axiom of reducibility. *Ibid.* **16**, 35–42.

1951a. Towards a consistent set theory. *Ibid.*, 130–136.

1952. The hypothesis that all classes are nameable. *Acad. U.S.A.* **38**, 979–981.

1952a. Some philosophical implications of mathematical logic. *Rev. of Metaphysics* **6**, 169–198. (Cf. V. E. Beneš: *Philos. Studies* **4**, 56–58; 1953; Myhill: *ibid.* **6**, 47–48, 1954.)

1953. Symposium: On the ontological significance of the Löwenheim-Skolem theorem. *Academic Freedom, Logic, and Religion.* Philadelphia. 57–70.

1967. Notes toward an axiomatization of intuitionistic analysis. *Logique et Analyse* **35**, 280–297.

1968. Formal systems of intuitionistic analysis I. Van Rootselaar-Staal 68, 161–178.

MYHILL, J. and SCOTT, D.

1971. Ordinal definability. Scott 71, 271–278.

NAGEL, E. and NEWMAN, J. R.

1958. Gödel's proof. New York. 118 pp. German ed. Wien, 1964.

NAGEL, E., SUPPES, P. and TARSKI, A.

1962 (editors). Logic, methodology and philosophy of science. Proc. of the 1960 intern. congr. Stanford, California. 661 pp.

NELSON, D.

1947. Recursive functions and intuitionistic number theory. *Tr.A.M.S.* **61**, 307–368. (Cf. N. A. Šanin: *Doklady Akad. Nauk SSSR* **93**, 779–782, 1953.)

1949. Constructible falsity. *J.S.L.* **14**, 16–26.

VON NEUMANN, J.

1923. Zur Einführung der transfiniten Zahlen. *Acta Szeged* **1**, 199–208; also in **61**, 24–33. English translation in van Heijenoort 67, 346–354.

1925. Eine Axiomatisierung der Mengenlehre. *J.f. Math.* **154**, 219–240; corrections, *ibid.* **155**, 128, 1926. Also in **61**, 34–56. English translation in van Heijenoort 67, 393–413.

1928. Die Axiomatisierung der Mengenlehre. *Math. Ztschr.* **27**, 669–752. Also in **61**, 339–422.

1928a. Über die Definition durch transfinite Induktion und verwandte Fragen der allgemeinen Mengenlehre. *Math. Annalen* **99**, 373–391; also in **61**, 320–338. (Cf. A. Fraenkel, *ibid.*, 392–393.)

1929. Über eine Widerspruchsfreiheitsfrage in der axiomatischen Mengenlehre. *J.f. Math.* **160**, 227–241; also in **61**, 494–508.

1929a. Zur allgemeinen Theorie des Masses. *Fund. Math.* **13**, 73–116; also in **61**, 599–643. (Cf. W. Sierpiński, *ibid.* **33**, 228–234, 1945.)

1961. Collected Works, vol. I. Oxford. 654 pp.

NOVAK, ILSE L.

1951. A construction for models of consistent systems. *Fund. Math.* **37**, 87–110.

NOVIKOV, P. S.

1955. On the algorithmic unsolvability of the word problem of group theory (Russian). *Trudy Mat. Inst. imeni V.A. Steklova* **44**, 1–143.

OBERSCHELP, A.

1964. Eigentliche Klassen als Urelemente in der Mengenlehre. *Math. Annalen* **157**, 234–260.

ONO, K.

1957. Set theory founded on a unique generating principle. *Nagoya Math. J.* **12**, 151–159.

OREY, S.

1955. Formal development of ordinal number theory. *J.S.L.* **20**, 95–104.

1956. On the relative consistency of set theory. *Ibid.*, 280–290.

1964. New foundations and the axiom of counting. *Duke Math. J.* **31**, 655–660.

PAP, A.

1954. The linguistic hierarchy and the vicious-circle-principle. *Philos. Studies* **5**, 49–53.

PEANO, G.

1890. Démonstration de l'intégrabilité des équations différentielles ordinaires. *Math. Annalen* **37**, 182–228; also in 57–59 I, 119–170. (Cf. D. van Dantzig: *Proc. Amsterdam* **45**, 367–373, 1942.)

1903. Les définitions mathématiques. Bibliothèque du Congrès International de Philosophie, Paris, 1900, III. Paris 1903. 279–288. Also in 57–59 II, 362–368. (*Cf. Bolletino della "Mathesis", N.S.* 7, 106–120, 1915; *Periodico di Mat.* (4) **1**, 175–189, 1921; W. Wilkosz: *Ann. de la Soc. Polon. de Math.* **16**, 176–178, 1938.)

1957–59. Opere Scelte. Roma. vol. I, 1957, 530 pp. vol. II, 1958, 518 pp. vol. III, 1959, 470 pp.

PEIRCE, C. S.

1933. Collected papers. Ed. by C. Hartshorne and P. Weiss. Vol. III. Cambridge, Mass. 433 pp.

PÉTER, RÓZSA

1951. Rekursive Funktionen. Budapest. 206 pp.

PIERPONT, J.

1928. Mathematical rigor, past and present. *Bull. A.M.S.* **34**, 23–53.

PINCUS, D.

1971. Zermelo-Fraenkel consistency results by Fraenkel-Mostowski methods. *J.S.L.*

POINCARÉ, H.

1902. Du rôle de l'intuition et de la logique en mathématiques. *C.R. du IIme Congr. Intern. des Math., Paris* 1900, 200–202.

1903. La science et l'hypothèse. Paris. 40me mille (Biblioth. de Philos. Scient.), 1925. English ed. 1905. German ed. *(Wissenschaft und Hypothese 1)*, 2nd ed., Leipzig, 1906.

1908. Science et méthode. Paris. 311 pp. English ed. in the book: The foundations of science, New York 1913, 553 pp.; German ed., Leipzig 1914.

1910. Über transfinite Zahlen. Sechs Vorträge über ausgewahlte Gegenstände aus der reinen Mathematik und mathematischen Physik (Leipzig & Berlin). 43–48.

1913. Dernières pensées. Paris. 258 pp. German ed., 1913.

POPPER, K. R.

1954. Self-reference and meaning in ordinary language. *Mind* **63**, 162–169.

POST, E. L.

1921. Introduction to a general theory of elementary propositions. *Am. J. of Math.* **43**, 163–185; also in van Heijenoort 67, 264–283.

1936. Finite combinatory processes – formulation 1. *J.S.L.* **1**, 103–105.

1943. Formal reductions of the general combinatorial decision problem. *Am. J. of Math.* **65**, 197–215.

1944. Recursively enumerable sets of positive integers and their decision problems. *Bull. A.M.S.* **50**, 284–316.

1947. Recursive unsolvability of a problem of Thue. *J.S.L.* **12**, 1–11.

PRAWITZ, D.

1965. Natural deduction; a proof-theoretical study. Stockholm. 113 pp.

PRESBURGER, M.

1930. Über die Vollständigkeit eines gewissen Systems der Arithmetik ganzer Zahlen, in welchem die Addition als einzige Operation hervortritt. C.R. du Ier Congr. des Math. des Pays Slaves, Warszawa 1929. 92–101, 395.

PRIOR, A. N.

1953. Three-valued logic and future contingents. *Philos. Quarterly* **3**, 317–326.

1955. Curry's paradox and 3-valued logic. *Australasian J. of Philos.* **33**, 177–182.

PUTNAM, H.

1957. Axioms of class existence. Summaries of talks at the Summer Institute for Symbolic Logic (mimeographed). Ithaca, N.Y. 2nd ed.: Princeton, 1960. 271–274.

QUINE, W. V.

1936. Set-theoretic foundations for logic. *J.S.L.* **1**, 45–57.

1937. New foundations for mathematical logic. *Am. Math. Monthly* **44**, 70–80. A revised and expanded version is in **53**, 80–101.

1950. Methods of logic. New York. 264 pp. (Revised ed. 1959.)

1951. Mathematical logic. Revised edition. Cambridge, Mass. 346 pp. (First edition, 1940.)

1951a. On the consistency of "New Foundations". *Acad. U.S.A.* 37, 538–540.

1952. On an application of Tarski's theory of truth. *Ibid.* **38**, 430–433.

1953. From a logical point of view. Cambridge, Mass. 184 pp.

1955. On Frege's way out. *Mind* **64**, 145–159.

1956. Unification of universes in set theory. *J.S.L.* **21**, 267–279.

1963. Set theory and its logic. Cambridge, Mass. 359 pp. (Revised ed.: 1969, 361 pp.)

RABIN, M. O.

1958. Recursive unsolvability of group-theoretic properties. *Annals of Math.* **67**, 172–194.

1960. Computable algebra, general theory and theory of computable fields. *Tr. A.M.S.* **95**, 341–360.

1965. A simple method for undecidability proofs and some applications. Bar-Hillel 65, 58–68.

1969. Decidability of second-order theories and automata on infinite trees. *Tr. A.M.S.* **141**, 1–35.

1970. Weakly definable relations and special automata. Bar-Hillel 70a, 1–23.

RAMSEY, F. P.

1926. The foundation of mathematics. *Proc. of the London Math. Soc.* (2)**25**, 338–384. Reprinted in 31.

1927. Mathematical logic. *Matn. Gazette* **13**, 185–194.

1931. The foundations of mathematics and other logical essays. Ed. by R. B. Braithwaite. New York & London.

RASIOWA, HELENA

1954. Constructive theories. *Bull. Acad. Polon. Sc.,* **2**, 121–124.

RASIOWA, HELENA and SIKORSKI, R.

1963. The mathematics of metamathematics. Warsaw. 519 pp.

REICHENBACH, H.

1944. Bertrand Russell's logic. Schilpp 44, 21–54.

REIDEMEISTER, K.

1949. Das exakte Denken der Griechen. Hamburg. 108 pp.

REINHARDT, W. N.

1970. Ackermann's set theory equals ZF. *Ann. of Math. Logic* **2**, 189–249.

REINHARDT, W. N. and SOLOVAY, R. M.

∞ Strong axioms of infinity and elementary embeddings. To appear.

RICE, H. G.

1954. Recursive real numbers. *Proc. A.M.S.* **5**, 784–791. (Cf. J. R. Myhill: *J.S.L.* **18**, 7–10, 1953.)

RICHARD, J.

1905. Les principes de mathématiques et le problème des ensembles. *Rev. Gen. des*

Sc. **16**, 541; also in *Acta Math.* **30**, 295–296, 1906. English translation in van Heijenoort 67, 142–144. (Cf. L. Olivier: *Rev. Gen. des Sc.* **16**, 542–543.)

1907. Sur un paradoxe de la théorie des ensembles et sur l'axiome de Zermelo. *L'Ens. Math.* **9**, 94–98. (Cf. *ibid.*, 39–44.)

1929. Sur l'axiome de Zermelo. *Bull. des Sc. Math.* (2)**53**, 106–109.

RIEGER, L.

1949. On the lattice theory of Brouwerian propositional logic. *Act. Fac. Rerum Nat. Univ. Carolinae,* No. **189**. Praha. 40 pp. (Cf. *Časopis* 74, 55–61, 1949.)

1957. A contribution to Gödel's axiomatic set theory I. *Czechoslovak Math. J.* 7 (82), 323–357.

1963. On the consistency of the generalized continuum hypothesis. Rozprawy Mat. 31. Warszawa. 45 pp.

ROBINSON, A.

1939. On the independence of the axioms of definiteness *(Axiome der Bestimmtheit). J.S.L.* **4**, 69–72.

1956. A result on consistency and its application to the theory of definition. *Proc. Amsterdam* **A59** (= *I.M.* **18**), 47–58.

1957. Some problems of definability in the lower predicate calculus. *Fund. Math.* **44**, 309–329. (Cf. G. Kreisel: *Bull. A.M.S.* **63**, 99–100, 1957.)

1959. Solution of a problem of Tarski. *Fund. Math.* **47**, 179–204.

1963. Introduction to model theory and to the metamathematics of algebra. Amsterdam. 284 pp.

ROBINSON, JULIA

1952. Existential definability in arithmetic. *Tr. A.M.S.* **72**, 437–449.

1969. Unsolvable diophantine problems. *Proc. A.M.S.* **22**, 534–538.

ROBINSON, R. M.

1937. The theory of classes. A modification of von Neumann's system. *J.S.L.* **2**, 69–72.

1945. On finite sequences of classes. *J.S.L.* **10**, 125–126.

1947. On the decomposition of spheres. *Fund. Math.* **34**, 246–260.

ROGERS, H.

1967. Theory of recursive functions and effective computability. New York. 482 pp.

VAN ROOTSELAAR, B.

1952. Un problème de M. Dijkman. *Proc. Amsterdam* **A55** (= *I.M.* **14**), 405–407.

1954. Generalization of the Brouwer integral. Thesis. Univ. of Amsterdam. (Cf. *Proc. Amsterdam* **A59** (= *I.M.* **18**), 579–580, 1956.)

VAN ROOTSELAAR, B. and STAAL, J. F.

1968 (editors). Logic, methodology and philosophy of science III; Proc. of the intern. congr. for logic, methodology and philos. of sc., Amsterdam, 1967. Amsterdam. 553 pp.

ROSENBLOOM, P. C.

1945. An elementary constructive proof of the fundamental theorem of algebra. *Am. Math. Monthly* **52**, 562–570. (Cf. Proc. of the 2nd Intern. Congr. of the Intern. Union for the Philos. of Sc., Zurich 1954, II, 135–137, 1955.)

ROSSER, J. B.

1936. Constructibility as a criterion for existence. *J.S.L.* **1**, 36–39.

1936a. Extensions of some theorems of Gödel and Church. *Ibid.*, 87–91.

1937. Gödel theorems for non-constructive logics. *Ibid.* **2**, 129–137. (Errata, *ibid.*, p. IV.)

1939. An informal exposition of proofs of Gödel's theorem and Church's theorems. *Ibid.* **4**, 53–60. (Errata, *ibid.*, p. IV.)

1942. The Burali-Forti paradox. *J.S.L.* 7, 1–17.

1952. The axiom of infinity in Quine's New Foundations. *Ibid.* **17**, 238–242.

1953. Logic for mathematicians. New York. 540 pp.

1955. Deux esquisses de logique. Paris. 65 pp.

1969. Simplified independence proofs. New York. 217 pp.

ROSSER, J. B. and TURQUETTE, A. R.

1952. Many-valued logics. Amsterdam. 124 pp.

ROSSER, J. B. and WANG, H.

1950. Non-standard models for formal logics. *J.S.L.* **15**, 113–129. (Errata, *ibid.*, p. IV.)

ROUGIER, L.

1920. La philosophie géométrique de Henri Poincaré. Biblioth. de Philos. Contemp. Paris.

RUBIN, H. and RUBIN, JEAN E.

1963. Equivalents of the axiom of choice. Amsterdam. 134 pp.

RUBIN, H. and SCOTT, D.

1954. Some topological theorems equivalent to the Boolean prime ideal theorem (abstract). *Bull. A.M.S.* **60**, 389.

RUDIN, MARY ELLEN

1969. Souslin's conjecture. *Am. Math. Monthly* **76**, 1113–1119.

RUSSELL, B.

1903. The principles of mathematics. I. London. 2nd ed.: London 1937 & New York 1938. (With a new introduction.) 534 pp. Reprinted 1950.

1906. On some difficulties in the theory of transfinite numbers and order types. *Proc. of the London Math. Soc.* (2)4, 29–53.

1908. Mathematical logic as based on the theory of types. *Am. J. of Math.* **30**, 222–262. Also in: Logic and knowledge, London, 1956, 59–102, and in van Heijenoort 67, 150–182. French translation in *R.M.M.* **18**, 263–301, 1910.

1919. Introduction to mathematical philosophy. London & New York. 206 pp. 2nd ed., 1920. German ed., 1923; French ed., 1928.

1931. Review of Ramsey 31. *Mind,* N.S. **40**, 476–482.

1944. Reply to criticisms. Schilpp 44, 679–741.

RYLL-NARDZEWSKI, C.

1953. The role of the axiom of induction in elementary arithmetic. *Fund. Math.* **39**, 239–263.

SAATY, L. S.

1965 (editor). Lectures on modern mathematics. Vol. III. New York. 321 pp.

SACKS, G. E.

1963. Degrees of unsolvability. Ann. of Math. Studies 55, Princeton. 174 pp.

1969. Measure-theoretic uniformity in recursion theory and set theory. *Tr. A.M.S.* **142**, 381–420.

ŠANIN, N. A.

1962. Constructive real numbers and constructive function spaces. *Trudy Mat. Inst. Steklov* **67**, 15–294. English translation: Translations Math. Monographs A.M.S. **21**, 1968.

SCARPELLINI, B.

1966. On a family of models of Zermelo-Fraenkel set theory. *Ztschr. f. math. Logik* **12**, 191–204.

SCHILLER, F. C. S.

1928. The infinite regress of proof. *Mind*, N.S. **37**, 353–354.

SCHILPP, P. A.

1944 (editor). The philosophy of Bertrand Russell. New York. 829 pp.

SCHLICK, M.

1926. Erleben, Erkennen, Metaphysik. *Kantstudien* **31**, 146–158.

1931. Die Kausalität in der gegenwärtigen Physik. *Naturwissenschaften* **19**, 145–162.

SCHMIDT, H. A.

1951. Die Zulässigkeit der Behandlung mehrsortiger Theorien mittels der üblichen einsortigen Prädikatenlogik. *Math. Annalen* **123**, 187–200.

1960. Mathematische Gesetze der Logik. Berlin. 555 pp.

SCHMIDT, H. A., SCHÜTTE, K. and THIELE, H. J.

1968 (editors). Contributions to mathematical logic, Proc. of the logic colloq., Hannover, 1966. Amsterdam. 298 pp.

SCHOENFLIES, A.

1900–07. Die Entwicklung der Lehre von den Punktmannigfaltigkeiten. I. Teil. (*Jahresb. D.M.V.* **8**, part 2.) Leipzig, 1900. 251 pp. II. Teil (2. *Ergänzungsband zu den Jahresb. D.M.V.*). Leipzig, 1907. 331 pp.

1911. Über die Stellung der Definition in der Axiomatik. *Jahresb. D.M.V.* **20**, 222–255.

1921. Zur Axiomatik der Mengenlehre. *Math. Annalen* **83**, 173–200. (Appeared already in *Proc. Amsterdam* **22**, 1920.)

1928. Die Krisis in Cantors mathematischem Schaffen. *Acta Math.* **50**, 1–23.

SCHOLZ, H.

1928. Review of O. Becker 27. *Deutsche Literaturzeitung* **49**, 679–690.

1930. Die Axiomatik der Alten. *Blätter f. deutsche Philos.* **4**, 259–278.

1950. Review of Weyl 49. *J.S.L.* **15**, 206–208.

SCHÖNFINKEL, M.

1924. Über die Bausteine der mathematischen Logik. *Math. Annalen* **92**, 305–316. English translation in van Heijenoort 67, 355–366.

SCHRÖTER, K.

1956. Theorie des bestimmten Artikels. *Ztschr. f. math. Logik* **2**, 37–56.

1956a. Über den Zusammenhang der in den Implikationsaxiomen vollständigen Axiomensysteme des zweiwertigen mit denen des intuitionistischen Aussagenkalküls. *Ibid.*, 173–176.

1957. Eine Umformung des Heytingschen Axiomensystems für den intuitionistischen Aussagenkalkül. *Ibid.* **3**, 18–29.

SCHÜTTE, K.

1950. Schlussweisen-Kalküle der Prädikatenlogik. *Math. Annalen* **122**, 47–65.

1951. Beweistheoretische Erfassung der unendlichen Induktion in der Zahlentheorie. *Math. Annalen* **122**, 369–389.

1952. Beweistheoretische Untersuchung der verzweigten Analysis. *Ibid.* **124**, 123–144.

1960. Beweistheorie. Berlin. 355 pp.

1968. Vollständige Systeme modaler und intuitionistischer Logik. Berlin.

SCHWABHÄUSER, W.

1956. Über die Vollständigkeit der elementaren Euklidischen Geometrie. *Ztschr. f. math. Logik* **2**, 137–165.

SCOTT, D.

1954. Prime ideal theorems for rings, lattices, and Boolean algebras (abstract). *Bull. A.M.S.* **60**, 390.

1955. Definitions by abstraction in axiomatic set theory (abstract). *Ibid.* **61**, 442.

1961. More on the axiom of extensionality. Bar-Hillel et al. 61, 115–131.

1961a. Measurable cardinals and constructible sets. *Bull. Acad. Polon. Sc.* **9**, 521–524.

1962. Quine's individuals. Nagel et al. 62, 111–115.

1967. A proof of the independence of the continuum hypothesis. *Math. Systems Theory* **1**, 89–111.

1968. Extending the topological interpretation to intuitionistic analysis. I. *Compositio Math.* **20**, 194–210.

1970. Extending the topological interpretation to intuitionistic analysis. II. Kino et al. 70, 235–257.

1971 (editor). Axiomatic set theory. Proceedings of symposia in pure math. **13**. Providence, R.I.

SCOTT, D. and SOLOVAY, R. M.

∞ Boolean-valued models for set theory. To appear.

SEIDENBERG, A.

1954. A new decision method for elementary algebra. *Annals of Math.* **60**, 365–374.

SHAW-KWEI, M.

1954. Logical paradoxes for many-valued systems. *J.S.L.* **19**, 37–40.

SHELAH, S.

1970. Solution of Łoś conjecture for uncountable languages (abstract). *Notices A.M.S.* **17**, 968.

SHEPHERDSON, J. C.

1951–53. Inner models for set theory. *J.S.L.* **16**, 161–190, 1951; **17**, 225–237, 1952; **18**, 145–167, 1953.

1965. Machine configuration and word problems of given degree of unsolvability. Bar-Hillel 65, 69–90.

SHOENFIELD, J. R.

1954. A relative consistency proof. *J.S.L.* **19**, 21–28.

1954a. Models of formal systems (abstract). *Bull. A.M.S.* **60**, 350.

1955. The independence of the axiom of choice (abstract). *Ibid.* **20**, p. 202.

1959. On the independence of the axiom of constructibility. *Am. J. of Math.* **81**, 537–540.

1967. Mathematical logic. Reading, Mass. 344 pp.

1971. Unramified forcing. Scott 71, 357–382.

1971a. Measurable cardinals. Gandy-Yates 71, 19–49.

SIERPIŃSKI, W.

1919. L'axiome de M. Zermelo et son rôle dans la théorie des ensembles et l'analyse. *Bull. de l'Acad. des Sc. de Cracovie, Cl. des Sc. Math. et Nat., Série A*, 1918, 97–152. (Cf. also *C.R. Varsovie* **24**, 184–190, 1932.)

1921. Les exemples effectifs et l'axiome du choix. *Fund. Math.* **2**, 112–118.

1927. Sur un probleme conduisant à un ensemble non-mesurable. *Ibid.* **10**, 177–179.

1934. Sur les ensembles partout de deuxième catégorie. *Ibid.* **22**, 1–3.

1941. L'axiome du choix et l'hypothèse du continu. *Les Entretiens de Zurich* 125–134. (Cf. *ibid.*, 134–143.)

1956. Hypothèse du continu. 2nd ed. New York. 274 pp.

1958. Cardinal and ordinal numbers. Warsaw. 487 pp. 2nd ed., 1965.

SIERPIŃSKI, W. and TARSKI, A.

1930. Sur une propriété caractéristique des nombres inaccessibles. *Fund. Math.* **15**, 292–300.

SIKORSKI, R.

1964. Boolean algebras. 2nd ed. Berlin. 237 pp.

SILVER, J.

1971. The consistency of the GCH with the existence of a measurable cardinal. Scott 71, 391–396.

1971a. The independence of Kurepa's conjecture and two-cardinal conjectures in model theory. Scott 71, 383–390.

SKOLEM, T.

1923. Einige Bermerkungen zur axiomatischen Begründung der Mengenlehre. Wiss. Vorträge gehalten auf dem 5. Kongress der skandinav. Mathematiker in Helsingfors 1922, 217–232; also in 70, 137–152. English translation in van Heijenoort 67, 290–301.

1923a. Begründung der elementaren Arithmetik durch die rekurrierende Denkweise ohne Anwendung scheinbarer Veränderlichen mit unendlichem Ausdehnungsbereich. *Skrifter uitgit av Videnskapsselskapet i Kristiania, I. Mat. Nat. Kl.*, No. **6**, 1–38; also in 70, 153–188. English translation in van Heijenoort 67, 302–333.

1924. Ein Verfahren zu beliebig angenäherter Bestimmung einer Wurzel einer beliebigen algebraischen Gleichung. *Norsk Mat. Forenings Skrifter* (1) **15**, 14 pp.

1929. Über einige Grundlagenfragen der Mathematik. *Skrifter utgit av det Norske Vid.-Akad. i Oslo*, I, No. **4**, 1–49; also in 70, 275–280.

1930. Einige Bemerkungen zu der Abhandlung von E. Zermelo *"Über die Definitheit in der Axiomatik". Fund. Math.* **15**, 337–341; also in 70, 355–366.

1934. Über die Nicht-Charakterisierbarkeit der Zahlenreihe mittels endlich oder abzählbar unendlich vieler Aussagen mit ausschliesslich Zahlenvariablen. *Fund. Math.* **23**, 150–161; also in 70, 455–482.

1941. Sur la portée du théorème de Löwenheim-Skolem. *Les Entretiens de Zurich,* 25–47; also in 70, 455–482. (Discussion, 47–52.)

1952. Some remarks on the foundation of set theory. Congr. Cambridge Mass. 1950, I, 695–704; also in 70, 519–528.

1955. Peano's axioms and models of arithmetic. Math. interpretations of formal systems. Amsterdam. 1–14; also in 70, 587–600.

1957. Two remarks on set theory. *Math. Scand.* **5**, 40–46; also in 70, 607–614.

1970. Selected works in logic. Oslo. 732 pp.

SŁUPECKI, J.

1953. St. Leśniewski's protothetics. *Studia Logica* **1**, 44–112.

1955. St. Leśniewski's calculus of classes. *Ibid.* **3**, 7–71. (Cf. Errata, *ibid.*, p. 299.)

SMULLYAN, R. M.

1957. Languages in which self-reference is possible. *J.S.L.* **22**, 55–67.

SOBOCIŃSKI, B.

1949–50. L'analyse de l'antinomie russellienne par Leśniewski. *Methodos* **1**, 94–107, 220–228, 308–316; **2**, 237–257.

SOLOVAY, R. M.

1964. The measure problem (abstract). *J.S.L.* **29**, 227–228.

1966. On the consistency of Takeuti's TT (abstract). *J.S.L.* **31**, 691.

1967. A nonconstructible Δ_3^1 set of integers. *Tr. A.M.S.* **127**, 50–75.

1969. On the cardinality of Σ_2^1 sets of reals. Bulloff et al. 69, 58–73.

1970. A model of set theory in which every set of reals is Lebesgue measurable. *Ann. of Math.* (2) **92**, 1–56.

1971. Real-valued measurable cardinals. Scott 71, 397–428.

SOLOVAY, R. M. and TENNENBAUM, S.

1971. Iterated Cohen extensions and Souslin's problem. *Ann. of Math.* (2) **94**, 201–245.

SPECKER, E.

1953. The axiom of choice in Quine's *New foundations for mathematical logic. Acad. U.S.A.* **39**, 972–975.

1954. Verallgemeinerte Kontinuumshypothese und Auswahlaxiom. *Archiv der Math.* **5**, 332–337.

1957. Zur Axiomatik der Mengenlehre (Fundierungs- und Auswahlaxiom). *Ztschr. f. math. Logik* **3**, 173–210.

1958. Dualität. *Dialectica* **12**, 451–465.

1962. Typical ambiguity. Nagel et al. 62, 116–124.

SPECTOR, C.

1962. Provably recursive functionals of analysis; a consistency proof of analysis by an extension of principles formulated in current intuitionistic mathematics. Recursive number theory. Proc. symposia pure math. *A.M.S.* **5**. 247 pp. 1–27.

STEGMÜLLER. W.

1956–57. Das Universalienproblem einst und jetzt. *Archiv f. Philos.* **6**, 192–225; **7**, 45–81.

1962. Eine Axiomatisierung der Mengenlehre beruhend auf den Systemen von Bernays und Quine. Logik und Logikkalkül (Käsbauer et al., editors). Freiburg. 57–103.

STEINITZ, E.

1910. Algebraische Theorie der Körper. *J. f. Math.* **137**, 167–309. New. ed. by R. Baer und H. Hasse, Berlin, 1930, 155 pp. + 27 pp. notes. Reprinted: New York, 1950.

STONE, M.

1937. Topological representations of distributive lattices and Brouwerian logics. *Časopis* **67**, 1–25.

DE SUE, F. See under D.

SUETUNA, Z.

1951–53. Über die Grundlagen der Mathematik. I–III. *J. of the Math. Soc. of Japan* **3**, 59–68, 1951; *Proc. of the Japan Acad.* **27**, 389–392, 1951; **29**, 91–95, 1953.

SUPPES, P.

1960. Axiomatic set theory. Princeton. 265 pp.

SUSZKO, R.

1951. Canonic axiomatic systems. *Studia Philos.* **4**, 301–330.

SZMIELEW, WANDA

1947. On choices from finite sets. *Fund. Math.* **34**, 75–80.

1955. Elementary properties of Abelian groups. *Fund. Math.* **41**, 203–271.

SZPILRAJN, E.

1930. Sur l'expansion de l'ordre partiel. *Fund. Math.* **16**, 386–389.

TAKAHASHI, M.

1967. On imcomparable cardinals. *Comm. Math. Univ. St. Paul* **16**, 129–142.

TAKEUTI, G.

1961. Axioms of infinity of set theory. *J. Math. Soc. Japan* **13**, 220–233.

1965. Transcendence of cardinals. *J.S.L.* **30**, 1–7.

1965a. Recursive functions and arithmetical functions of ordinal numbers. Bar-Hillel 65, 179–196.

1969. The universe of set theory. Bulloff et al. 69, 74–128.

TAKEUTI, G. and KINO, A.

1962. On hierarchies of predicates of ordinal numbers. *J. Math. Soc. Japan* **14**, 199–232.

TARSKI, A.

1924. Sur quelques théorèmes qui équivalent à l'axiome du choix. *Fund. Math.* **5**, 147–154.

1925. Sur les ensembles finis. *Fund. Math.* **6**, 45–95. (Cf. C. Kuratowski: *ibid.* **1**, 131–132, 1920.)

1931. Sur les ensembles définissables de nombres réels I. *Fund. Math.* **17**, 210–239. Revised English ed. in 56, 110–142.

1933. Einige Betrachtungen über die Begriffe der ω-Widerspruchsfreiheit und der ω-Vollständigkeit. *Monatsh. Math. Ph.* **40**, 97–112. English ed. in 56, 279–295

1936. Der Wahrheitsbegriff in den formalisierten Sprachen. *Studia Philos.* **1**, 261–405. English ed. in 56, 156–278.

1938. Über unerreichbare Kardinalzahlen. *Fund. Math.* **30**, 68–89.

1938a. Der Aussagenkalkül und die Topologie. *Ibid.* **31**, 103–134. English ed. in 56, 421–454.

1939. On undecidable statements in enlarged systems of logic and the concept of truth. *J.S.L.* **4**, 105–112.

1939a. On well-ordered subsets of any set. *Fund. Math.* **32**, 176–183.

1944. The semantic conception of truth and the foundations of semantics. *Philos. and Phenomen. Research* **4**, 341–376. (Reprinted in H. Feigl and W. Sellars, eds.: Readings in philosophical analysis, New York, 1949 and in L. Linsky, ed.: Semantics and the philosophy of language, Urbana III., 1952.)

1948. Axiomatic and algebraic aspects of two theorems on sums of cardinals. *Fund. Math.* **35**, 79–104.

1951 (with the assistance of J.C.C. McKinsey). A decision method for elementary algebra and geometry. 2nd ed. Berkeley & Los Angeles. 63 pp.

1954. Prime ideal theorems for Boolean algebras and the axiom of choice; Prime ideal theorems for set algebras and ordering principles; Prime ideal theorems for set algebras and the axiom of choice (3 abstracts). *Bull. A.M.S.* **60**, 390–391.

1955. General principles of induction and recursion (abstract). *Bull. A.M.S.* **61**, 442–443.

1956. Logic, semantics, metamathematics. Oxford. 471 pp.

1956a. Notions or proper models for set theories (abstract). *Bull. A.M.S.* **62**, 601.

TARSKI, A., MOSTOWSKI, A. and ROBINSON, R. M.

1953. Undecidable theories. Amsterdam. 98 pp.

TARSKI, A. and VAUGHT. R. L.

1957. Arithmetical extensions of relational systems. *Compositio Math.* **18**, 81–102.

TENNENBAUM, S.

1968. Souslin's problem. *Acad. U.S.A.* **59**, 60–63.

THARP, L. H.

1966. Bernays' set theory and the continuum hypothesis (abstract). *Notices A.M.S.* **13**, 138.

1967. On a set theory of Bernays. *J.S.L.* **32**, 319–321.

THIELE, E. J.

1955. Ein axiomatisches System der Mengenlehre nach Zermelo und Fraenkel. *Ztschr. f. math. Logik* **1**, 173–195.

TITGEMEIER, R.

1961. Über einen Widerspruch in Cogans Darstellung der Mengenlehre. *Ztschr. f. math. Logik* 7, 161–163.

TOMS, E.

1941. The law of excluded middle. *Philos. of Sc.* **8**, 33–38.

TROELSTRA, A. S.

1966. Intuitionistic general topology. Doctoral thesis. Amsterdam.

1968. New sets of postulates for intuitionistic topology. *Compositio Math.* **20**, 211–222.

1969. Principles of intuitionism. Lecture Notes in Math. 95. Berlin. 111 pp.

1970. Informal theory of choice sequences. *Studia Logica* **25**, 31–52.

1971. Notions of realizability for intuitionistic arithmetic and intuitionistic arithmetic in all finite types. Proc. of the second Scandinavian logic symposium, 1970 (J.E. Fenstad, ed.) Amsterdam. 369–405.

TUGUÉ, T.

1964. On the partial recursive functions of ordinal numbers. *J. Math. Soc. Japan* **16**, 1–31.

TURING, A. M.

1936. On computable numbers, with an application to the *Entscheidungsproblem. Proc. of the London Math. Soc.* (2)**42**, 230–265. (Cf. *ibid.* **43**, 544–546, 1937.)

1937. Computability and λ-definability. *J.S.L.* **2**, 153–163.

1950. The word problem in semi-groups with cancellation. *Annals of Math.* (2)**52**, 491–505.

1954. Solvable and unsolvable problems. Science News (Penguin Books), No. 31, 7–23.

TURQUETTE, A. R.

1950. Gödel and the synthetic a priori. *J. of Philos.* **47**, 125–129. (Cf. I.M. Copi: *Ibid.* **46**, 243–245, 1949.)

VAN DALEN, D. See Dalen, D. van

VAN DANTZIG, P. See Dantzig, P. van

VAN DER CORPUT, J. G. See Corput, J. G. van der

VANDIVER, H. S.

1934–35. On the foundations of a constructive theory of discrete commutative algebra. *Acad. U.S.A.* **20**, 579–165. (Cf. *Annals of Math.* (2)37, 1–6, 1936.)

VAN HEIJENOORT, J. See Heijenoort, J. van

VAN ROOTSELAAR, B. See Rootselaar, B. van

VAUGHT, R. L.

1954. Applications of the Löwenheim-Skolem-Tarski theorem to problems of completeness and decidability. *Proc. Amsterdam* A57 (= *I.M.* **16**), 467–472.

1967. Axiomatizability by a schema. *J.S.L.* **32**, 473–479.

VON NEUMANN, J. See Neumann, J. von

VAN VLECK, E. B.

1908. On non-measurable sets of points, with an example. *Tr. A.M.S.* **9**, 237–244.

VOPĚNKA, P.

1962. Die Konstruktion von Modellen der Mengenlehre nach der Methode der Ultraprodukte (Russian with a German summary). *Ztschr. f. math. Logik* **8**, 293–304.

1964. The independence of the continuum hypothesis. (Russian with an English summary). *Comm. Math. Univ. Carolinae*, supplement to vol. 5. 48 pp.

1965–67. The limits of sheaves and applications on constructions of models. *Bull. Acad. Polon. Sc.* **13**,189–192, 1965. On ∇-model of set theory. *Ibid.*, 267–272. Properties of ∇-model. *Ibid.*, 441–444. ∇-models in which the generalized continuum hypothesis does not hold. *Ibid.*, **14**, 95–99, 1966. The limits of sheaves over extremally disconnected compact Hausdorff spaces. *Ibid.* **15**, 1–4, 1967.

VOPĚNKA, P. and HÁJEK, P.

1965–67. Permutation submodels of the model ∇. *Bull. Acad. Polon. Sci.* **13**, 611–614, 1965. Some permutation submodels of the model ∇. *Ibid.* **14**, 95–99, 1966. Concerning the ∇-models of set theory. *Ibid.* **15**, 113–117, 1967.

VOPĚNKA, P. and HRBÁČEK, K.

1966. On strongly measurable cardinals. *Bull. Acad. Polon. Sc.* **14**, 587–591.

VREDENDUIN, P. G. J.

1954. The logic of negationless mathematics. *Compositio Math.* **11**, 204–270.

VUYSJE, D.

1953. Significs, its tendency, methodology, and applications. *Proc. of the Am. Acad. of Arts and Sc.* **80**, 223–270.

VAN DER WAERDEN, B. L.

1930. Eine Bemerkung über die Unzerlegbarkeit von Polynomen. *Math. Annalen* **102**, 738–739.

1954. Science awakening. Groningen. 306 pp.

1955. Algebra, 4th ed. Berlin.

WAJSBERG, M.

1938. Untersuchungen über den Aussagenkalkül von A. Heyting. *Wiadomości Mat.* **46**, 45–101.

WANG, H.

1949. On Zermelo's and von Neumann's axioms for set theory. *Acad, U.S.A.* **35**, 150–155.

1950. Remarks on the comparison of axiom systems. *Ibid.* **36**, 448–453.

1950a. A formal system of logic. *J.S.L.* **15**, 25–32.

1952. The irreducibility of impredicative principles. *Math. Annalen* **125**, 56–66. (Cf. 63, 432–442.)

1952a. Logic of many-sorted theories. *J.S.L.* **17**, 105–116. (Cf. 63, 322–333.)

1952b. Truth definitions and consistency proofs. *Tr. A.M.S.* 73, 243–275; also in **63**, 443–477.

1953. Quelques notions d'axiomatique. *Rev. Philos. de Louvain* **51**, 409–443. (Cf. 63, 1–33.)

1953a. The categoricity question of certain grand logics. *Math. Ztschr.* **59**, 47–56.
1954. The formalization of mathematics. *J.S.L.* **19**, 241–266; also in **63**, 559–584.
1955. On denumerable bases of formal systems. Math. interpretations of formal systems. Amsterdam. 57–84.
1955a. On formalization. *Mind* **64**, 226–238; also in **63**, 57–67.
1955b. Undecidable sentences generated by semantical paradoxes. *J.S.L.* **20**, 31–43; also in **63**, 546–558.
1957. The axiomatization of arithmetic. *J.S.L.* **22**, 145–158; also in **63**, 68–81.
1963. A survey of mathematical logic. Amsterdam. 651 pp.

WANG, H. and McNAUGHTON, R.
1953. Les systèmes axiomatiques de la théorie des ensembles. Paris. 55 pp.

WARD, L. E.
1962. The weak Tychonoff theorem and the axiom of choice. *Proc. A.M.S.* **13**, 757–758.

WAVRE, R.
1926. Sur le principe du tiers exclu. *R.M.M.* **33**, 425–430.

WEGEL, H.
1956. Axiomatische Mengenlehre ohne Elemente von Mengen. *Math. Annalen* **131**, 435–462.

WEISS, P.
1928. The theory of types. *Mind* **37**, 338–348.

WEYL, H.
1918. Das Kontinuum. Kritische Untersuchungen über die Grundlagen der Analysis. Leipzig. 83 pp. Reprinted 1932.
1919. Der circulus vitiosus in der heutigen Begründung der Mathematik. *Jahresb D.M.V.* **28**, 85–92.
1921. Über die neue Grundlagenkrise der Mathematik. *Math. Ztschr.* **10**, 39–79. Reprinted, with *Nachtrag Juni 1955,* in Selecta Hermann Weyl (Basel & Stuttgart 1956), 211–248.
1924. Randbemerkungen zu Hauptproblemen der Mathematik. *Ibid.* **20**, 131–150.
1926. Philosophie der Mathematik und Naturwissenschaft. Teil I. (Handbuch der Philos. IIA.) München. 64 pp. Reprinted 1950. English ed. in 49.
1931. Die Stufen des Unendlichen. Jena. 19 pp.
1946. Mathematics and logic. A brief survey serving as preface to a review of "The philosophy of Bertrand Russell". *Am. Math. Monthly* **53**, 2–13.
1949. Philosophy of mathematics and natural science. Princeton. 311 pp.

WHITEHEAD, A. N. and RUSSELL. B.
1910–13. Principia Mathematica. 3 vols. Cambridge, 1910, 1912, 1913. 666+772+491 pp. – 2nd. ed.: 1925, 1927, 1927, 674+742+491 pp., besides an Introduction to the 2nd ed. and Appendices. A, B, C in vol. I: 34+15+9+8 pp.

WILDER, R. L.
1952. Introduction to the foundations of mathematics. New York. 305 pp.

WITTGENSTEIN, L.
1922. Tractatus logico-philosophicus (German with English translation). London. 189 pp. 4th impression 1949. In German already in: *Annalen der Naturphilos.* **14**, 185–262, 1921.

1956. Remarks on the foundations of mathematics (German with English translation). Edited by G. H. von Wright, R. Rees and G. E. M. Anscombe. Oxford. 400 pp.

YATES, C. E. M.

1970. Initial segments of the degrees of unsolvability. Bar-Hillel 70a, 63–83.

ZERMELO, E.

1904. Beweis, dass jede Menge wohlgeordnet werden kann. *Math. Annalen* **59**, 514–516. English translation in van Heijenoort 67, 139–141. (Cf. A. Schoenflies, E. Borel, P. E. B. Jourdain: *Math. Annalen* **60**, 181–186, 194–195, 465–470, 1905.)

1908. Neuer Beweis für die Wohlordnung. *Math. Annalen* **65**, 107–128. English translation in van Heijenoort 67, 181–198.

1908a. Untersuchungen über die Grundlagen der Mengenlehre I. *Math. Annalen* **65**, 261–281. English translation in van Heijenoort 67, 199–215.

1909. Sur les ensembles finis et le principe de l'induction complète. *Acta Math.* **32**, 185–193.

1929. Über den Begriff der Definitheit in der Axiomatik. *Fund. Math.* **14**, 339–344.

1930. Über Grenzzahlen und Mengenbereiche. *Ibid.* **16**, 29–47.

1935. Grundlagen einer allgemeinen Theorie der mathematischen Satzsysteme. *Fund. Math.* **25**, 136–146.

ZICH, O. V.

1948. Příspěvek k teorii celých čísel a jednojednoznačného zobrazeni (A contribution to the theory of whole numbers and of one-to-one transformations). *Třídy České Akad.*, Rozprawy II, vol. **58**, No. 11. 12 pp.

ZLOT, W. L.

1960. The principle of choice in pre-axiomatic set theory. *Scripta Math.* **25**, 105–123.

ZORN, M.

1944. Idempotency of infinite cardinals. Univ. of California Publ. in Math., N.S. 2, 9–12.

INDEX OF PERSONS

INDEX OF SYMBOLS

logical

special

SUBJECT INDEX

(see also Table of Contents: numbers refer to pages)